Plants, Lichens, Fungi, and Algae Extracts and Derivatives with Antimicrobial Properties for Nutrition and Health

Plants, Lichens, Fungi, and Algae Extracts and Derivatives with Antimicrobial Properties for Nutrition and Health

Guest Editors

Violeta Popovici
Emma Adriana Ozon
Andreea Letitia Arsene
Verginica Schroder

MDPI

Basel • Beijing • Wuhan • Barcelona • Belgrade • Novi Sad • Cluj • Manchester

Guest Editors

Violeta Popovici
Department of Integrated Mountain Territorial Development
"Costin C. Kiriţescu" National Institute of Economic Research
Vatra Dornei
Romania

Emma Adriana Ozon
Department of Pharmaceutical Technology and Biopharmacy
"Carol Davila" University of Medicine and Pharmacy
Bucharest
Romania

Andreea Letitia Arsene
Department of Pharmaceutical Microbiology
University of Medicine and Pharmacy "Carol Davila"
Bucharest
Romania

Verginica Schroder
Department of Cellular and Molecular Biology
Ovidius University of Constanta
Constanta
Romania

Editorial Office
MDPI AG
Grosspeteranlage 5
4052 Basel, Switzerland

This is a reprint of the Special Issue, published open access by the journal *Antibiotics* (ISSN 2079-6382), freely accessible at: https://www.mdpi.com/journal/antibiotics/special_issues/Z11GRXL7WY.

For citation purposes, cite each article independently as indicated on the article page online and as indicated below:

Lastname, A.A.; Lastname, B.B. Article Title. *Journal Name* **Year**, *Volume Number*, Page Range.

ISBN 978-3-7258-7817-8 (Hbk)
ISBN 978-3-7258-7818-5 (PDF)
https://doi.org/10.3390/books978-3-7258-7818-5

Contents

About the Editors

Violeta Popovici

Violeta Popovici is a Senior Clinical Pharmacist, Ph.D. in Dental Medicine, and Senior Researcher III at the Mountain Economy Center (CE-MONT), Vatra-Dornei. She finished her Post-Doc Studies at Ovidius University of Constanta. Her interdisciplinary research over 6 years focused on the lichen Usnea barbata in the Călimani Mountains, Romania. The phytochemical profile and pharmacological properties of U. barbata extract in different solvents were explored. This was completed by incorporating the most active U. barbata extracts in mucoadhesive oral films, with promising effects in oral cancer and various oral infections with opportunistic germs. She currently collaborates with research teams across the medical field. Her research interests include plant/lichen secondary metabolites, identification and pharmacological potential, antibiotics, antimicrobial activity, in vitro anticancer activity, oral cancer, ROS, oxidative stress, antioxidants/pro-oxidants, food, diets, and case studies.

Emma Adriana Ozon

Emma Adriana Ozon is a Doctor of Pharmacy and Professor in the Department of Pharmaceutical Technology and Biopharmacy at Carol Davila University of Medicine and Pharmacy in Bucharest, Romania. Her work focuses on pharmaceutical formulation, preformulation studies, and physicochemical characterization of various dosage forms. She has significant experience enhancing drug solubility and bioavailability, particularly for poorly soluble drugs, using methods such as cyclodextrin inclusion complexes and lyophilization. Key research areas include the development of oral solid dosage forms, mouth-dissolving films, topical gel-creams, and cosmetic preparations, as well as innovative methods such as 3D printing and hot-melt extrusion in drug manufacturing.

Andreea Letitia Arsene

Andreea Letiția Arsene is a Professor in the Discipline of General Microbiology and Pharmaceuticals, Faculty of Pharmacy, UMF Carol Davila, Bucharest. She obtained a bachelor's degree at the UMF Carol Davila Bucharest, Faculty of Pharmacy in 2001 and a Ph.D. degree at the Carol Davila Bucharest, Faculty of Pharmacy in 2008. Since 2002, she has worked at UMF Carol Davila Bucharest, serving as a University Trainer (2002–2004), a University Assistant (2004–2007), a Supervisor (2007–2013), a Lecturer (2013–2019), and a Professor (2019–Present). She is a member of the European Federation for Pharmaceutical Sciences (EUFEPS), the European Association for the Study of Diabetes (EASD), the International Society of History of Pharmacy (IGGP), and the Catalan Society of the Friends of History of Pharmacy and Pharmaceutical Sciences (SAHCFC-Barcelona).

Verginica Schroder

Verginica Schroder is Associate Professor and Director of the Department of Pharmaceutical Sciences II, and Member of the Council of the Faculty of Pharmacy and the Bioethics Commission of the University. She has a PhD in Biology, and her research focuses on medicinal plants and natural compounds with therapeutic potential. She contributes to prestigious international journals and participates in interdisciplinary research projects, including 3D technologies for heritage conservation.

antibiotics

MDPI

Editorial

Plants, Lichens, Fungi, and Algae Extracts and Derivatives with Antimicrobial Properties for Nutrition and Health

Violeta Popovici [1,*], Emma Adriana Ozon [2], Andreea Letiția Arsene [3] and Verginica Schröder [4]

1 Center for Mountain Economics, "Costin C. Kirițescu" National Institute of Economic Research (INCE-CEMONT), Romanian Academy, 725700 Vatra-Dornei, Romania
2 Department of Pharmaceutical Technology and Biopharmacy, Faculty of Pharmacy, "Carol Davila" University of Medicine and Pharmacy, 020956 Bucharest, Romania; emma.budura@umfcd.ro
3 Department of Pharmaceutical Microbiology, Faculty of Pharmacy, "Carol Davila" University of Medicine and Pharmacy, 020956 Bucharest, Romania; andreea.arsene@umfcd.ro
4 Department of Cellular and Molecular Biology, Faculty of Pharmacy, Ovidius University of Constanta, 6 Capitan Al. Serbanescu Street, 900001 Constanta, Romania; verginica.schroder@univ-ovidius.ro
* Correspondence: violeta.popovici@ce-mont.ro

Plants are a rich source of bioactive compounds with a wide range of nutritional and therapeutic properties [1]. Studies have demonstrated their potential to prevent and treat various infectious diseases, highlighting the importance of their phytochemicals as antioxidants and antimicrobial agents. Plant-derived compounds present a promising opportunity to combat a wide range of bacterial pathogens, including those prioritized by the World Health Organization (WHO) [2]. Their unique chemical structures, often difficult or impossible to replicate with synthetic methods, highlight the potential of natural compounds to address bacterial resistance [3]. By leveraging various mechanisms to target microbial pathogens, these compounds showcase a multifaceted approach to anti-infectious therapy [4]. Furthermore, their structural diversity, combined with their safety, underscores the importance of further exploring natural substances in the development of effective antimicrobial therapies [5].

This Special Issue focuses on the above-mentioned topics by presenting the most recent research on the antimicrobial activity of various plant species and plant-derived products. New and valuable data have been added to this field by the authors of the nine original studies in this Special Issue.

Plant extracts are widely studied for their inhibitory activity against various pathogens. Unlike conventional antibiotics, which often target a single site (leading to resistance), plant extracts exhibit multiple mechanisms, depending on the phytochemicals extracted and the solvents used (mainly water or ethanol) [6]. Lipophilic compounds disrupt cell membrane integrity, leading to the leakage of cellular contents [7]. Phenolic compounds bind to bacterial enzymes and cell wall proteins, thereby neutralizing microbial function and replication [8]. Lectins are valuable agents in preventing the attachment of viruses and bacteria to host cells [9–11]. Other plant extracts can alter the cytoplasmic pH of microorganisms, thereby disrupting their metabolic processes [12]. These mechanisms offer a promising avenue for further exploration in microbial management and the development of treatment protocols.

Three original studies examined the phytochemical composition and pharmacological potential of different plant extracts in ethanol.

First, an interdisciplinary Romanian research team conducted a complex analysis of the phenolic profile and selected biological properties of ethanolic extracts of *Agave amica* (Medik.) Thiede & Govaerts cultivated in Romania [13]. This research is a comparative

 https://doi.org/10.3390/antibiotics15010111

study of the polyphenolic profile and the antioxidant, antimicrobial, antibiofilm, and cytotoxic/antiproliferative properties of 70% ethanolic extracts from the aerial parts and bulbs of *A. amica*. Both ethanolic extracts had high polyphenol content, with caffeic acid identified as a key compound. The aerial part extract shows promising results, with significantly higher concentrations of total polyphenols, tannins, flavonoids, and caffeic acid derivatives, as well as greater antioxidant activity than the bulb extract ($p < 0.001$). Both extracts exhibited significant inhibition of *Candida albicans* growth and moderate efficacy against *Listeria monocytogenes* and *Staphylococcus aureus*, affecting both planktonic cells and biofilms. Furthermore, promising evidence of dose-dependent cytotoxicity against the colorectal adenocarcinoma cell line suggests pathways for further research on potential therapeutic applications [13].

In the second paper, researchers from Thailand investigated the ethanolic extracts of Thai medicinal flowers—*Mesua ferrea* L. (Bunnak), *Mammea siamensis* T. Anderson (Saraphi), and *Clitoria ternatea* (Anchan)—known for their use in traditional medicine. The phytochemical profile determined by HPLC indicated that the *M. ferrea* extract contained the highest level of gallic acid. In contrast, the *M. siamensis* extract reported the highest quercetin concentration. LC-MS analysis identified fifteen key phenolic metabolites in all Thai flower extracts. The extract from *C. ternatea* had the highest total flavonoid content and showed substantial antioxidant activity. Antibacterial activity against enteric pathogens (*E. coli*, *E. coli* O157:H7, *Salmonella typhi*, *Shigella dysenteriae*, and *Vibrio cholerae*) revealed that M. ferrea exhibited the most significant inhibitory effects. All extracts exhibited time-dependent antibacterial activity, significantly prevented biofilm formation, disrupted established biofilms, reduced bacterial adhesion to intestinal cells, and induced membrane damage in *E. coli* O157:H7, leading to leakage of cellular components (DNA, RNA, and proteins). The second study results recommend Thai flower extracts as plant-based alternatives for preventing or controlling enteric bacterial infections [14].

The third paper reported the antiviral (anti-HIV), antifungal (anti-*C. albicans*), and antibacterial (anti-*E. coli*) activities of a green ultrasound-assisted extract from different parts of *Momordica charantia* fruits cultivated in India and Saudi Arabia. The authors identified and quantified, using LC-MS/MS analysis, three bioactive cucurbitane triterpenoid glycosides (karavilosides) in the *M. charantia* extracts tested, which are responsible for their pharmaceutical potential, and supported their results with molecular docking [15].

Incorporating plant extracts into nanoparticles (NPs) is an innovative approach to enhancing antibacterial activity. By combining bioactive plant phenolic metabolites with the advantageous properties of nanoparticles—such as their large surface area and targeted delivery capabilities—potent antimicrobial agents can be developed. These nanoparticles, which often include silver, zinc oxide, or copper oxide, show great promise for combating drug-resistant bacteria such as *E. coli* and *S. aureus*. They have effective antibacterial mechanisms (disrupting bacterial cells and generating reactive oxygen species (ROS)) and could be a cost-effective alternative to traditional antibiotics [16]. Thus, Jeon et al. synthesized silver nanoparticles (AgNPs) using an extract of *Morus nigra* L. [17]. The green-synthesized nanoparticles exhibited high antioxidant, cytotoxic, and anti-inflammatory activities, and significantly increased inhibitory effects against *E. coli* and *S. aureus* [17].

Propolis is a natural product derived from three primary sources: plant resins collected by bees, substances produced through their metabolic processes (including wax), and additional materials incorporated during propolis processing. The chemical composition of propolis is strongly determined by the specific plant sources of the resins, making it a unique and valuable plant-derived product with antimicrobial properties [18]. Thus, Oliveira et al. evaluated the inhibitory effects of hydroethanolic extracts of Portuguese propolis harvested from the Gerês apiary against a panel of three Gram-positive (*Bacillus subtilis*,

 https://doi.org/10.3390/antibiotics15010111

methicillin-sensitive *S. aureus* (MSSA), and methicillin-resistant *S. aureus* (MRSA)) and one Gram-negative bacterium (*E. coli*), as well as two yeasts (*C. albicans* and *Saccharomyces cerevisiae*). The authors compared the antibacterial activity of propolis extracts with that of three conventional antibiotics (erythromycin, vancomycin, and amoxicillin/clavulanate). They found that propolis exhibited greater antimicrobial activity against *B. subtilis* than vancomycin. Moreover, MRSA strains were resistant to all three broad-spectrum antibiotics and susceptible to propolis extracts [19].

Essential oils exhibit antimicrobial activity through various mechanisms that contribute to their effectiveness. These include disrupting the structure of cell walls, which enhances permeability, and altering the flow of electrons and protons. Additionally, they promote active transport, facilitate the coagulation of cell contents, damage cytoplasmic membranes, and hydrolyze ATP. By understanding these mechanisms, we can better harness the antimicrobial potential of essential oils, which can synergize with conventional antibiotics [20,21]. Four original studies that analyzed the inhibitory activity of various essential oils against food-borne and other clinically significant pathogens were published in our Special Issue. The results obtained by research teams from Italy, Romania, Poland, Turkey, and Portugal could enrich the scientific database with valuable information regarding the antimicrobial activity and potential synergism of the essential oils investigated in the contributions listed below.

1. Iseppi, R.; Truzzi, E.; Sabia, C.; Messi, P. Efficacy and Synergistic Potential of Cinnamon (*Cinnamomum zeylanicum*) and Clove (*Syzygium aromaticum* L. Merr. & Perry) Essential Oils to Control Food-Borne Pathogens in Fresh-Cut Fruits. *Antibiotics* **2024**, *13*, 319. https://doi.org/10.3390/antibiotics13040319
2. Neagu, R.; Popovici, V.; Ionescu, L.-E.; Ordeanu, V.; Biță, A.; Popescu, D.M.; Ozon, E.A.; Gîrd, C.E. Phytochemical Screening and Antibacterial Activity of Commercially Available Essential Oils Combinations with Conventional Antibiotics against Gram-Positive and Gram-Negative Bacteria. *Antibiotics* **2024**, *13*, 478. https://doi.org/10.3390/antibiotics13060478
3. Luca, S.V.; Wojtanowski, K.; Korona-Głowniak, I.; Skalicka-Woźniak, K.; Minceva, M.; Trifan, A. Spent Material Extractives from Hemp Hydrodistillation as an Underexplored Source of Antimicrobial Cannabinoids. *Antibiotics* **2024**, *13*, 485. https://doi.org/10.3390/antibiotics13060485
4. Kılıç, C.S.; Catarino, I.; Alves-Silva, J.; Demirci, B.; Kırcı, D.; Salgueiro, L.; Zuzarte, M. Unlocking the Skin-Protective Effects of *Ferulago* spp. Essential Oils: A Focus on Antidermatophytic and Wound-Healing Potential. *Antibiotics* **2025**, *14*, 343. https://doi.org/10.3390/antibiotics14040343

Author Contributions: As Academic Editors, all authors equally contributed to this editorial. All authors have read and agreed to the published version of the manuscript.

Funding: This editorial received no external funding.

Acknowledgments: The Academic Editors of this Special Issue are grateful to all Contributors and Editors of the *Antibiotics* MDPI journal for their excellent collaboration and support.

Conflicts of Interest: The authors declare no conflicts of interest.

References

1. Hossain, M.S.; Wazed, M.A.; Asha, S.; Amin, M.R.; Shimul, I.M. Dietary Phytochemicals in Health and Disease: Mechanisms, Clinical Evidence, and Applications—A Comprehensive Review. *Food Sci. Nutr.* **2025**, *13*, e70101. [CrossRef]

2. SeyedAlinaghi, S.; Mehraeen, E.; Mirzapour, P.; Yarmohammadi, S.; Dehghani, S.; Zare, S.; Gholami, S.; Attarian, N.; Abiri, A.; Farahani Rad, F.; et al. A Systematic Review on Natural Products with Antimicrobial Potential against WHO's Priority Pathogens. *Eur. J. Med. Res.* **2025**, *30*, 525. [CrossRef] [PubMed]
3. Ślusarczyk, J.; Adamska, E.; Czerwik-Marcinkowska, J. Fungi and Algae as Sources of Medicinal and Other Biologically Active Compounds: A Review. *Nutrients* **2021**, *13*, 3178. [CrossRef] [PubMed]
4. Vaou, N.; Stavropoulou, E.; Voidarou, C.; Tsigalou, C.; Bezirtzoglou, E. Towards Advances in Medicinal Plant Antimicrobial Activity: A Review Study on Challenges and Future Perspectives. *Microorganisms* **2021**, *9*, 2041. [CrossRef] [PubMed]
5. Angelini, P. Plant-Derived Antimicrobials and Their Crucial Role in Combating Antimicrobial Resistance. *Antibiotics* **2024**, *13*, 746. [CrossRef]
6. Cowan, M.M. Plant Products as Antimicrobial Agents. *Clin. Microbiol. Rev.* **1999**, *12*, 564–582. [CrossRef]
7. Valdez-Lara, A.G.; Jaramillo-Granada, Á.M.; Ortega-Zambrano, D.; García-Marquez, E.; García-Fajardo, J.A.; Mercado-Uribe, H.; Ruiz-Suárez, J.C. Disruption of Biological Membranes by Hydrophobic Molecules: A Way to Inhibit Bacterial Growth. *Front. Microbiol.* **2025**, *15*, 1478519. [CrossRef]
8. Lobiuc, A.; Pavăl, N.-E.; Mangalagiu, I.I.; Gheorghiță, R.; Teliban, G.-C.; Amăriucăi-Mantu, D.; Stoleru, V. Future Antimicrobials: Natural and Functionalized Phenolics. *Molecules* **2023**, *28*, 1114. [CrossRef]
9. Hartmann, M.; Lindhorst, T.K. The Bacterial Lectin FimH, a Target for Drug Discovery—Carbohydrate Inhibitors of Type 1 Fimbriae-Mediated Bacterial Adhesion. *Eur. J. Org. Chem.* **2011**, *2011*, 3583–3609. [CrossRef]
10. Carter, M.D.; Tran, T.M.; Cope-Arguello, M.L.; Weinstein, S.; Li, H.; Hendrich, C.G.; Prom, J.L.; Li, J.; Chu, L.T.; Bui, L.; et al. Lectins and Polysaccharide EPS I Have Flow-Responsive Roles in the Attachment and Biofilm Mechanics of Plant Pathogenic Ralstonia. *PLoS Pathog.* **2024**, *20*, e1012358. [CrossRef]
11. Omidvar, R.; Ayala, Y.A.; Brandel, A.; Hasenclever, L.; Helmstädter, M.; Rohrbach, A.; Römer, W.; Madl, J. Quantification of Nanoscale Forces in Lectin-Mediated Bacterial Attachment and Uptake into Giant Liposomes. *Nanoscale* **2021**, *13*, 4016–4028. [CrossRef]
12. Gonelimali, F.D.; Lin, J.; Miao, W.; Xuan, J.; Charles, F.; Chen, M.; Hatab, S.R. Antimicrobial Properties and Mechanism of Action of Some Plant Extracts Against Food Pathogens and Spoilage Microorganisms. *Front. Microbiol.* **2018**, *9*, 1639. [CrossRef] [PubMed]
13. Niculae, M.; Hanganu, D.; Oniga, I.; Burcă, S.-A.; Tiperciuc, B.; Ielciu, I.; Pall, E.; Bab, T.; Burtescu, R.F.; Sava, M.A.; et al. Agave Amica (Medik.) Thiede & Govaerts (Asparagaceae)—Insights into Its Valuable Phenolic Profile and In Vitro Antimicrobial, Antibiofilm, Antioxidative, and Antiproliferative Properties. *Antibiotics* **2025**, *14*, 638. [CrossRef] [PubMed]
14. Suriyaprom, S.; Cheepchirasuk, N.; Ngamsaard, P.; Intachaisri, V.; Inta, A.; Tragoolpua, Y. Thai Medicinal Flowers as Natural Antioxidants and Antibacterial Agents Against Pathogenic Enteric Bacteria: A Comparative Study of Mesua Ferrea, Mammea Siamensis, and Clitoria Ternatea. *Antibiotics* **2025**, *14*, 1038. [CrossRef] [PubMed]
15. Aldholmi, M.; Ahmad, R.; Shaikh, M.H.; Salem, A.M.; Alqurashi, M.; Alturki, M. Anti-Infective Activity of Momordica Charantia Extract with Molecular Docking of Its Triterpenoid Glycosides. *Antibiotics* **2024**, *13*, 544. [CrossRef]
16. Reda, M.; Ashames, A.; Edis, Z.; Bloukh, S.; Bhandare, R.; Abu Sara, H. Green Synthesis of Potent Antimicrobial Silver Nanoparticles Using Different Plant Extracts and Their Mixtures. *Processes* **2019**, *7*, 510. [CrossRef]
17. Jeon, Y.-N.; Ryu, S.-J.; Lee, H.-Y.; Kim, J.-O.; Baek, J.-S. Green Synthesis of Silver Nanoparticle Using Black Mulberry and Characterization, Phytochemical, and Bioactivity. *Antibiotics* **2024**, *13*, 686. [CrossRef]
18. Dezmirean, D.S.; Paşca, C.; Moise, A.R.; Bobiş, O. Plant Sources Responsible for the Chemical Composition and Main Bioactive Properties of Poplar-Type Propolis. *Plants* **2020**, *10*, 22. [CrossRef]
19. Oliveira, R.D.; Araújo, C.; Almeida-Aguiar, C. In Vitro Antimicrobial Potential of Portuguese Propolis Extracts from Gerês against Pathogenic Microorganisms. *Antibiotics* **2024**, *13*, 655. [CrossRef]
20. Christaki, S.; Moschakis, T.; Kyriakoudi, A.; Biliaderis, C.G.; Mourtzinos, I. Recent Advances in Plant Essential Oils and Extracts: Delivery Systems and Potential Uses as Preservatives and Antioxidants in Cheese. *Trends Food Sci. Technol.* **2021**, *116*, 264–278. [CrossRef]
21. Bolouri, P.; Salami, R.; Kouhi, S.; Kordi, M.; Asgari Lajayer, B.; Hadian, J.; Astatkie, T. Applications of Essential Oils and Plant Extracts in Different Industries. *Molecules* **2022**, *27*, 8999. [CrossRef]

 https://doi.org/10.3390/antibiotics15010111

antibiotics

MDPI

Article

Efficacy and Synergistic Potential of Cinnamon (*Cinnamomum zeylanicum*) and Clove (*Syzygium aromaticum* L. Merr. & Perry) Essential Oils to Control Food-Borne Pathogens in Fresh-Cut Fruits

Ramona Iseppi [1], Eleonora Truzzi [2], Carla Sabia [1] and Patrizia Messi [1,*]

1 Department of Life Sciences, University of Modena and Reggio Emilia, Via G. Campi 287, 41125 Modena, Italy; ramona.iseppi@unimore.it (R.I.); carla.sabia@unimore.it (C.S.)
2 Department of Chemical and Geological Sciences, University of Modena and Reggio Emilia, Via G. Campi 103, 41125 Modena, Italy; eleonora.truzzi@unimore.it
* Correspondence: patrizia.messi@unimore.it; Tel.: +39-059-205-5615

Abstract: The presence of microbial pathogens in ready-to-eat produce represents a serious health problem. The antibacterial activity of cinnamon (*Cinnamomum zeylanicum*) and clove (*Syzygium aromaticum* L. Merr. & Perry) essential oils (EOs) was determined toward food-borne pathogens by agar disk diffusion and minimum inhibitory concentration (MIC) assays. The growth kinetics of all strains, both in a buffer suspension assay and "on food" in artificially contaminated samples, were also investigated. The two EOs demonstrated a good antibacterial effect both alone and in combination (EO/EO). The use of EO/EO led to a synergistic antibacterial effect, also confirmed by the growth kinetics studies, where the EOs were active after 10 h of incubation ($p < 0.0001$) at significantly lower concentrations than those when alone. In the "on food" studies performed on artificially contaminated fruit samples stored at 4 °C for 8 days, the greatest killing activity was observed at the end of the trial (8 days) with a reduction of up to 7 log CFU/g compared to the control. These results confirm the good antibacterial activity of the EOs, which were more effective when used in combination. Data from the "on food" studies suggest cinnamon and clove essential oils, traditionally used in the food industry, as a possible natural alternative to chemical additives.

Keywords: cinnamon and clove essential oils; synergistic activity; food-borne pathogens; fresh-cut fruit; bio-preservation

1. Introduction

The Mediterranean diet is known throughout the world for its balanced nutrient content, and in this context, ready-to-eat (RTE) fruits and vegetables offer a range of minerals, vitamins and phytochemicals essential for human health [1]. Adequate consumption of fresh produce plays an important role in the prevention of chronic pathologies such as cardiovascular diseases, cancer, hypertension, diabetes and obesity [2]. Given that the beneficial health value of these foods is widely recognized by the scientific community and consumers, the ready-to-eat (RTE) fresh produce market is one of the most rapidly expanding sectors of the food industry [3].

It is well known that the mild treatments undergone by RTE produce favor more rapid physiological deterioration, biochemical changes and microbial degradation, mainly in fruits for their high water and sugar contents [4]. This can cause both economic loss through spoilage bacteria and lead to outbreaks, resulting in illness and death due to the presence of food-borne pathogens. Among RTE produce, fresh-cut fruits could act as a vehicle for transmitting pathogens of major concern like *Listeria monocytogenes*, *Staphylococcus aureus*, pathogenic *Escherichia coli* mainly O157:H7 and *Salmonella* spp. [5,6].

Several listeriosis outbreaks have been linked to fresh produce contamination around the world, and many studies have detected *L. monocytogenes* in fresh minimally processed fruits [7–9] linked to melons, caramel apples, peaches, plums and nectarines (stone fruits) [10–12].

S. aureus has been detected on fresh produce and ready-to-eat vegetable salads, and it proved capable of growing on peeled Hamlin oranges stored at 24 °C (75.2 °F) and survived up to 14 days when stored at 4 to 8 °C (39.2 to 46.4 °F) [13].

E. coli O157:H7 is recognized as an important food-borne pathogen, and it has been identified as a causative agent for cantaloupe outbreaks in 1993, 1997 and 2004 [14,15].

Multistate outbreaks of *Salmonella enteritidis* and *Salmonella* Sundsvall linked to the consumption of peaches and imported Mexican cantaloupes, respectively, were reported by the U.S. Food and Drug Administration (FDA) in 2021 and 2024 [16,17], and a *Salmonella* Newport outbreak in ready-to-eat (RTE) cut watermelons was reported in the United Kingdom (UK) in 2012 [18].

Y. enterocolitica is a ubiquitous pathogen frequently isolated from soil and water from fruit and vegetables, thus being a potential source of yersiniosis [19,20]. In 2021, berries and juices and other derived produce were implicated in outbreaks caused by unspecified *Salmonella* and *Y. enterocolitica* [21].

Fresh-cut produce, due to its high water content, can favor the development of pathogens and spoilage bacteria [22]. To counteract their multiplication in food, chemical preservatives have always been used, but consumer demand is increasingly oriented towards a safe product with a long shelf-life obtained with the use of natural preservatives. Recent exploitation of natural products to control decay and extend the shelf-life of perishables has received more and more attention by researchers, including essential oils (EOs), natural compounds isolated from aromatic plants and generally recognized as safe (GRAS) for the environment and human health. Although originally added to change or improve taste, essential oils have antimicrobial activity that makes them an attractive choice to replace synthetic preservatives [23,24]. Many studies have demonstrated the antimicrobial activity of several spices toward pathogenic and spoilage bacteria and fungi, such as cinnamon and clove. *Cinnamomum zeylanicum* (cinnamon) is a very common spice obtained from the bark and leaves of trees of the *Cinnamomum* genus, widely used as a flavoring component in the food industry. The main components of cinnamon EO extracted from bark are *trans*-cinnamaldehyde, eugenol and linalool. Based on several studies, cinnamon essential oil exhibits significant inhibitory effects on both Gram-negative and Gram-positive bacteria like *Salmonella* enterica, *E. coli*, *S. aureus* and *L. monocytogenes* [25]. *Syzygium aromaticum* (clove) EO is considered the most useful and beneficial among spices that have been used in the food field. Clove contains flavonoids and phenolic molecules, including eugenol, which represents the main bioactive molecule of this EO [26]. With its main component of eugenol, clove EO has high potential not only for pharmaceutical applications but also for agricultural and food applications. The antimicrobial activities of clove EO appear to surpass those of other spices, showing extraordinary antibacterial activity against a great number of microorganisms [27].

The use of essential oils in the food industry is an interesting approach, but their intense aroma limits their use in the field of food preservation. An approach to overcome this problem could be to create combinations of different essential oils that, due to their synergistic effects, could allow reducing the concentration and, consequently, the negative sensorial impact without compromising the antimicrobial activity [28].

The aim of the present investigation was to assess the efficacy and synergistic potential of two essential oils traditionally used in the food industry to control food-borne pathogens in fresh-cut fruits. In this study, cinnamon and clove essential oils (EOs) were used alone and in combination to find a safe, effective and natural method to improve the safety and the shelf life of highly perishable produce.

2. Results

2.1. Chemical Composition of the EOs

C. zeylanicum (cinnamon) and *S. aromaticum* EOs were phytochemically characterized in terms of chemical composition via GC–MS and GC–FID. More than 95% of the total composition was determined for each EO. As reported in Table 1, cinnamon EO was mainly rich in *trans*-cinnamaldehyde, which accounted for 69% of the total composition. Relevant amounts of β-caryophyllene, linalool, eugenol, limonene, cinnamyl acetate, γ-cadinene, *p*-cymene and α-humulene were also detected. Regarding clove EO, eugenol and its ester derivative represented more than 91% of the total composition.

Table 1. Chemical percent composition of *Cinnamomum zeylanicum* (cinnamon) bark essential oil and *Syzygium aromaticum* (clove) flower bud essential oil.

Compound	Lit. LRI	Exp. LRI	*Cinnamomum zeylanicum*	*Syzygium aromaticum*
α-thujene	928	926	0.10	-
α-pinene	936	932	0.47	-
camphene	950	947	0.19	-
sabinene	973	972	0.21	-
α-phellandrene	1004	1002	0.37	-
α-terpinene	1017	1014	0.18	-
p-cymene	1024	1022	1.44	-
limonene	1029	1026	2.23	0.26
linalool	1099	1100	3.36	0.71
fenchol	1112	1116	-	0.14
camphor	1143	1143	-	0.24
terpinen-4-ol	1177	1176	0.21	-
α-terpineol	1190	1190	0.42	-
trans-cinnamaldehyde	1277	1278	68.96	-
bornyl acetate	1287	1287	0.30	-
thymol	1292	1292	0.13	-
citronellyl acetate	1357	1362	3.37	-
eugenol	1378	1378	0.74	79.61
α-copaene	1376	1381	-	0.34
β-caryophyllene	1420	1424	5.64	3.05
α-bergamotene	1444	1445	0.26	-
cinnamyl acetate	1445	1449	2.07	-
α-humulene	1453	1458	1.15	0.51
γ-cadinene	1523	1522	1.95	-
δ-cadinene	1527	1527	0.19	-
eugenyl acetate	1526	1535	-	11.47
caryophyllene oxide	1590	1594	1.06	1.64
Total			95.00	97.97

Experimental retention indices and literature retention indices (HP-5 column) according to NIST 14 (National Institute of Standards and Technology, USA; 14th edition) library database [29].

2.2. Agar Disk Diffusion Assay

Table 2 shows the antibacterial activity of cinnamon and clove EOs using the agar disk diffusion assay as a preliminary screening. Both Gram-negative and Gram-positive bacteria were sensitive to the EOs. Cinnamon and clove EOs displayed a bacterial growth inhibition zone ranging from 27 to 32 mm and 30 to 35 mm, respectively, with similar inhibition zones observed for both Gram-positive and Gram-negative strains.

Table 2. Agar disk diffusion assay of cinnamon and clove essential oils. The inhibition zones were measured in mm.

Indicator Strains	Cinnamon Essential Oil	Clove Essential Oil
Yersinia enterocolitica ATCC 23715	32 ± 1.1	35 ± 0.6
Escherichia coli ATCC 25922	29 ± 1.0	31 ± 1.4
Salmonella Typhimurium ATCC 19585	27 ± 0.9	30 ± 1.3
Listeria monocytogenes NCTC 10888	29 ± 0.5	32 ± 1.6
Staphylococcus aureus ATCC 6538	27 ± 0.9	31 ± 1.6

2.3. Minimum Inhibitory Concentration (MIC) and Fractional Inhibitory Concentration Index (FIC Index) Determination

The results of the minimum inhibitory concentration (MIC) of both EOs, performed on the same strains, confirmed the good antibacterial activity against both Gram-positive and Gram-negative microorganisms, as revealed by the disk diffusion test (Table 3). Cinnamon EO showed an MIC value ranging from 1 to 8 µg/mL, whereas clove EO displayed a higher MIC value (from 4 to 64 µg/mL), especially toward *S. typhimurium* and *L. monocytogenes* (64 µg/mL for both strains).

Table 3. Synergistic activity of EO/EO combination, by FIC index calculation. MIC values are expressed in µg/mL.

Strains	EO	MIC EOs (µg/mL)	MIC EO/EO (µg/mL)	FIC Index
Yersinia enterocolitica ATCC 23715	Cinnamon	4	1	0.5
	Clove	1	0.25	
Escherichia coli ATCC 25922	Cinnamon	8	2	0.5
	Clove	4	1	
Salmonella Typhimurium ATCC 19585	Cinnamon	64	16	0.5
	Clove	8	2	
Listeria monocytogenes NCTC 10888	Cinnamon	64	16	0.5
	Clove	8	2	
Staphylococcus aureus ATCC 6538	Cinnamon	64	16	0.5
	Clove	8	2	

Regarding the effectiveness of EOs used in combination (cinnamon EO/clove EO), a synergistic effect against all tested strains was observed. The combination of EOs led to an increase in the antibacterial effect, with a consequent decrease in the concentrations used (four times) (Table 3).

2.4. Growth Kinetics Study

Cinnamon and clove EOs showed activity against viable cells of all tested strains after 10 h of incubation, especially for *E. coli* (Figures S1–S4). Both EOs and the EO/EO combination were active ($p < 0.0001$) at the end of the experiments toward Gram-negative and Gram-positive microorganisms, notably against *L. monocytogenes* (Figure 1). Also, in this case, the mixture of EOs allowed a significant reduction in the concentrations of the individual EOs.

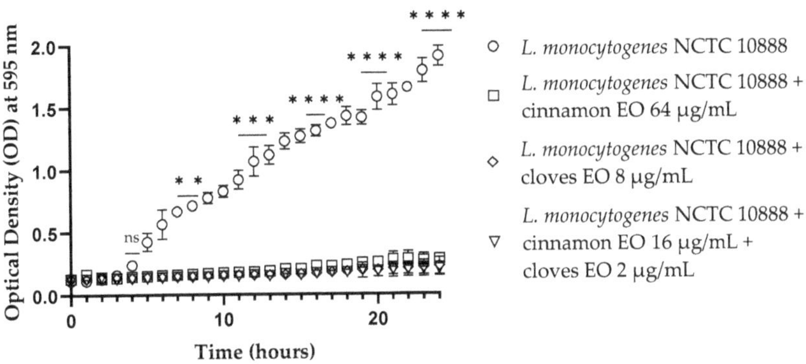

Figure 1. Time–kill studies of cinnamon and clove EOs alone and in combination against *Listeria monocytogenes* NCTC 10888 viable cells. *p*-values of <0.01 (**), $p < 0.001$ (***) and $p < 0.0001$ (****) were considered significant by *t*-test and ANOVA with Bonferroni correction. Results are expressed as mean ± SD of the three determinations (error bar = S.D.; $n = 3$).

2.5. Evaluation of the Antibacterial Activity by "On Food" Studies

The trend observed for *Y. enterocolitica* ATCC 23715 in fresh-cut fruits added with cinnamon and clove EOs, alone or in combination, and stored at refrigeration temperature showed a viable cell decrease after 24 h. After 4 days, for both the cinnamon and clove EOs and the EO mixture, a further reduction of about 2.00 log CFU/g compared to the control ($p = 0.0007$, $p = 0.0001$ and $p = 0.0002$, respectively) was observed. After 6 days, both EOs resulted in a reduction of approximately 5 log CFU/g ($p < 0.0007$ for cinnamon and clove EOs), and the EO/EO combination displayed a reduction of 6 log CFU/g ($p < 0.0001$). At the end of the trial (8 days), cinnamon and clove EOs showed a viable cell reduction of 6 log CFU/g and 6.5 log CFU/g, respectively ($p < 0.0001$), and for the EO/EO combination, a 7 log CFU/g decrease was observed ($p < 0.0001$) (Figure 2).

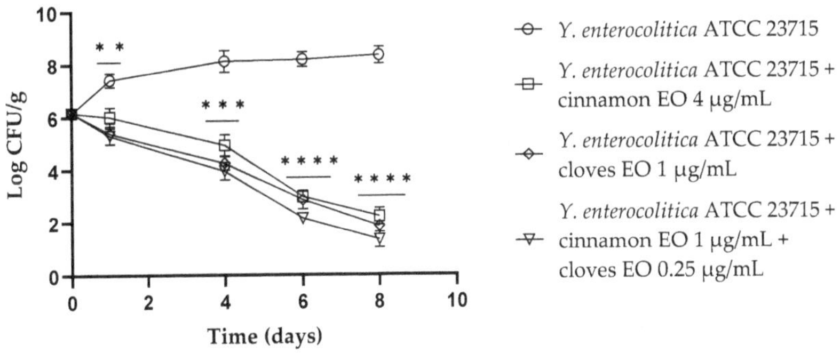

Figure 2. *Yersinia enterocolitica* viable counts (log CFU/g) observed in the contaminated fresh-cut fruits. *p*-values of <0.01 (**), $p < 0.001$ (***) and $p < 0.0001$ (****) were considered significant by *t*-test and ANOVA with Bonferroni correction. ns stands for not statistically significant. Results are expressed as mean ± SD of the three determinations (error bar = S.D.; $n = 3$).

Regarding *E. coli* ATCC 25922, both single EOs and the EO/EO combination showed a reduction in viable cells of about 2 log CFU/g ($p < 0.01$) after 24 h of the experiment. After 4 days, clove EO and the EO/EO combination displayed the best results ($p < 0.0001$) compared to the control with a reduction of about 5 log CFU/g, and the cinnamon EO showed a reduction of 4 log CFU/g ($p < 0.0001$). After 6 days of the trial, clove EO and the combination still confirmed the best results with a reduction of 6 log CFU/g ($p < 0.0001$) of viable *E. coli* cells. At the end of the trial (8 days), the EO/EO combination had the best results, showing 7.7 log CFU/g *E. coli* viable cells (Figure 3).

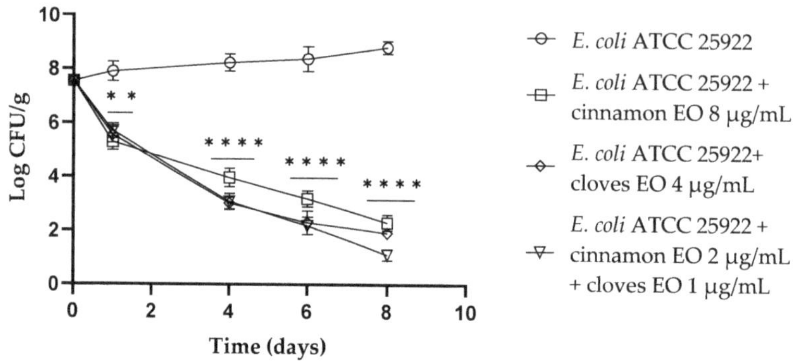

Figure 3. *Escherichia coli* viable counts (log CFU/g) observed in the contaminated fresh-cut fruits. p-values of <0.01 (**) and $p < 0.0001$ (****) were considered significant by t-test and ANOVA with Bonferroni correction. ns stands for not statistically significant. Results are expressed as mean ± SD of the three determinations (error bar = S.D.; $n = 3$).

Regarding *S. typhimurium* ATCC 19585, at 24 h of the experiment, both EOs displayed a reduction of about 1.5 log CFU/g in viable cells, and a better result emerged using the EO/EO combination, with a reduction of 2.5 log CFU/g viable cells ($p = 0.0007$). After 4 days, cinnamon and clove EOs alone and the EO/EO combination exhibited the same activity with a reduction of about 4 log CFU/g ($p < 0.001$) viable cells. From day 6 until the end of the experiment, clove EO and the EO/EO combination demonstrated the best activity against *S. typhimurium* compared to the control ($p < 0.0001$) (Figure 4).

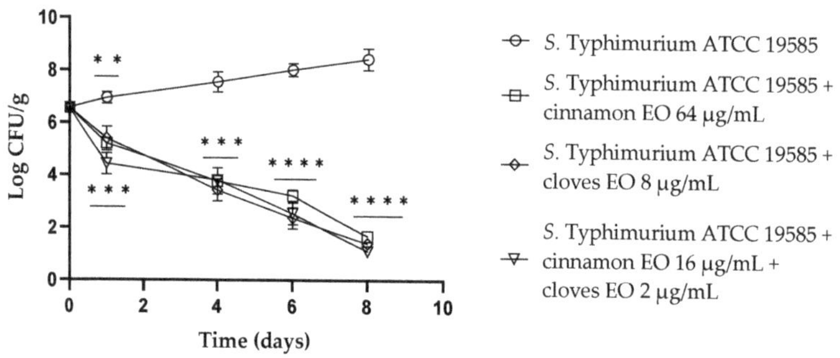

Figure 4. *Salmonella* Typhimurium ATCC 19585 viable counts (log CFU/g) observed in the contaminated fresh-cut fruits. p-values of <0.01 (**), $p < 0.001$ (***) and $p < 0.0001$ (****) were considered significant by t-test and ANOVA with Bonferroni correction. ns stands for not statistically significant. Results are expressed as mean ± SD of the three determinations (error bar = S.D.; $n = 3$).

Figure 5 shows the anti-listeria activity of cinnamon and clove EOs and of the EO/EO combination throughout the experiment time (8 days). The single EOs and the EO/EO combination displayed the same activity after 24 h ($p < 0.01$), and after 4 days, the EO/EO combination showed the best activity with a reduction of about 4.5 log CFU/g compared to the control ($p = 0.0002$). From day 6 until the end of the experiment, all samples (EOs alone and the EO/EO combination) demonstrated remarkable activity against *L. monocytogenes* compared to the control ($p < 0.0001$).

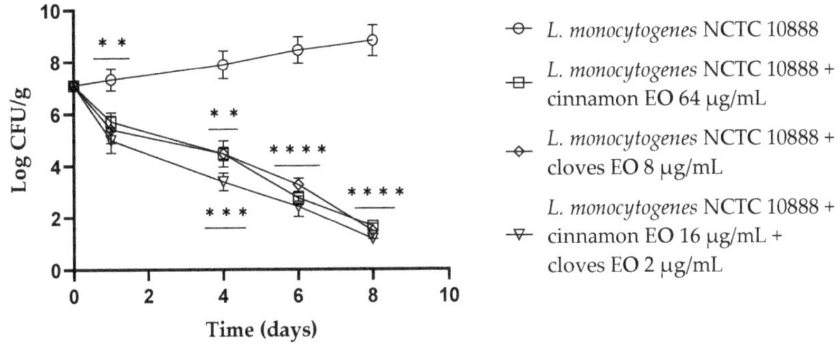

Figure 5. *Listeria monocytogenes* NCTC 10888 viable counts (log CFU/g) observed in the contaminated fresh-cut fruits. *p*-values of <0.01 (**), $p < 0.001$ (***) and $p < 0.0001$ (****) were considered significant by *t*-test and ANOVA with Bonferroni correction. ns stands for not statistically significant. Results are expressed as mean ± SD of the three determinations (error bar = S.D.; $n = 3$).

Concerning *S. aureus*, both the single EOs showed a reduction in viable cells of about 1.50 log CFU/g ($p = 0.006$) after 24 h of the experiment, and the EO/EO combination displayed a reduction of 2.5 log CFU/g ($p = 0.0011$). After 4 days, clove EO demonstrated the best results ($p = 0.0002$) compared to the control with a reduction of about 4 log CFU/g, and the cinnamon EO and the EO/EO combination showed a reduction of 3.7 log CFU/g ($p = 0.0005$). Clove EO and the EO/EO combination showed the best results with a reduction of about 5.5 log CFU/g ($p < 0.0001$) viable *S. aureus* cells, even after 6 days of the experiment. At the end of the trial (8 days), the EO/EO combination presented the best result with a 7.6 log CFU/g reduction of *S. aureus* viable cells (Figure 6).

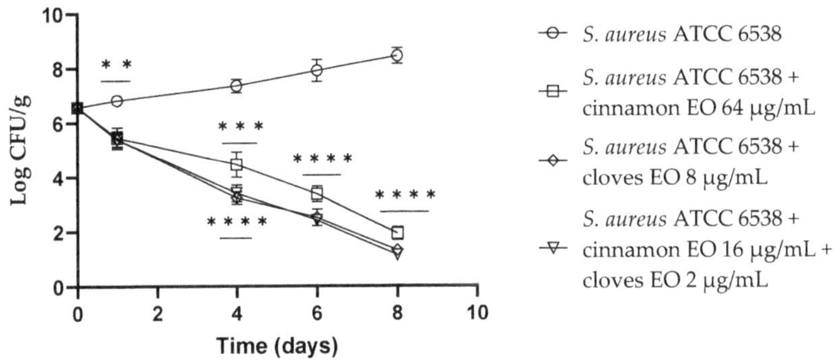

Figure 6. *Staphylococcus aureus* viable counts (log CFU/g) observed in the contaminated fresh-cut fruits. *p*-values of <0.01 (**), $p < 0.001$ (***) and $p < 0.0001$ (****) were considered significant by *t*-test and ANOVA with Bonferroni correction. ns stands for not statistically significant. Results are expressed as mean ± SD of the three determinations (error bar = S.D.; $n = 3$).

3. Discussion

Today's busy lifestyles negatively affect eating habits, and the limited time to prepare meals, even simple ones, creates the necessity to purchase ready-to-eat produce that constitutes a suitable meal, as it does not require extra preparation, is easy to consume and represents a valid alternative to maintain a balanced diet [30]. Natural barriers such as skin and rinds as well as a naturally acidic pH prevent or retard the growth of pathogenic bacteria in fruits. However, some fully matured fruits have pH values approaching 7, and once cut, they expose the internal flesh to environmental contaminants and can serve as substrates for the growth of bacteria [14]. Cantaloupe and watermelon are among these fruits and are not exempt from the list of foods known to be vehicles of food-borne illness [31]. A critical aspect of RTE foods like this is microbial contamination, and the primary objective of the manufacturer must be not only to obtain an increasingly prolonged shelf life but also to guarantee the safety of the product. In the present investigation, the EOs showed antimicrobial efficacy against all the food-borne pathogens, both alone and in combination. Clove EO has already shown its effective antimicrobial efficacy against some food-borne pathogens [32]. The main active component of clove EO is eugenol [33,34], which can deteriorate the cell wall and determine cell lysis [35]. *Trans*-cinnamaldehyde is the main compound in cinnamon EO [36]. Cinnamaldehyde has antimicrobial effects, as it inhibits cell wall biosynthesis, membrane function and certain enzymatic activities [37]. Also, in the present investigation, clove and cinnamon EOs were found to be abundantly effective against both Gram-negative and Gram-positive strains. Clove EO and cinnamon EO showed the same anti-listeria activity throughout the experiment. The single EOs displayed the same activity against *Y. enterocolitica* and *S. typhimurium*, although clove was more effective than cinnamon at 4 and 6 days of experimentation for *Yersinia* and at 6 days for *Salmonella*. Concerning *E. coli* and *S. aureus*, clove EO exhibited better activity than cinnamon EO at 4 and 6 days of experimentation. Clove EO reduced viable cells of *E. coli* by 5 log and 6 log at 4 and 6 days, whereas cinnamon EO showed a reduction of 4 log and 5 log, respectively. Other authors have demonstrated the better activity of clove EO compared to other essential oils such as cinnamon, cardamom and oregano against Gram-negative and Gram-positive bacteria such as *S. typhimurium*, *E. coli*, *Bacillus cereus* and *Listeria innocua* [38,39]. Furthermore, Liang et al. [40] demonstrated that clove EO has better antimicrobial activities against spoilage microorganisms in apple cider than those of other spices tested. Other studies have also highlighted the antibacterial activity of cinnamon EO in food-borne and spoilage bacteria such as *B. cereus*, *E. coli*, *Salmonella enterica*, *Y. enterocolitica*, *L. monocytogenes*, *S. aureus* and *P. aeruginosa* [41,42]. Cinnamon EO displays prominent activity against fungi that are more sensitive compared to bacteria [26,43]. Moreover, cinnamon EO exhibits antibacterial and anti-biofilm activity against methicillin-resistant *S. aureus* (MRSA) [44]. *S. aureus* can acquire antimicrobial resistance, complicating the problem of food-borne illnesses. Furthermore, ready-to-eat (RTE) foods, which do not require heat treatment before consumption, represent a vehicle for the spread of antibiotic-resistant *S. aureus* and of heat-resistant staphylococcal enterotoxins [45]. The presence of MRSA strains in RTE fresh produce may pose a further threat to public health, and therefore, the use of EOs as natural disinfectants can improve the safety of these foods. Last, regarding the effectiveness of the EO mixture, a synergistic effect against all tested strains was observed. The EO/EO combination led to a significant cell viable count decrease in all tested strains, with a reduction of 2.5 log cfu/g (p = 0.0007) in artificially contaminated RTE fruits with *S. typhimurium* ATCC 19585 after 24 h of exposure. This increased activity was maintained throughout the experiment, with a decrease in viable cells of the tested bacteria of at least 7 logs after 8 days. Regarding *E. coli* and *L. monocytogenes*, the decrease in viable cells was 7.7 log at the end of the experiment. The synergy between cinnamon EO/clove EO against food-borne pathogens and spoilage bacteria (*S. aureus*, *L. monocytogenes*, *S. typhimurium* and *P. aeruginosa*) was also reported by Purkait et al. [46]. The EO/EO combination led to an increase in the antibacterial effect, with a decrease in their employed concentrations. An important aspect to evaluate, before any use of essential oils as preservatives, is the alteration of flavor. The olfactory contribution that even small amounts of these compounds bring to food could be remarkable, and not all consumers are sure to like it. Such a negative aspect can be overcome by combining two essential oils, as also revealed in the present investigation. In our "on food" tests, we mixed essential oils derived from different plants, and the encouraging results obtained support the possibility of their use in fresh-cut fruit, similar to what happens for vegetables, where these natural additives are already used as flavors. Even a careful selection of EOs, based on food taste characteristics, could contribute to obtaining a minimal impact on the organoleptic properties of the product. Last, seeing the rapid emergence of drug-resistant pathogens, the antimicrobial efficacy of EOs could be exploited in depth. The emergence of antibiotic-resistant pathogens in the food chain is considered a cross-sectoral problem, and fresh-cut fruits could also represent a favorable environment for exchange through the conjugation mechanism of genes responsible for antibiotic resistance and a vehicle of difficult-to-treat infections. As revealed in other investigations, EOs are natural substances capable of positively modulating the sensitivity of antibiotic-resistant pathogens [47], even when they are organized in biofilms [48]. Thus, from today's perspective of One Health approaches to infectious diseases, EOs also represent a valid alternative as a means of dealing with the problem of antibiotic-resistant pathogens.

4. Materials and Methods

4.1. Microbial Strains and Essential Oils

Five food pathogen classified strains were used in this study, including *Yersinia enterocolitica* ATCC 23715, *Escherichia coli* ATCC 25922, *Salmonella typhimurium* ATCC 19585, *Listeria monocytogenes* NCTC 10888 and *Staphylococcus aureus* ATCC 6538.

All strains were confirmed by matrix-assisted laser desorption ionization (MALDI) time-of-flight mass spectrometry (TOF/MS) and maintained in Tryptic Soy Broth (TSB, Oxoid S.p.A, Milan, Italy) supplemented with 20% (*vol./vol.*) glycerine at −80 °C until use.

Cinnamon (*Cinnamomum zeylanicum*) and clove (*Syzygium aromaticum* L. Merr. & Perry) essential oils (EOs) were purchased from a local herbalist shop. EOs were stored at a low temperature (4 °C) and protected from light and humidity until use.

4.2. Chemical Characterization of the EOs

4.2.1. Gas Chromatography–Mass Spectrometry (GC–MS) Analysis

Analyses were performed on a 7890A gas chromatograph coupled with a 5975C network mass spectrometer (GC–MS) (Agilent Technologies, Milan, Italy). Compounds were separated on an Agilent Technologies HP-5 MS cross-linked poly-5% diphenyl–95% dimethyl polysiloxane (30 m × 0.25 mm i.d., 0.25 μm film thickness) capillary column. The column temperature was initially set at 45 °C, increased at a rate of 2 °C/min up to 100 °C, raised to 250 °C at a rate of 5 °C/min and finally held for 5 min. The injection volume was 0.1 μL with a split ratio of 1:20. Helium was used as the carrier gas at a flow rate of 0.7 mL/min. The injector, transfer line and ion source temperatures were 250, 280 and 230 °C, respectively. MS detection was performed with electron ionization at 70 eV, operating in the full-scan acquisition mode in the *m/z* range 40–400. The EOs were diluted 1:20 (*v*/*v*) with *n*-hexane before GC–MS analysis.

4.2.2. Gas Chromatography–Flame Ionization Detection (GC–FID) Analysis

Chromatographic characterization of EOs was performed on a 7820 gas chromatograph (Agilent Technologies, Milan, Italy) with a flame ionization detector (FID). EOs and the mixture of aliphatic hydrocarbons (C_8–C_{40}) were diluted 1:20 (*v*/*v*) with *n*-hexane before GC–FID analysis. Helium was used as a carrier gas at a flow rate of 1 mL/min. The injector and detector temperatures were set at 250 and 300 °C, respectively. EO components were separated on an Agilent Technologies HP-5 crosslinked poly-5% diphenyl–95% dimethylsiloxane (30 m × 0.32 mm i.d., 0.25 μm film thickness) capillary column. The column temperature was initially set at 45 °C, increased at a rate of 2 °C/min up to 100 °C, raised to 250 °C at a rate of 5 °C/min and finally maintained for 5 min. The injection volume was 1 μL with a split ratio of 1:20.

Compounds were identified by comparing the retention times of the chromatographic peaks with those of authentic reference standards run under the same conditions and by comparing the linear retention indices (*LRIs*) relative to C_8–C_{40} n-alkanes obtained on the HP-5 column under the above-mentioned conditions according to the literature [49]. Peak enrichment by co-injection with authentic reference compounds was also carried out. A comparison of the MS fragmentation pattern of the target analytes with those of pure components was performed by using the National Institute of Standards and Technology (NIST version 2.0d, 2005) mass spectral database.

The percentage of the relative number of individual components was expressed as the percent peak area relative to the total peak area obtained by the GC–FID analysis. Semi-quantitative data were acquired from the mean of two analyses.

Data acquisition and processing were performed using OpenLab CDS C.01.04 (Agilent Technologies, Santa Clara, CA, USA) software.

4.3. Agar Disk Diffusion Assay

The preliminary determination of the antibacterial activity of cinnamon and clove EOs against all tested strains was carried out by using the agar disk diffusion assay according to the standard procedure of the Clinical and Laboratory Standards Institute [50]. Sterile disks of 6 mm in diameter containing 10 μL of each EO were placed on Tryptic Soy Agar (TSA, Oxoid S.p.A, Milan, Italy) plates previously seeded with 100 μL of 10^6 CFU/mL of each cell suspension. After incubation at 37 °C for 24 h, the antagonistic activity of the EOs was quantified by a clear zone of inhibition of the bacterial growth around the disks, and the diameters in millimeters of these zones were measured [51].

4.4. Minimum Inhibitory Concentration (MIC)

According to the Clinical Laboratory Standards Institute (CLSI) guidelines (2019) [52], the MIC of EOs was determined against all microorganisms using the broth microdilution method in 96-well microplates. Briefly, in each well of a sterile 96-well microplate, 95 μL of Tryptic Soy Broth (TSB, Oxoid S.p.A, Milan, Italy) and 5 μL of bacterial suspension were added to a final inoculum concentration of 10^6 CFU/mL. Then, 100 μL of EO serial dilutions was added to obtain concentrations ranging from 512 to 0.125 μg/mL. Negative control wells consisted of bacteria in TSB without EOs. The plates were incubated at 37 °C for 24 h on a plate shaker at 150 rpm. The MIC was defined as the lowest concentration of EOs that inhibited visible growth of the tested microorganisms when the optical density (OD) was measured at 570 nm using a microtiter plate reader. All experiments were conducted in triplicate, and the results were expressed as the arithmetic mean of the three determinations.

4.5. Determination of the Fractional Inhibitory Concentration Index (FIC Index)

The checkerboard method [53] was carried out to check the synergistic antibacterial activity of the combined EO/EO by using the microdilution method in the same way as previously described for the MIC evaluation. The FIC index value was calculated by comparing the value of the MIC of each EO alone with the combination-derived MIC. An FIC index value of $\leq$0.5 reveals synergism, $\leq$0.5 to $\geq$1 reveals an additive effect, 1 to 4 reveals indifference, and >4 reveals antagonism.

4.6. Growth Kinetics Study

The growth of all test strains was determined in the presence of single EOs and the EO/EO (cinnamon–clove) combination added at the MIC and FIC index values. In a 96-well sterile microplate, 100 μL of sterile nutrient broth was mixed with 50 μL of single EOs or the EO/EO combination and with 50 μL of each microbial strain from a stock previously diluted to obtain a bacteria cell density of about 10^5 CFU/mL. Measurements were obtained in an automatic micro plate reader (Tecan Sunrise™) at an optical density (OD) of 595 nm with orbital shaking at 150 rpm for a total of 26 h at 1-h intervals. The experiments were replicated three times, and the results were expressed as the arithmetic mean of the three determinations.

4.7. Evaluation of the Antibacterial Activity by "On Food" Studies

The antibacterial activity of the EOs, both individually and in combination, was assessed in individually packaged samples of fresh-cut fruits (a mixed pool of watermelons, pineapples, strawberries and peaches) purchased from a local supermarket on the first day of their shelf life (indicated by the expiration date on the package). Before starting the study, the microbial contamination of the fresh-cut fruits was determined. On average, the samples exhibited a microbial load of 18 CFU/g. The presence of bacteria used in this study was excluded.

The samples were artificially contaminated by a 10 min immersion in 50 mL of each bacterial suspension (10^8 CFU/mL) in sterile saline solution (NaCl 0.85%). After drying, the contaminated samples were submerged for 10 min in 20 mL of a solution (sterile water and sucrose 1%) of single EOs and the EO/EO combination (EO/EO), added to the values of the MIC and FIC index. Contaminated samples without immersion in a solution containing EOs were used as controls.

Subsequently, portions of fresh-cut fruit were packed in 10 g servings and stored at refrigeration temperature (4 °C). At regular intervals (0 h, 24 h, 4 days, 6 days and 8 days), the viable load of each strain was determined by direct counting in selective plates. Individual portions (10 g) were opened and collected in sterile plastic bags, supplemented with 90 mL of Buffered Peptone Water (Oxoid, Milan, Italy) and homogenized for 1 min in Stomacher (Stomacher Lab Blender, Seward Medical, London, UK). Serial tenfold dilutions of the obtained suspensions were spread in triplicate on selective plates and incubated aerobically at 37 °C for 48 h. Viable cells of tested bacteria were enumerated, and the results were expressed as CFU/g. The experiments were replicated three times, and the results were expressed as the arithmetic mean of the three determinations.

4.8. Statistical Analysis

The statistical significance was determined by a *t*-test and an ANOVA using statistical program GraphPad Prism 9.2.0. (San Diego, CA, USA). The analysis was followed by Bonferroni's post hoc test. The statistical analysis of kinetic data was performed following the "GraphPad guide to comparing dose–response or kinetic curves" [54]. For each kinetic curve obtained in this study, the Area Under the Curve (AUC) was calculated to summarize the curve into a single value. The statistical analysis was performed on the AUC values of each experimental group using an unpaired *t*-test. The *p*-values were declared significant at ≤ 0.05. To verify the reproducibility of the results, each experiment was replicated three times under the same conditions.

5. Conclusions

The results of the present study provide encouraging information concerning the effects of two natural antimicrobial agents on the safety of minimally processed fruits. The results also highlight the importance of the synergistic effect of the EO/EO combination based on the FIC index, time–kill assay and "on food" studies. This synergistic effect is shown by the efficacy in inhibiting the growth and survival of pathogenic bacteria at low concentrations, even when applied in fresh-cut produce. Thus, the application of this EO mixture in fruit produce would also be acceptable to consumers not only for its microbial safety obtained with the addition of natural preservatives like EOs but also for its organoleptic characteristics, which are not altered by the low concentrations of use. These encouraging data can suggest the EO mixture as a new antimicrobial strategy for the correct preservation of perishable products like fresh-cut fruits.

To confirm the effectiveness of the low EO concentrations of the synergistic mixture, further studies will be performed on its ability to damage the structural integrity of the cell membrane of the pathogenic strains, both in planktonic and in biofilm states. Studies on the toxicity, safety and interaction of EOs at the cellular level still need an in-depth investigation.

Supplementary Materials: The following supporting information can be downloaded at https://www.mdpi.com/article/10.3390/antibiotics13040319/s1: Figure S1: Time–kill studies of cinnamon and clove EOs alone and in combination against *Escherichia coli* ATCC 25922 viable cells. *p*-values of <0.05 (*), $p < 0.01$ (**), $p < 0.001$ (***) and $p < 0.0001$ (****) were considered significant by *t*-test and

ANOVA with Bonferroni correction. Results are expressed as mean ± SD of the three determinations (error bar = S.D.; $n = 3$). Figure S2: Time–kill studies of cinnamon and clove EOs alone and in combination against *Yersinia enterocolitica* ATCC 23715 viable cells. p-values of <0.05 (*), $p < 0.01$ (**), $p < 0.001$ (***) and $p < 0.0001$ (****) were considered significant by t-test and ANOVA with Bonferroni correction. Results are expressed as mean ± SD of the three determinations (error bar = S.D.; $n = 3$). Figure S3: Time–kill studies of cinnamon and clove EOs alone and in combination against *Salmonella* Typhimurium ATCC 19585 viable cells. p-values of <0.05 (*), $p < 0.01$ (**), $p < 0.001$ (***) and $p < 0.0001$ (****) were considered significant by t-test and ANOVA with Bonferroni correction. Results are expressed as mean ± SD of the three determinations (error bar = S.D.; $n = 3$). Figure S4: Time–kill studies of cinnamon and clove EOs alone and in combination against *Staphylococcus aureus* ATCC 6538 viable cells. p-values of <0.05 (*), $p < 0.01$ (**), $p < 0.001$ (***) and $p < 0.0001$ (****) were considered significant by t-test and ANOVA with Bonferroni correction. Results are expressed as mean ± SD of the three determinations (error bar = S.D.; $n = 3$).

Author Contributions: Conceptualization, R.I. and P.M.; methodology, C.S., E.T. and R.I.; validation, P.M. and C.S.; investigation, R.I., E.T. and P.M.; resources, P.M.; data curation, R.I. and C.S.; writing—original draft preparation, R.I., E.T. and P.M.; writing—review and editing, P.M. and R.I.; visualization, R.I. and C.S.; supervision, P.M.; project administration, P.M. and RI. All authors have read and agreed to the published version of the manuscript.

Funding: This research received no external funding.

Institutional Review Board Statement: Not applicable.

Conflicts of Interest: The authors declare no conflicts of interest.

References

1. Ramos, B.; Miller, F.; Brandão, T.R.; Teixeira, P.; Silva, C.L. Fresh fruits and vegetables—An overview on applied methodologies to improve its quality and safety. *Innov. Food Sci. Emerg. Technol.* **2013**, *20*, 1–15. [CrossRef]
2. World Health Organization (WHO). *Diet, Nutrition and the Prevention of Chronic Diseases*; Report of a joint WHO/FAO; WHO: Geneva, Switzerland, 2002; WHO Technical Report Series No 916. Available online: https://apps.who.int/iris/bitstream/handle/10665/42665/WHO_TRS_916.pdf (accessed on 11 May 2019).
3. Santos, M.I.; Grácio, M.; Silva, M.C.; Pedroso, L.; Lima, A. One health perspectives on food safety in minimally processed vegetables and fruits: From farm to fork. *Microorganisms* **2023**, *11*, 2990. [CrossRef] [PubMed]
4. Balali, G.I.; Yar, D.D.; Afua Dela, V.G.; Adjei-Kusi, P. Microbial contamination, an increasing threat to the consumption of fresh fruits and vegetables in today's world. *Int. J. Microbiol.* **2020**, *2020*, 3029295. [CrossRef] [PubMed]
5. Carstens, C.K.; Salazar, J.K.; Darkoh, C. Multistate outbreaks of foodborne illness in the United States associated with fresh produce from 2010 to 2017. *Front. Microbiol.* **2019**, *10*, 2667. [CrossRef] [PubMed]
6. Wadamori, Y.; Gooneratne, R.; Hussain, M.A. Outbreaks and factors influencing microbiological contamination of fresh produce. *J. Sci. Food Agric.* **2017**, *97*, 1396–1403. [CrossRef] [PubMed]
7. Melo, J.; Quintas, C. Minimally processed fruits as vehicles for foodborne pathogens. *AIMS Microbiol.* **2023**, *9*, 1–19. [CrossRef] [PubMed]
8. Machado-Moreira, B.; Richards, K.; Brennan, F.; Abram, F.; Burgess, C.M. Microbial contamination of freshp: What, where, and how? *Compr. Rev. Food Sci. Food Saf.* **2019**, *18*, 1727–1750. [CrossRef] [PubMed]
9. Luciano, W.A.; Griffin, S.; Targino de Souza Pedrosa, G.; Alvarenga, V.; Valdramidis, V.; Magnani, M. Growth behavior of low populations of *Listeria monocytogenes* on fresh-cut mango, melon and papaya under different storage temperatures. *Food Microbiol.* **2022**, *102*, 103930. [CrossRef] [PubMed]
10. Centers for Disease Control and Prevention (CDC). Multistate Outbreak of Listeriosis Linked to Whole Cantaloupes from Jensen farms, Colorado. 2011. Available online: https://www.cdc.gov/listeria/outbreaks/cantaloupes-jensen-farms/index.html (accessed on 27 August 2012).
11. Centers for Disease Control and Prevention (CDC). Multistate Outbreak of Listeriosis Linked to Commercially Produced, Prepackaged Caramel Apples. 2015. Available online: https://www.cdc.gov/listeria/outbreaks/caramel-apples-12-14/index.html (accessed on 12 February 2015).
12. Food and Drug Administration (FDA). Outbreak Investigation of Listeria monocytogenes: Peaches, Plums, & Nectarines. 2023. Available online: https://www.fda.gov/food/outbreaks-foodborne-illness/outbreak-investigation-listeria-monocytogenes-peaches-plums-nectarines-november-2023 (accessed on 30 January 2024).
13. Harris, L.J.; Farber, J.N.; Beuchat, L.R.; Parish, M.E.; Suslow, T.V.; Garrett, E.H.; Busta, F.F. Outbreaks associated with fresh produce: Incidence, growth, and survival of pathogens in fresh and fresh-cut produce. *Compr. Rev. Food Sci. Food Saf.* **2003**, *2* (Suppl. S1), 78–141. [CrossRef]

14. Del Rosario, B.A.; Beuchat, L.R. Survival and growth of enterohemorrhagic *Escherichia coli* O157:H7 in cantaloupe and watermelon. *J. Food Prot.* **1995**, *58*, 105–107. [CrossRef]
15. Castillo, A.; Martínez-Téllez, M.A.; Rodríguez-García, M.O. *Produce Contamination Problem: Causes and Solutions*; Melons (Chapter 10); Matthews, K.R., Sapers, G.M., Doyle, M.P., Eds.; Academic Press: New York, NY, USA, 2014; pp. 207–236.
16. Food and Drug Administration (FDA). Outbreak Investigation of *Salmonella enteritidis*: Peaches. 2020. Available online: https://www.fda.gov/food/outbreaks-foodborne-illness/outbreak-investigation-salmonella-enteritidis-peaches-august-2020 (accessed on 6 June 2021).
17. Centers for Disease Control and Prevention (CDC). Salmonella Outbreak Linked to Cantaloupes. 2024. Available online: https://www.cdc.gov/salmonella/sundsvall-11-23/index.html (accessed on 19 January 2024).
18. Byrne, L.; Fisher, I.; Peters, T.; Mather, A.; Thomson, N.; Rosner, B.; Bernard, H.; McKeown, P.; Cormican, M.; Cowden, J.; et al. International outbreak control team. A multi-country outbreak of *Salmonella* Newport gastroenteritis in Europe associated with watermelon from Brazil, confirmed by whole genome sequencing: October 2011 to January 2012. *Eurosurveillance* **2014**, *19*, 6–13. [CrossRef] [PubMed]
19. Bari, M.L.; Hossain, M.A.; Isshiki, K.; Ukuku, D. Behavior of *Yersinia enterocolitica* in foods. *J. Pathog.* **2011**, *2011*, 420732. [CrossRef]
20. Quaglio, P.; Aggazzotti, G.; Fabio, A.; Messi, P.; Mantovani, G. Presenza di batteri appartenenti al genere *Yersinia* in acque superficiali e di falda Presence of bacteria belonging to the genus *Yersinia* in surface and ground water. *Ann. Ig.* **1989**, *1*, 157–164.
21. European Food Safety Authority; European Centre for Disease Prevention and Control. European Food Safety Authority and European Centre for Disease Prevention and Control. The European Union One Health 2021 Zoonoses Report. *EFSA J.* **2022**, *20*, 7666. [CrossRef]
22. Xu, F.; Liu, S.; Liu, Y.; Wang, S. Effect of mechanical vibration on postharvest quality and volatile compounds of blueberry fruit. *Food Chem.* **2021**, *349*, 129216. [CrossRef]
23. Iseppi, R.; Camellini, S.; Sabia, C.; Messi, P. Combined antimicrobial use of essential oils and bacteriocin bacLP17 as seafood biopreservative to control *Listeria monocytogenes* both in planktonic and in sessile forms. *Res. Microbiol.* **2020**, *171*, 351–356. [CrossRef] [PubMed]
24. Falleh, H.; Ben Jemaa, M.; Saada, M.; Ksouri, R. Essential oils: A promising eco-friendly food preservative. *Food Chem.* **2020**, *330*, 127268. [CrossRef] [PubMed]
25. Zhang, L.; Chen, F.; Lai, S.; Wang, H.; Yang, H. Impact of soybean protein isolate-chitosan edible coating on the softening of apricot fruit during storage. *LWT* **2018**, *96*, 604–611. [CrossRef]
26. Liu, Q.; Meng, X.; Li, Y.; Zhao, C.N.; Tang, G.Y.; Li, H.B. Antibacterial and antifungal activities of spices. *Int. J. Mol. Sci.* **2017**, *18*, 1283. [CrossRef]
27. Mohamed, S.H.; Zaky, W.M.; Kassem, J.M.; Abbas, H.M.; Salem, M.M.E.; Said-Al Ahl, H.A.H. Impact of antimicrobial properties of some essential oils on cheese yoghurt quality. *World Appl. Sci. J.* **2013**, *27*, 497–507.
28. Hyldgaard, M.; Mygind, T.; Meyer, R.L. Essential oils in food preservation: Mode of action, synergies, and interactions with food matrix components. *Front. Microbiol.* **2012**, *3*, 12. [CrossRef]
29. National Institute of Standards and Technology Library Database. Available online: https://webbook.nist.gov/ (accessed on 11 March 2024).
30. Jabs, J.; Devine, C.M. Time scarcity and food choices: An overview. *Appetite* **2006**, *47*, 196–204. [CrossRef]
31. Walsh, K.A.; Bennett, S.D.; Mahovic, M.; Gould, L.H. Outbreaks associated with cantaloupe, watermelon, and honeydew in the United States, 1973–2011. *Foodborne Pathog. Dis.* **2014**, *11*, 945–952. [CrossRef] [PubMed]
32. Xu, J.G.; Liu, T.; Hu, Q.P.; Cao, X.M. Chemical composition, antibacterial properties and mechanism of action of essential oil from clove buds against *Staphylococcus aureus*. *Molecules* **2016**, *21*, 1194. [CrossRef]
33. Chaieb, K.; Hajlaoui, H.; Zmantar, T.; Kahla-Nakbi, A.B.; Rouabhia, M.; Mahdouani, K.; Bakhrouf, A. The chemical composition and biological activity of clove essential oil, *Eugenia caryophyllata* (*Syzigium aromaticum* L. Myrtaceae): A short review. *Phytother. Res.* **2007**, *21*, 501–506. [CrossRef]
34. Lomarat, P.; Phanthong, P.; Wongsariya, K.; Chomnawang, M.T.; Bunyapraphatsara, N. Bioautography-guided isolation of antibacterial compounds of essential oils from Thai spices against histamine-producing bacteria. *Pak. J. Pharm. Sci.* **2013**, *26*, 473–477. [PubMed]
35. Burt, S. Essential oils: Their antibacterial properties and potential applications in foods—A review. *Int. J. Food Microbiol.* **2004**, *94*, 223–253. [CrossRef] [PubMed]
36. Rao, P.V.; Gan, S.H. Cinnamon: A multifaceted medicinal plant. *Evid. Based Complement. Alternat. Med.* **2014**, *2014*, 642942. [CrossRef]
37. Shreaz, S.; Wani, W.A.; Behbehani, J.M.; Raja, V.; Irshad, M.; Karched, M.; Ali, I.; Siddiqi, W.A.; Hun, L.T. Cinnamaldehyde and its derivatives, a novel class of antifungal agents. *Fitoterapia* **2016**, *112*, 116–131. [CrossRef]
38. Badei, A.; Faheld, S.; El-Akel, A.; Mahmoud, B. Application of some spices in flavoring and preservation of cookies: 2-antimicrobial and sensory properties of cardamom, cinnamon and clove. *Dtsch. Lebensm. Rundsch.* **2002**, *98*, 261–265.
39. Angienda, P.O.; Onyango, D.M.; Hill, D.J. Potential application of plant essential oils at sub-lethal concentrations under extrinsic conditions that enhance their antimicrobial effectiveness against pathogenic bacteria. *Afr. J. Microbiol. Res.* **2010**, *4*, 1678–1684.
40. Liang, Z.W.; Cheng, Z.H.; Mittal, G.S. Inactivation of microorganisms in apple cider using spice powders, extracts and oils as antimicrobials with and without low-energy pulsed electric field. *J. Food Agric. Environ.* **2003**, *1*, 28–33.

41. Nanasombat, S.; Wimuttigosol, P. Antimicrobial and antioxidant activity of spice essential oils. *Food Sci. Biotechnol.* **2011**, *20*, 45–53. [CrossRef]
42. Trajano, V.N.; Lima, E.D.; de Souza, E.L.; Travassos, A. Antibacterial property of spice essential oils on food contaminating bacteria. *Ciencia Tecnol. Aliment.* **2009**, *29*, 542–545. [CrossRef]
43. Mvuemba, H.N.; Green, S.E.; Tsoomo, A.; Avis, T.J. Antimicrobial efficacy of cinnamon, ginger, horseradish and nutmeg extracts against spoilage pathogens. *Phytoprotection* **2009**, *90*, 65–70. [CrossRef]
44. Cui, H.; Li, W.; Li, C.; Vittayapadung, S.; Lin, L. Liposome containing cinnamon oil with antibacterial activity against methicillin-resistant *Staphylococcus aureus* biofilm. *Biofouling* **2016**, *32*, 215–225. [CrossRef] [PubMed]
45. Mikhaylova, Y.; Shelenkov, A.; Chernyshkov, A.; Tyumentseva, M.; Saenko, S.; Egorova, A.; Manzeniuk, I.; Akimkin, V. Whole-genome analysis of *Staphylococcus aureus* isolates from ready-to-eat food in Russia. *Foods* **2022**, *11*, 2574. [CrossRef] [PubMed]
46. Purkait, S.; Bhattacharya, A.; Bag, A.; Chattopadhyay, R.R. Synergistic antibacterial, antifungal and antioxidant efficacy of cinnamon and clove essential oils in combination. *Arch. Microbiol.* **2020**, *202*, 1439–1448. [CrossRef]
47. Iseppi, R.; Condò, C.; Messi, P. Synergistic inhibition of methicillin-resistant *Staphylococcus aureus* (MRSA) by *Melaleuca alternifolia* Chell (Tea Tree) and *Eucalyptus globulus* Labill. essential oils in association with oxacillin. *Antibiotics* **2023**, *12*, 846. [CrossRef]
48. Iseppi, R.; Mariani, M.; Condò, C.; Sabia, C.; Messi, P. Essential oils: A natural weapon against antibiotic-resistant bacteria responsible for nosocomial infections. *Antibiotics* **2021**, *10*, 417. [CrossRef]
49. Adams, R. *Identification of Essential Oil Components by Gas Chromatography/Mass Spectrometry*, 4th ed.; Allured Pub Corp: Carol Stream, IL, USA, 2007.
50. *CLSI document M02eA11*; Performance Standards for Antimicrobial Disk Susceptibility Tests: Approved Standard, 11th ed. Clinical and Laboratory Standards Institute: Wayne, PA, USA, 2012.
51. Klancnik, A.; Piskernik, S.; Jersek, B.; Mozina, S.S. Evaluation of diffusion and dilution methods to determine the antibacterial activity of plant extracts. *J. Microbiol. Methods.* **2010**, *81*, 121–126. [CrossRef] [PubMed]
52. CLSI. *Performance Standards for Antimicrobial Susceptibility Testing*, 29th ed.; Clinical and Laboratory Standards Institute: Wayne, PA, USA, 2019; Volume 39.
53. Hemaiswaryaa, S.; Kruthiventib, A.K.; Doblea, M. Synergism between natural products and antibiotics against infectious diseases. *Phytomedicine* **2008**, *15*, 639–652. [CrossRef] [PubMed]
54. Motulsky, H. *The GraphPad Guide to Comparing Dose-Response or Kinetic Curves*; GraphPad Software: San Diego, CA, USA, 1998.

antibiotics

MDPI

Article

Phytochemical Screening and Antibacterial Activity of Commercially Available Essential Oils Combinations with Conventional Antibiotics against Gram-Positive and Gram-Negative Bacteria

Răzvan Neagu [1,2], Violeta Popovici [3,*], Lucia-Elena Ionescu [4,*], Viorel Ordeanu [5], Andrei Biță [6], Diana Mihaela Popescu [2], Emma Adriana Ozon [7] and Cerasela Elena Gîrd [1]

1 Department of Pharmacognosy, Phytochemistry, and Phytotherapy, Faculty of Pharmacy, Carol Davila University of Medicine and Pharmacy, 6 Traian Vuia Street, 020956 Bucharest, Romania; razvan.neagu@drd.umfcd.ro (R.N.); cerasela.gird@umfcd.ro (C.E.G.)
2 Regenerative Medicine Laboratory, "Cantacuzino" National Military Medical Institute for Research and Development, 103 Spl. Independentei, 050096 Bucharest, Romania; popescu.diana@cantacuzino.ro
3 Center for Mountain Economics, "Costin C. Kirițescu" National Institute of Economic Research (INCE-CEMONT), Romanian Academy, 725700 Vatra-Dornei, Romania
4 Experimental Microbiology Laboratory, "Cantacuzino" National Military Medical Institute for Research and Development, 103 Spl. Independentei, 050096 Bucharest, Romania
5 Faculty of Pharmacy, "Titu Maiorescu" University, 16 Sincai, 040314 Bucharest, Romania; viorel.ordeanu@prof.utm.ro
6 Department of Pharmacognosy & Phytotherapy, Faculty of Pharmacy, University of Medicine and Pharmacy of Craiova, 200349 Craiova, Romania; andrei.bita@umfcv.ro
7 Department of Pharmaceutical Technology and Biopharmacy, Faculty of Pharmacy, Carol Davila University of Medicine and Pharmacy, 6 Traian Vuia Street, 020956 Bucharest, Romania; emma.budura@umfcd.ro
* Correspondence: violeta.popovici@ce-mont.ro (V.P.); ionescu.lucia@cantacuzino.ro (L.-E.I.)

Abstract: The present study aims to evaluate the antibacterial activity of five commercially available essential oils (EOs), Lavender (LEO), Clove (CEO), Oregano (OEO), Eucalyptus (EEO), and Peppermint (PEO), against the most-known MDR Gram-positive and Gram-negative bacteria—*Staphylococcus aureus* (ATCC 25923), *Escherichia coli* (ATCC 25922), and *Pseudomonas aeruginosa* (ATCC 27853)—alone and in various combinations. Gas Chromatography–Mass Spectrometry (GC-MS) analysis established their complex compositions. Then, their antibacterial activity—expressed as the inhibition zone diameter (IZD) value (mm)—was investigated in vitro by the diffusimetric antibiogram method, using sterile cellulose discs with Ø 6 mm impregnated with 10 μL of sample and sterile borosilicate glass cylinders loaded with 100 μL; the minimum inhibitory concentration (MIC) value (μg/mL) for each EO was calculated from the IZD values (mm) measured after 24 h. The following EO combinations were evaluated: OEO+CEO, CEO+EEO, CEO+PEO, LEO+EEO, and EEO+PEO. Then, the influence of each dual combination on the activity of three conventional antibacterial drugs—Neomycin (NEO), Tetracycline (TET), and Bacitracin (BAC)—was investigated. The most active EOs against *S. aureus* and *E. coli* were LEO and OEO (IZD = 40 mm). They were followed by CEO and EEO (IZD = 20–27 mm); PEO exhibited the lowest antibacterial activity (IZD = 15–20 mm). EEO alone showed the highest inhibitory activity on *P. aeruginosa* (IZD = 25–35 mm). It was followed by CEO, LEO, and EEO (IZD = 7–11 mm), while PEO proved no antibacterial action against it (IZD = 0 mm). Only one synergic action was recorded (OEO+CEO against *P. aeruginosa*); EEO+PEO revealed partial synergism against *S. aureus* and CEO+PEO showed additive behavior against *E. coli*. Two triple associations with TET showed partial synergism against *E. coli*, and the other two (with NEO and TET) evidenced the same behavior against *S. aureus*; all contained EEO+PEO or CEO+PEO. Most combinations reported indifference. However, numerous cases involved antagonism between the constituents included in the double and triple combinations, and the EOs with the strongest antibacterial activities belonged to the highest antagonistic combinations. A consistent statistical analysis supported our results, showing that the EOs with moderate antibacterial activities could generate combinations with higher inhibitory effects based on synergistic or additive interactions.

Keywords: essential oils; GC-MS; antibacterial activity; Gram-positive and Gram-negative bacteria; diffusimetric antibiogram; antibacterial drugs; combinations; interactions

1. Introduction

Currently, treating bacterial diseases is a challenge for modern medicine due to the expansion of antibiotic resistance. It is well known that numerous infections are induced by bacterial species present in many people without producing disease. Generally, these strains are non-pathogenic; however, these bacteria could become pathogenic when found in a place different from where they are commonly located or when their number significantly increases due to immunosuppression. According to the Federal Office of Public Health, the most common conditions are connected with the following bacterial species: *S. aureus*, *S. pneumoniae*, *E. coli*, *K. pneumoniae*, *A. baumani*, and *P. aeruginosa* [1]. All mentioned bacteria can quickly develop resistance to numerous conventional antibiotics [1]. Medical research focuses on finding and developing new inhibitory agents for multidrug-resistant (MDR) bacteria. Another potential strategy is to combine classical antibiotic drugs with plant-derived antimicrobials: various plant extracts, phytochemicals, and essential oils (EOs).

The EO chemical composition [2] includes phenolic compounds, terpenes, terpenoids [3], phenylpropanoids, and other aliphatic and aromatic constituents with strong lipophilic properties [4]. Their content varies depending on seasonal variations, climate, sub-species, and even the oil extraction method [5,6], which can lead to consequences for their antibacterial activity [7]. Investigating the EO antibacterial action is essential, considering the interactions between the main constituents and various mechanisms [8]. For example, OH phenolic from terpenoids is responsible for antibacterial activity and dual redox behavior. Thus, Eugenol, Linalool, and Linalyl-acetate block extracellular enzyme synthesis, inhibit histidine decarboxylase's and ATPase's enzymatic activity, and induce cell membrane depletion, losing ATP and K [9,10]. Other major constituents of various EOs—1,8-cineol and menthol derivatives (isomenthol and neomenthol)—exhibit significant antibacterial activity through bacterial surface charge alteration. They increase intracellular oxidative stress, inducing membrane disruption and bacterial damage [11,12]. Thymol induces extensive membrane destruction, inhibiting bacterial growth, decreasing intracellular ATP, cell membrane depolarization, and membrane damage [13]. Other EOs' specific constituents, terpenes (p-Cymene, α-Pinene, γ-Terpinene, and Limonene), exhibit very low or no antibacterial activity; they can increase the antibacterial potential of other compounds. For example, p-cymene has a strong affinity for bacterial membranes, which changes their potential. This could affect the membrane permeability and decrease cellular motility [14].

Depending on each EO phytochemical profile, the antibacterial effects could be due to cellular membrane disruption through protein damage, cell content leakage, motive proton force depletion, and cytoplasm coagulation [15–17]. The initial negative impact of EOs on the bacterial cell membrane influences the spectrum of bacterial metabolic processes affected by the responsible EO phytocompounds (ATPase, lipase, and coagulase activity inhibition, leakage of cell ions, citrate metabolic pathway disruption, efflux pump blockage, and fluidization of membrane lipids) [18–21]. Moreover, they can inhibit bacterial toxins production (α-toxin, enterotoxins A and B, and listeriolisin), diminishing their virulence [22,23]. Various scientists have combined different EOs to identify their synergistic associations [24] for therapeutic or food preservative purposes [20]. Moreover, based on differences in antibacterial mechanisms, EOs were coupled with conventional antibiotics [16], because most antibacterial drugs act by a cell wall and protein synthesis inhibition [25]. These combinations were tested against MDR bacterial strains to determine their usefulness. The most studied synergies imply EOs of *Melaleuca alternifolia*, *Coriandrum sativum*, *Lippia sidoides*, *Thymus maroccanus*, *Cinnamomum zeylanicum*, *Syzygium aromaticum*, *Mentha piperita*, *Origanum vulgare*, and *Rosmarinus officinalis* combined with antibiotic drugs from various classes (beta-lactams, quinolones, aminoglycozides, chloramphenicol, tetracyclin, and polypeptides) against MDR Gram-positive and Gram-negative bacteria [26].

In this context, the present study aims to investigate the antibacterial activity and potential interactions of five commercially available EOs—Lavender (LEO), Clove (CEO), Oregano (OEO), Eucalyptus (EEO), and Peppermint (PEO)—in binary combinations and triple associations with antibiotic drugs—Tetracycline (TET), Neomycin (NEO), and Bacitracin (BAC) against *S. aureus*, *E coli*, and *P. aeruginosa*. A comparative GC-MS analysis of autochthonous EOs and standard ones, rarely used antibiotic drugs, and selected EO combinations investigated through two different techniques of diffussimetric antibiogram, by using paper discs (disc diffusion method—DDM) and borosilicate glass cylinders (cylinder technique—CT), represents the novelty of our research.

2. Results

2.1. Gas Chromatography–Mass Spectrometry Analysis

The GC-MS chromatograms of the tested EOs are displayed in Figure 1. The main quantified constituents, expressed as percentages (%), are registered in Table 1. For each EO, both its sample and standard compositions are displayed. The GC-MS chromatograms of the EOs used as standards are in the Supplementary Materials, Figure S1. Moreover, all constituents identified and quantified in all EOs are in Table S1 from the Supplementary Materials.

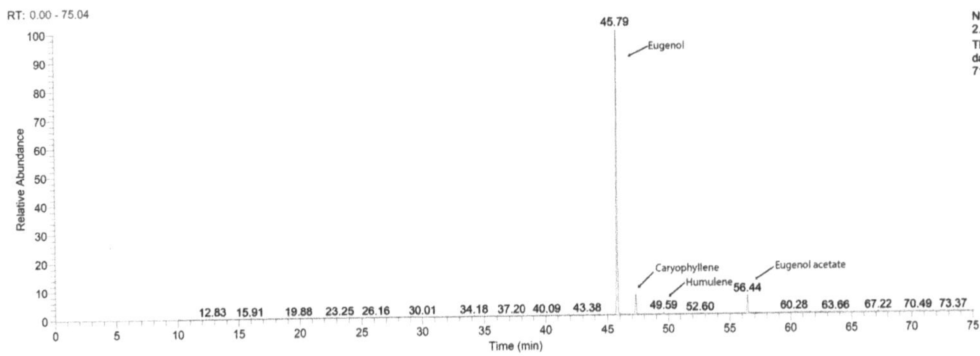

(A)

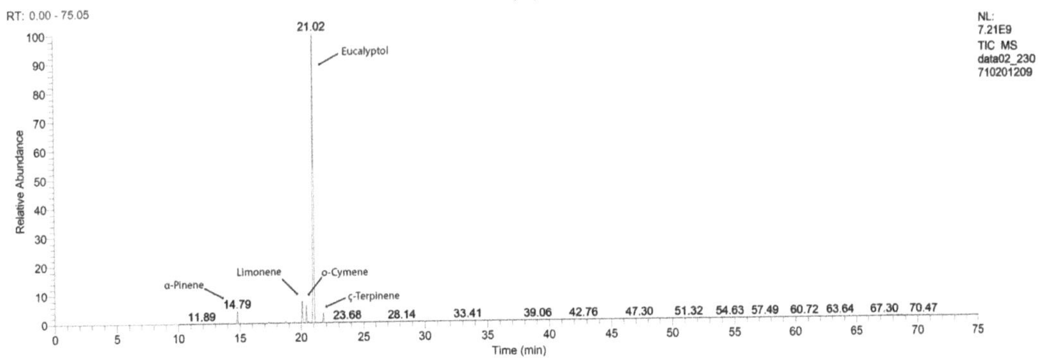

(B)

Figure 1. *Cont.*

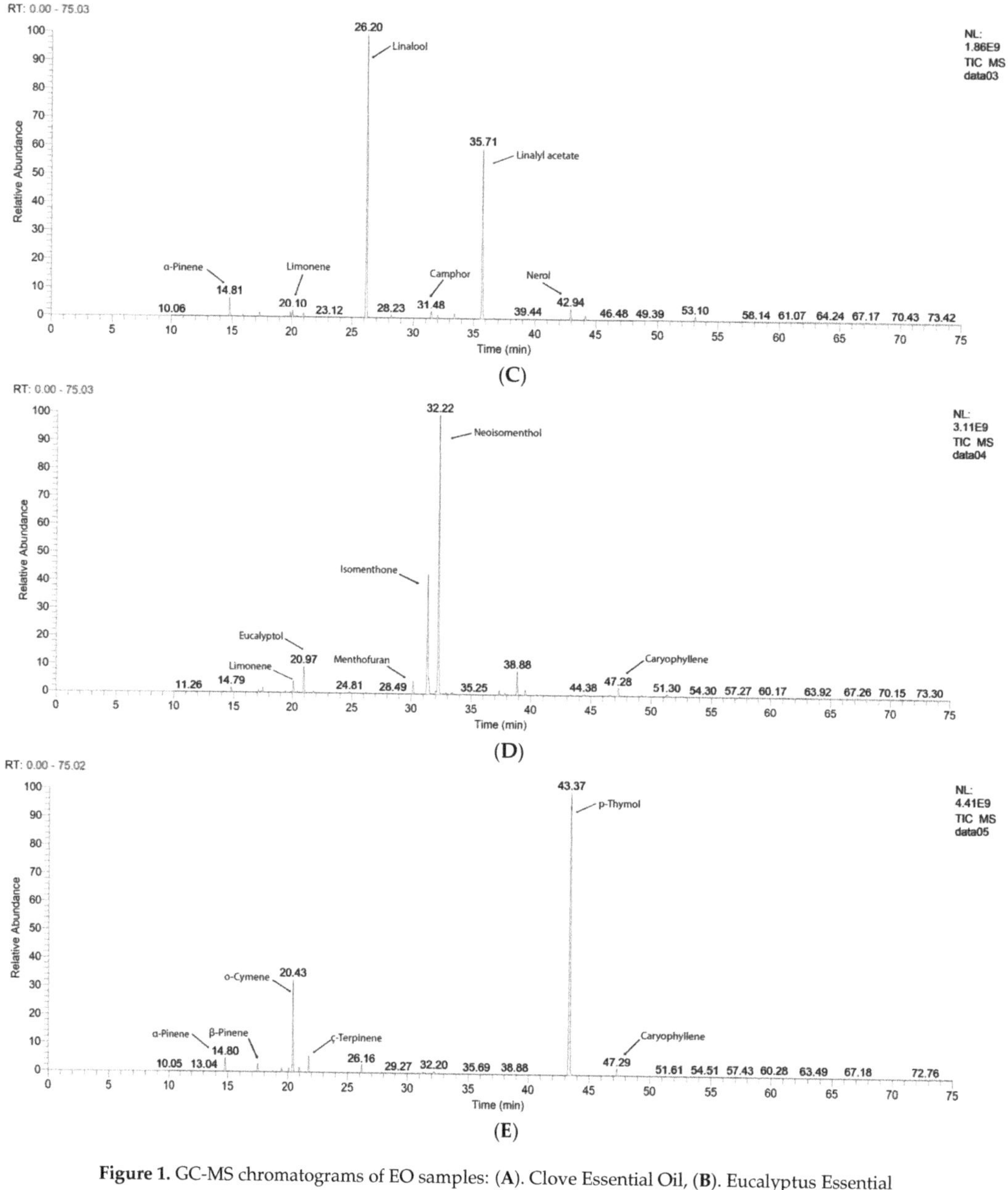

Figure 1. GC-MS chromatograms of EO samples: (**A**). Clove Essential Oil, (**B**). Eucalyptus Essential Oil, (**C**). Lavender Essential Oil, (**D**). Peppermint Essential Oil, and (**E**). Oregano Essential Oil.

The main constituents of each EO are as follows: Eugenol (86.22%), Caryophyllene (6.87%), and Eugenol acetate (5.75%) in CEO; Eucalyptol (83.74%), Limonene (5.45%), o-Cymene (4.13%), α-Pinene (2.81%), and γ-Terpinene (2.23%) in EEO; Linalool (52.93%), Linalyl acetate (32.31%), α-Pinene (3.16%), Nerol (2.11%), Camphor (1.34%), and Limonene (1.12%) in LEO; Neoisomenthol (55.09%), Isomenthone (26.52%), Eucalyptol (5.04%), Menthofuran (2.46%), Limonene (2.08%), and Caryophyllene (1.61%) in PEO; and p-Thymol

(72.08%), o-Cymene (16.26%), γ-Terpinene (2.98%), α-Pinene (2.39%), Caryophyllene (1.45%), Linalool (1.40%), and β-Pinene (1.39%) in OEO (Figure 1 and Table 1).

Table 1. The main constituents of EOs quantified by GC-MS analysis expressed as percentages and registered as a mean of 3 determinations ± SD.

RT [min]	Essential Oil	CEO	EEO	LEO	PEO	OEO
	Compound Name	**Quantity (%)**				
45.79	Eugenol	86.2272 ± 4.8956	-	-	-	-
47.29	Caryophyllene	6.8733 ± 0.4562	-	-	1.6170 ± 0.2035	1.4581 ± 0.2668
56.44	Eugenol acetate	5.7515 ± 0.6057	-	-	-	-
21.00	Eucalyptol	-	83.7478 ± 4.8364	0.7360 ± 0.1087	5.0452 ± 0.2877	0.8904 ± 0.0880
21.77	γ-Terpinene	-	2.2315 ± 0.2701	-	0.2864 ± 0.0166	2.9827 ± 0.3837
31.34	Isomenthone	-	-	-	26.5289 ± 1.5690	-
26.16	Linalool	-	0.0102 ± 0.0019	52.9348 ± 7.4084	0.0670 ± 0.0036	1.4072 ± 0.1055
31.48	Camphor	-	0.0138 ± 0.0009	1.3463 ± 0.2607	-	-
32.23	Neoisomenthol	-	-	-	55.0951 ± 7.7417	-
43.29	p-Thymol	-	-	-	0.0896 ± 0.0147	72.0862 ± 9.3764
35.69	Linalyl acetate	-	-	32.3146 ± 2.5124	-	0.0152 ± 0.0021
42.95	Nerol	-	-	2.1132 ± 0.1975	-	-
20.10	Limonene	-	5.4506 ± 1.0435	1.1219 ± 0.1812	2.0857 ± 0.2091	0.6877 ± 0.0737
20.42	o-Cymene	-	4.1323 ± 0.5402	0.2470 ± 0.0253	0.1803 ± 0.0320	16.2692 ± 1.4297
14.79	α-Pinene	-	2.8165 ± 0.2594	3.1612 ± 0.2142	0.8279 ± 0.1647	2.3971 ± 0.3353
17.46	β-Pinene	-	0.2631 ± 0.0227	0.0478 ± 0.0055	0.8802 ± 0.0767	1.3942 ± 0.2326
30.16	Menthofuran	-	-	-	2.4656 ± 0.3486	-

Eucalyptol, γ-Terpinene, Linalool, Terpinen 4-ol, Limonene, o-Cymene, α-Pinene, α-Myrcene, β-Pinene, and Camphene are common constituents in four EOs: EEO, LEO, PEO, and OEO. Caryophylene is found in CEO, PEO, and OEO. The specific compounds with the highest contents are Eugenol and Eugenyl acetate in CEO, Isomenthone, Neoisomenthone, Menthofuran, and Eucalyptol in PEO, p-Thymol, γ-Terpinene, and o-Cymene in OEO, Linalool, Linalyl-acetate, α-Pinene, and Nerol in LEO, and Eucalyptol, γ-Terpinene, Limonene, o-Cymene, and α-Pinene in EEO (Figure 1, Table S1 and Figure S1 from the Supplementary Materials.

In some EO samples, the main constituents' contents were higher than those in the standards (Eucalyptol in EEOs—83.7478 > 81.9587%; Linalool in LEOs—52.9348 > 35.8533%; Neoisomenthol in PEOs—55.0951 > 50.8339%; and Eugenol-acetate in CEOs—5.7515 > 1.5998%). Contrariwise, in other EO samples, they had lower contents than those in the standards (Eugenol and Caryophyllene in CEOs—86.2272 < 87.0884%; 6.8733 < 9.8462%; p-Thymol and γ-Terpinene in OEOs—72.0862 < 77.9238%; 2.9827 < 5.1776%, and Linalyl-acetate, o-Cymene, and α-Pinen in LEOs—32.3146 < 36.2963%, 0.2470 < 3.0292%, and 3.1612 < 5.6669%).

2.2. Antibacterial Activity on Gram-Positive and Gram-Negative Bacteria

Five binary EO combinations were performed based on the EO phytochemical profiles and our previous screening [27]. CEO's constituents differ in the highest measure from all other EOs tested (Figure 1, Table S1, and Figure S1 from the Supplementary Materials). Therefore, CEO, EEO, and PEO were combined in three pairs (CEO+EEO, CEO+PEO, and EEO+PEO); two other EOs were added in only two combinations: OEO+CEO and LEO+EEO. Each binary combination was combined with three conventional antibiotics: TET, NEO, and BAC, resulting in 15 triple combinations.

The selected antibiotic drugs belong to three different classes. Tetracycline is a broad-spectrum polypeptide antibiotic that exerts a bacteriostatic effect by reversibly binding bacterial 30S ribosomal subunit and blocking protein synthesis [28]. Neomycin is a broad-spectrum aminoglycoside [29] active against Gram-positive and Gram-negative bacteria by linking cellular ribosomes and inhibiting protein synthesis [30]. Bacitracin is a polypeptide antibacterial drug that acts against Gram-positive bacteria by inhibiting cell wall synthesis [30].

For an accurate evaluation of plant-derived products' antibacterial potential, Vorobets et al. recommend using at least two different techniques of diffusimetric antibiogram [31]. Therefore, we selected the disk diffusion method (DDM) [32] and cylinder technique (CT) [33]. The data obtained are presented synthetically in Tables 2–4.

The scale of measurement was as follows: a powerful inhibitory effect at IZD $\geq$ 35 mm, strong inhibitory effects at 35 > IZD $\geq$ 25 mm, a moderate inhibitory effect at 25 > IZD $\geq$15 mm, a mild inhibitory effect when 15 > IZD $\geq$ 10 mm, and a very low inhibitory effect at IZD < 10 mm [34]. Several similarities were observed between the IZD (inhibition zone diameter) was measured through both methods. However, these inconsistencies were due to the different diffusion levels of the samples' constituents in the culture medium, especially when they were tested in a large volume (100 μL).

2.2.1. Antibacterial Activity against *S. aureus*

S. aureus represents a part of skin flora in approximately 30% of humans without causing disease. However, *S. aureus* can induce skin, bone, and bloodstream infections in immunocompromised patients; it is the most frequent bacteria implied in post-surgical infections. [1]. The data from Table 2 show that LEO and OEO displayed a powerful inhibitory effect on *S. aureus* (IZD = 40 mm). Similar results can also be found in other studies [35,36]. PEO and EEO had the lowest effects measured by DDM; the same IZD range (15–24 mm) belongs to the values reported by other researchers [37,38]. However, their combination (EEO+PEO) tested by DDM was partially synergistic (FICI = 0.8), with a substantial inhibitory effect (IZD = 35 mm), significantly higher than each EO (Table 2). The same binary EWO combination, tested by CT, showed indifference (FICI = 1.4).

The corresponding *p*-values for each technique are displayed in Figure S2A,B in the Supplementary Materials.

All EOs and combinations exhibited inhibitory activity against *S. aureus*. The anti-staphylococcal effect of antibiotic drugs increased in the order of NEO < BAC < TET; combined with EEO+PEO, NEO and TET generated partially synergistic triple combinations when cylinder technique was used (Table 2).

Table 2. The antibacterial effects of EOs combinations and classical antibiotics on *S. aureus*, evaluated by both techniques of diffusimetric method, registered as a mean of 3 repetitions: IZD (mm), MIC (μg/mL), and FICI (for double and triple combinations).

Technique	Cellulose Disc Technique (DDM)						Cylinder Technique (CT)					
	S. aureus											
Sample	IZD (mm)		MIC (μg/mL)	FICI			IZD (mm)		MIC (μg/mL)	FICI		
	Range	Value		Range		Value	Range	Value		Range		Value
LEO	≥35	40	2.1	-		-	≥35	40	21.2	-		-
CEO	25–34	27	4.7	-		-	25–34	25	54.3	-		-
OEO	≥35	40	2.1	-		-	≥35	40	21.2	-		-
EEO	15–24	20	8.5	-		-	25–34	25	54.3	-		-
PEO	25–34	25	5.4	-		-	15–24	15	150.9	-		-
OEO+CEO	25–34	30	3.8	>1	≤4	2.60	15–24	15	150.9	>4	≤10	9.90
CEO+EEO	15–24	15	15.1	>4	≤10	5.00	25–34	27.5	44.9	>1	≤4	1.60

Table 2. *Cont.*

Technique	Cellulose Disc Technique (DDM)					Cylinder Technique (CT)				
	S. aureus									
Sample	IZD (mm)		MIC (µg/mL)	FICI		IZD (mm)		MIC (µg/mL)	FICI	
	Range	Value		Range	Value	Range	Value		Range	Value
CEO+PEO	15–24	15	15.1	>4 ≤10	6.00	15–24	20	84.9	>1 ≤4	2.20
LEO+EEO	≥35	40	2.1	>1 ≤4	1.20	25–34	30	37.7	>1 ≤4	2.50
EEO+PEO	≥35	35	2.8	>0.5 <1	0.80	25–34	25	54.3	>1 ≤4	1.40
NEO	<10	7	86.7	-	-	15–24	20	106.2	-	-
OEO+CEO+NEO	25–34	25	5.7	>1 ≤4	3.98	15–24	20	89.2	>4 ≤10	6.67
CEO+EEO+NEO	25–34	25	5.7	>1 ≤4	1.94	≥35	37.5	25.4	>1 ≤4	1.15
CEO+PEO+NEO	10–14	12.5	22.8	>4 ≤10	9.33	15–24	22	73.7	>1 ≤4	2.52
LEO+EEO+NEO	15–24	23.5	6.5	>1 ≤4	3.83	25–34	25	57.1	>4 ≤10	4.27
EEO+PEO+NEO	25–34	26	5.3	>1 ≤4	1.66	15–24	15	15.9	>0.5 <1	0.53
TET	25–34	25	6.8	-	-	25–34	29	50.5	-	-
OEO+CEO+TET	25–34	25	5.7	>4 ≤10	4.87	≥35	40	21.2	>1 ≤4	1.80
CEO+EEO+TET	15–24	22	7.4	>1 ≤4	3.52	15–24	15	150.9	>4 ≤10	8.52
CEO+PEO+TET	25–34	25	5.7	>1 ≤4	3.09	≥35	45	16.8	>0.5 <1	0.74
LEO+EEO+TET	25–34	26	5.3	>4 ≤10	4.83	25–34	30	37.7	>1 ≤4	3.20
EEO+PEO+TET	25–34	26	5.3	>1 ≤4	2.37	10–14	10	339.7	>15 ≤20	15.22
BAC	15–24	17	14.7	-	-	25–34	25	67.9	-	-
OEO+CEO+BAC	25–34	30	4.0	>1 ≤4	3.02	25–34	30	47.2	>1 ≤4	3.77
CEO+EEO+BAC	25–34	32.5	3.4	>1 ≤4	1.35	0	0	-	-	-
CEO+PEO+BAC	15–24	17.5	11.6	>4 ≤10	5.38	0	0	-	-	-
LEO+EEO+BAC	25–34	25	5.7	>1 ≤4	3.76	25–34	27	48.9	>4 ≤10	5.32
EEO+PEO+BAC	25–34	32.5	3.4	>1 ≤4	1.25	15–24	20	89.2	>1 ≤4	3.54

LEO—Lavender Essential Oil; CEO—Clove Essential Oil; OEO—Oregano Essential Oil; EEO—Eucalyptus Essential Oil; PEO—Peppermint Essential Oil; NEO—Neomicin; TET—Tetracyclin; and BAC—Bacitracin (https://microbiologie-clinique.com/antibiotic-family-abbreviation.html, accessed on 5 February 2024). IZD—inhibition zone diameter (mm); the scale of measurement was as follows: very strong inhibitory effect at IZD ≥ 35 mm, strong inhibitory effects at 35 > IZD ≥ 25, moderate inhibitory effect at 25 > IZD ≥ 15 mm, mild inhibitory effect when 15 > IZD ≥ 10 mm, and very low inhibitory effect at IZD < 10 mm. MIC—minimum inhibitory concentration (µg/mL), FICI—fractional inhibitory concentration index. If IZD = 0 mm, the substance has no effect; there is no MIC. FICI ≤ 0.5 indicates synergism (S). 0.5 < FICI < 1 means partial synergism (PS); FICI = 1 indicates additive effects (Add.); 1 < FICI ≤ 4—indifference (I); FICI > 4 is antagonism (Ant.) [39]; 4 < FICI ≤ 10—low antagonism; 10 < FICI ≤ 15—moderate antagonism; 15 < FICI ≤ 20—strong antagonism; FICI > 20—very strong antagonism.

2.2.2. Antibacterial Activity against *E. coli*

E. coli lives in the human gut, commonly without inducing disease. It can cause infections when it penetrates the urinary tract [1]; in human household contacts, dogs can be reservoirs of *E. coli* strains responsible for urinary tract infections [40]. Moreover, in immunosuppressed patients, *E. coli* can produce severe infections: otitis [41] and meningitis (in neonates and older adults [42]).

The data from Table 3 shows that OEO and LEO exhibited the highest antibacterial effect on *E. coli*. Conversely, PEO and BAC had no inhibitory activity against it, and CEO, EEO, TET, and NEO displayed moderate ones. In this context, using the DDM technique, CEO+PEO revealed additive antibacterial activity on *E coli*, and triple combinations with TET (CEO+PEO+TET and EEO+PEO+TET) acted partially synergistically. The corresponding *p*-values for each technique are displayed in Figure S2C,D in the Supplementary Materials.

Table 3. The antibacterial effect of essential oils (EOs) combinations and classical antibiotics on *E. coli* evaluated by both techniques of diffusimetric method, registered as a mean of 3 repetitions: IZD (mm), MIC (μg/mL), and FICI (for double and triple combinations).

Technique	Cellulose Disc Technique					Cylinder Technique				
	E. coli									
Sample	IZD (mm)		MIC (μg/mL)	FICI		IZD (mm)		MIC (μg/mL)	FICI	
	Range	Value		Range	Value	Range	Value		Range	Value
LEO	≥35	40	2.1	-	-	≥35	40	21.2	-	-
CEO	15–24	22	7.0	-	-	15–24	15	150.9	-	-
OEO	≥35	45	1.7	-	-	≥35	45	16.8	-	-
EEO	25–34	25	5.4	-	-	25–34	30	37.7	-	-
PEO	0	0	-	-	-	10–14	10	339.7	-	-
OEO+CEO	25–34	25	5.4	>1 ≤4	4.00	15–24	15	150.9	>4 ≤10	10.00
CEO+EEO	15–24	15	15.1	>4 ≤10	4.90	10–14	10	339.7	>10 ≤15	11.30
CEO+PEO	15–24	22	7.0	=1	1.00	0	0	-	-	-
LEO+EEO	10–14	12	23.6	>15 ≤20	15.60	15–24	15	150.9	>10 ≤15	11.10
EEO+PEO	10–14	13	20.1	>1 ≤4	3.70	10–14	12	235.9	>4 ≤10	7.00
NEO	15–24	15	18.9	-	-	25–34	25	67.9	-	-
OEO+CEO+NEO	25–34	30	4.0	>1 ≤4	3.70	0	0	-	-	-
CEO+EEO+NEO	15–24	15	15.9	>4 ≤10	6.05	25–34	27	48.9	>1 ≤4	2.33
CEO+PEO+NEO	25–34	25	5.7	>1 ≤4	1.11	15–24	20	89.2	>1 ≤4	2.16
LEO+EEO+NEO	15–24	15	15.9	>10 ≤15	11.35	25–34	25	57.1	>4 ≤10	5.04
EEO+PEO+NEO	15–24	15	15.9	>1 ≤4	3.78	15–24	20	89.2	>1 ≤4	3.93
TET	10–14	14	21.7	-	-	25–34	30	47.2	-	-
OEO+CEO+TET	25–34	25	5.7	>4 ≤10	4.42	25–34	25	57.1	>4 ≤10	4.96
CEO+EEO+TET	25–34	26	5.3	>1 ≤4	1.97	25–34	28	45.5	>1 ≤4	2.46
CEO+PEO+TET	25–34	27	4.9	>0.5 <1	0.92	25–34	26	52.8	>1 ≤4	1.59
LEO+EEO+TET	25–34	26	5.3	>1 ≤4	3.74	25–34	27.5	47.2	>4 ≤10	4.47
EEO+PEO+TET	25–34	30	4.0	>0.5 <1	0.92	25–34	30	39.6	>1 ≤4	1.99
BAC	0	0	-	-	-	0	0	-	-	-
OEO+CEO+BAC	25–34	25	5.7	>4 ≤10	4.16	15–24	15	158.5	>10 ≤15	10.48
CEO+EEO+BAC	10–14	13	21.1	>4 ≤10	6.91	<10	8	557.4	>15 20 ≤20	18.47
CEO+PEO+BAC	15–24	20	8.9	>1 ≤4	1.27	0	0	-	-	-
LEO+EEO+BAC	10–14	10	35.7	>20	23.61	10–14	10	356.7	>20	26.28
EEO+PEO+BAC	<10	9	44.0	>4 ≤10	8.14	10–14	10	356.7	>10 ≤15	10.51

LEO—Lavender Essential Oil; CEO—Clove Essential Oil; OEO—Oregano Essential Oil; EEO—Eucalyptus Essential Oil; PEO—Peppermint Essential Oil; NEO—Neomycin; TET—Tetracycline; BAC—Bacitracin; IZD—inhibition zone diameter (mm); the scale of measurement was as follows: very strong inhibitory effect at IZD $\geq$ 35 mm, strong inhibitory effects at 35 > IZD $\geq$ 25, moderate inhibitory effect at 25 > IZD $\geq$ 15 mm, mild inhibitory effect when 15 >IZD $\geq$ 10 mm, and very low inhibitory effect at IZD < 10 mm. MIC—minimum inhibitory concentration (μg/mL), FICI—Fractional inhibitory concentration; FICI $\leq$ 0.5 indicates synergism (S). 0.5 < FICI < 1 means partial synergism (PS); FICI = 1 indicates additive effects (Add.); 1 < FICI $\leq$ 4—indifference (I); FICI > 4 is antagonism (Ant.) [39]; 4 < FICI $\leq$ 10—low antagonism; 10 < FICI $\leq$ 15—moderate antagonism; 15 < FICI $\leq$ 20—strong antagonism; FICI > 20—very strong antagonism.

2.2.3. Antibacterial Activity against *P. aeruginosa*

P. aeruginosa is most known as a nosocomial pathogen, mainly determining severe infections within hospitals and care homes. Its resistance to carbapenems and polymyxins is continuously increasing. Surgical and medical devices (eq., endoscopes) potentially spread this MDR bacteria, leading to nosocomial outcomes [43].

Table 4 indicates that PEO and BAC had no inhibitory effects on *P. aeruginosa*; most EO triple combinations with BAC acted similarly through DDM. Moreover, all the other

EOs evidenced very low/low/moderate antibacterial activity on *P. aeruginosa*. One binary combination was synergistic—OEO+CEO, as evaluated by the cylinder technique. The corresponding *p*-values for each technique are displayed in Figure S2E,F in the Supplementary Materials.

Table 4 also shows the lowest IZD values from all three tables and the highest MIC ones, indicating the modest inhibitory activity of all EOs alone or combined.

Table 4. The antibacterial activity of EO combinations and classical antibiotics against *P. aeruginosa*, evaluated by both techniques of diffusimetric method, registered as a mean of 3 repetitions: IZD (mm), MIC (μg/mL) and FICI (for double and triple combinations).

Technique	Cellulose Disc Technique					Cylinder Technique				
	P. aeruginosa									
Sample	IZD (mm)		MIC (μg/mL)	FICI		IZD (mm)		MIC (μg/mL)	FICI	
	Range	Value		Range	Value	Range	Value		Range	Value
LEO	10–14	11	28.1	-	-	<10	7	693.2	-	-
CEO	10–14	10	33.9	-	-	<10	8	530.9	-	-
OEO	10–14	11	28.1	-	-	10–14	10	339.7	-	-
EEO	25–34	25	5.4	-	-	≥35	35	27.7	-	-
PEO	0	0	-	-	-	0	0	-	-	-
OEO+CEO	<10	7.5	60.6	>1 ≤4	3.90	25–34	30	37.7	≤0.5	0.20
CEO+EEO	<10	7.5	60.6	>10 ≤15	13.00	≥35	35	27.7	>1 ≤4	1.10
CEO+PEO	<10	7.5	60.6	>1 ≤4	1.80	0	0	-	-	-
LEO+EEO	<10	8	53.0	>10 ≤15	11.70	15–24	16	132.7	>4 ≤10	5.00
EEO+PEO	<10	8	53.0	>4 ≤10	9.80	15–24	17	117.5	>4 ≤10	4.20
NEO	10–14	10	42.4	-	-	15–24	20	106.2	-	-
OEO+CEO+NEO	0	0	-	-	-	15–24	22	73.7	>1 ≤4	1.16
CEO+EEO+NEO	10–14	12.5	22.8	>4 ≤10	5.56	15–24	15	158.5	>4 ≤10	7.50
CEO+PEO+NEO	15–24	15	15.9	>1 ≤4	1.29	0	0	-	-	-
LEO+EEO+NEO	<10	9	44.0	>10 ≤15	10.73	10–14	14	182.0	>1 ≤4	2.05
EEO+PEO+NEO	15–24	17	12.3	>1 ≤4	2.56	15–24	21	80.9	>1 ≤4	3.68
TET	15–24	15	18.9	-	-	25–34	25	67.9	-	-
OEO+CEO+TET	0	0	-	-	-	<10	7	727.9	>10 ≤15	14.23
CEO+EEO+TET	15–24	18	11.0	>1 ≤4	2.93	15–24	23	67.4	>1 ≤4	3.54
CEO+PEO+TET	10–14	14	18.2	>1 ≤4	1.49	15–24	15	158.5	>1 ≤4	2.62
LEO+EEO+TET	10–14	14	18.2	>4 ≤10	4.97	15–24	15	158.5	>4 ≤10	8.27
EEO+PEO+TET	15–24	15	15.9	>1 ≤4	3.78	15–24	22	73.7	>1 ≤4	3.74
BAC	0	0	-	-	-	0	0	-	-	-
OEO+CEO+BAC	0	0	-	-	-	0	0	-	-	-
CEO+EEO+BAC	0	0	-	-	-	15–24	22	73.7	>1 ≤4	2.79
CEO+PEO+BAC	0	0	-	-	-	0	0	-	-	-
LEO+EEO+BAC	0	0	-	-	-	15–24	21	80.9	>1 ≤4	3.03
EEO+PEO+BAC	<10	6	99.1	>15 ≤20	18.35	15–24	20	89.2	>1 ≤4	3.22

LEO—Lavender Essential Oil; CEO—Clove Essential Oil; OEO—Oregano Essential Oil; EEO—Eucalyptus Essential Oil; PEO—Peppermint Essential Oil; NEO—Neomycin; TET—Tetracyclin; and BAC—Bacitracin. IZD—inhibition zone diameter (mm); the scale of measurement was as follows: very strong inhibitory effect at IZD ≥ 35 mm, strong inhibitory effects at 35 > IZD ≥ 25, moderate inhibitory effect at 25 > IZD ≥ 15 mm, mild inhibitory effect when 15 > IZD ≥ 10 mm, and very low inhibitory effect at IZD < 10 mm. MIC—minimum inhibitory concentration (μg/mL), FICI—fractional inhibitory concentration. If IZD = 0 mm, the substance has no effect and no MIC; FICI ≤ 0.5—synergism (S). 0.5 < FICI < 1—partial synergism (PS); FICI = 1—additive effects (Add.); 1 < FICI ≤ 4—indifference (I); FICI > 4—antagonism (Ant.) [39]; 4 < FICI ≤ 10—low antagonism; 10 < FICI ≤ 15—moderate antagonism; 15 < FICI ≤ 20—strong antagonism; FICI > 20—very strong antagonism.

The data from Tables 3 and 4 show that the IZD sizes recorded on the Gram-negative bacteria (especially on *P. aeruginosa*) through the cylinder technique are higher than those measured by DDM. Moreover, Vorobets et al. noted that, in their study, the cylinder technique was the most sensitive, recording the highest IZD values [31].

Tables 1–3 report a few EO binary combinations (eq. CEO+EEO) with an IZD lower than that of each EO alone. These results suggest similar inhibitory mechanisms, with some compounds' competition for various binding sites.

Different data regarding antibacterial activity expressed as IZD (mm) for EOs, conventional antibiotics, and double and triple combinations are displayed in Figure 2 and the corresponding *p*-values are shown in Figure S2A–F in the Supplementary Materials. Thus, we can see that double EO combinations lead to an increasing antibacterial activity against *P. aeruginosa* in the cylinder technique (Figure 2). Generally, using the cylinder technique, NEO reduces the EO inhibitory activity on the bacteria tested (Figure 2, except on *P. aeruginosa*. On the other hand, triple combinations with TET have a higher antibacterial activity on all tested strains compared to the double EO combinations. Contrariwise, BAC mixed with double EO combinations had diminished IZD values (Figure 2).

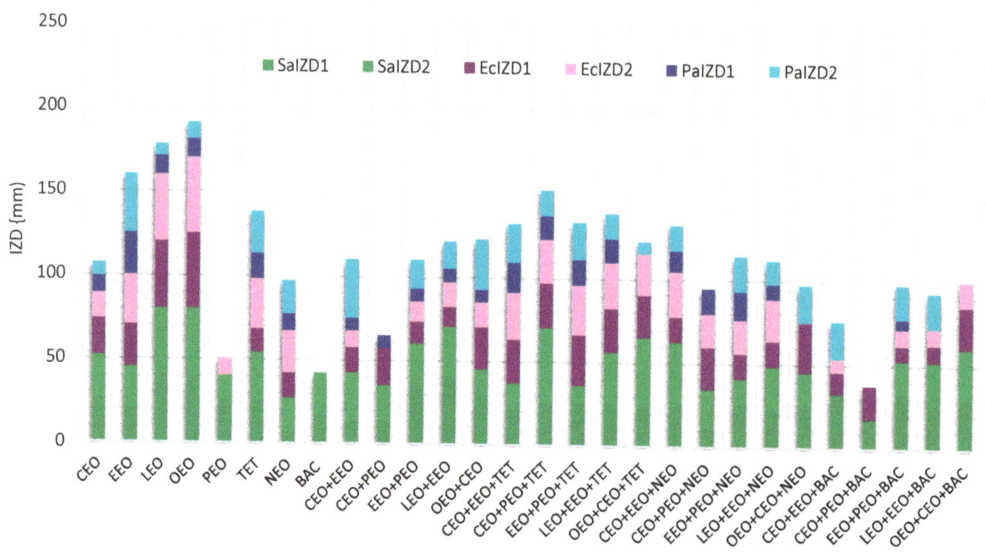

Figure 2. Stacked columns chart, as an overview of antibacterial activity against Gram-positive and Gram-negative bacteria, expressed as IZD (mm). LEO—Lavender Essential Oil; CEO—Clove Essential Oil; OEO—Oregano Essential Oil; EEO—Eucalyptus Essential Oil; PEO—Peppermint Essential Oil; TET—Tetracycline; NEO—Neomycin; BAC—Bacitracin; and IZD—inhibition zone diameter (mm), the scale of measurement was as follows: powerful inhibitory effect at IZD $\geq$ 35 mm, strong inhibitory effects at 35 > IZD $\geq$ 25 moderate inhibitory effect at 25 > IZD $\geq$ 15 mm, mild inhibitory effect when 15 > IZD $\geq$ 10 mm, and no inhibitory effect at IZD < 10 mm. Sa—*S. aureus*, Ec—*E. coli*, and Pa—*P. aeruginosa*.

Despite the different volumes of the samples used in both techniques (10 μ in DDM and 100 μ in CT), the IZD ranges were similar. Only the calculated MIC values recorded considerable differences, because the calculation formula considered the size of the sample's volume. The antibacterial potential was negatively correlated with MIC value. The DDM technique had the lowest MIC values and could be considered as better than CT.

The results of both techniques (DDM and CT, 1 and 2) of the diffusimetric method were compared using a Bland–Altman analysis. The results are displayed in the MS text, Figure 3, and Figure S3 from the Supplementary Materials. Similar results ($p < 0.05$) were obtained by both techniques against *S. aureus* ($p = 0.566$) and *E. coli* ($p = 0.460$). Significant differences between DDM and CT were recorded in terms of antibacterial activity against *P. aeruginosa* ($p = 0.002 < 0.05$), according to Figure 3E. The triple combination EOs+TET were responsible for this ($p = 0.036 < 0.05$).

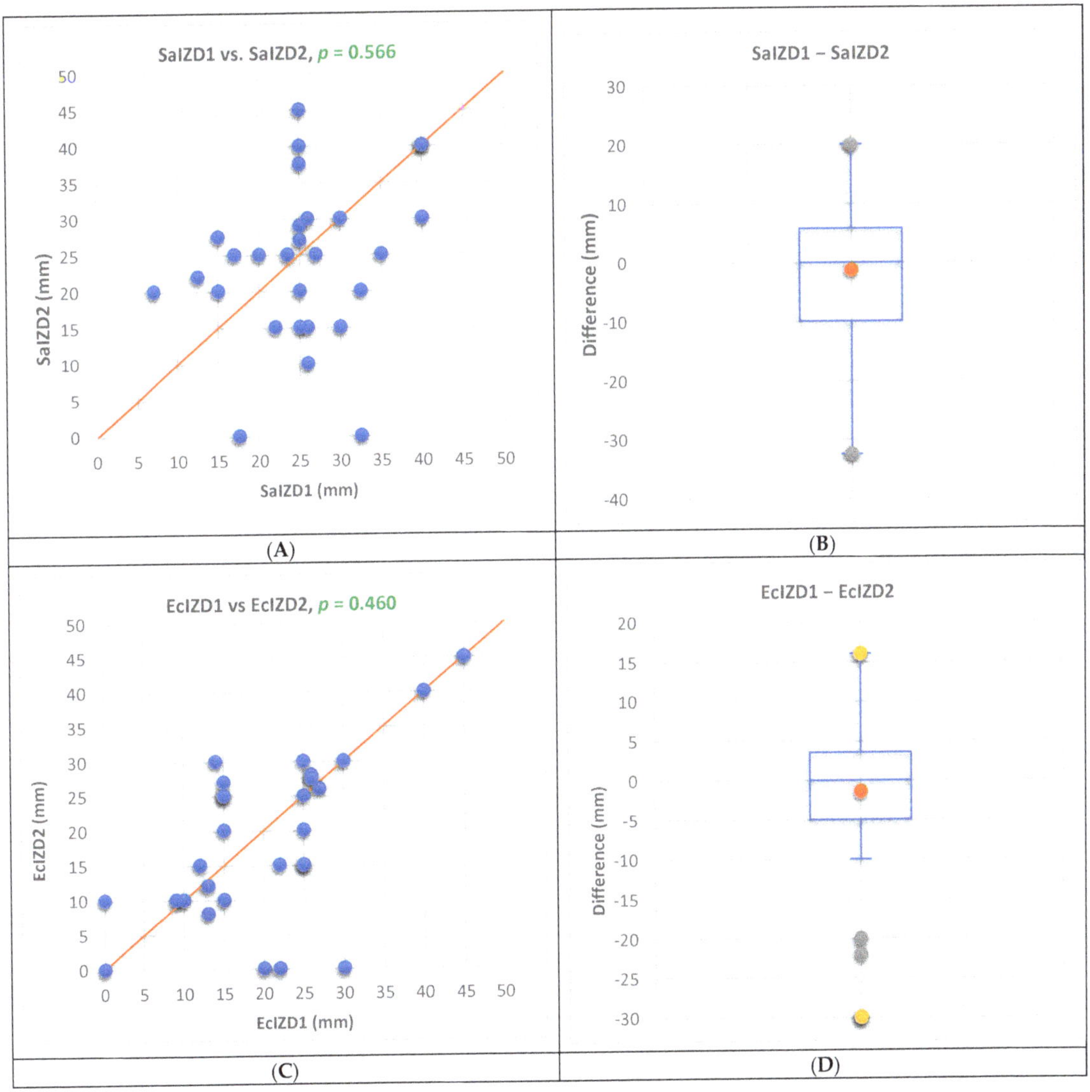

Figure 3. *Cont.*

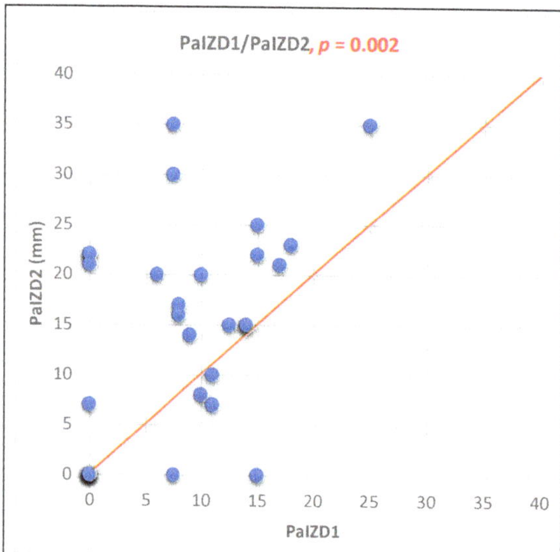

(E)

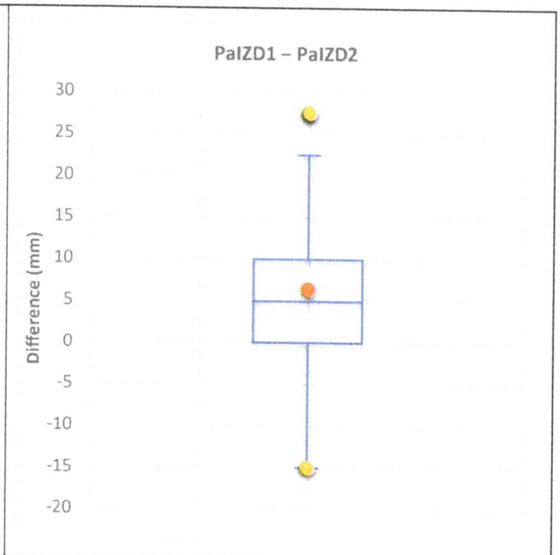

(F)

Figure 3. Bland–Altman comparison between both techniques of diffusimetric antibiogram for all bacteria tested: *S. aureus* (**A**,**B**), *E. coli* (**C**,**D**), and *P. aeruginosa* (**E**,**F**). The differences are statistically significant at $p < 0.05$.

In Figure 3B,D,F are evidenced the differences between both techniques.

The main constituents of the tested EOs are mono- and sesquiterpene hydrocarbons and their oxygenated derivatives; they are volatile compounds sensitive to heating, air, and humidity. Therefore, the antibacterial activity of the EOs could be reduced during both technical processes by losses due to evaporation [44–47] at 37 °C. This process occurs more in DDM, because small sample volumes (10 µL) are incorporated in filter paper discs placed on a culture medium. Therefore, the EO bioactive constituents only partially diffuse in culture media, inhibiting bacterial strain growth. In CT, 10× higher volumes (100 µL) progressively diffuse in the culture media, thus diminishing the volatile compounds' evaporation and increasing their biodisponibility.

On the other hand, some EO constituents absorbed in agar layers could suffer alteration processes during the incubation period of 24 h. Inouye et al. reported limonene and α-pinene oxidation [48]; in the present study, both constituents were found in four EOs tested (EEO, OEO, LEO, and PEO).

All these aspects could explain the different IZD sizes obtained through both techniques; depending on each EO phytochemical profile, one technique or the other could be more suitable, and the results more relevant and similar to those of previously published research. Thus, comparing the IZD values obtained through both methods, the most appropriate ones could be considered for the MIC and FICI calculation and interpretation.

Several studies reported that CEO is efficient against *S. aureus* and *E. coli* [49], OEO is active on all three bacteria, and PEO shows moderate or low activity on *S. aureus* and *P. aeruginosa* [50,51]. LEO and EEO exhibited inhibitory activity against *S. aureus* and *E. coli* [52,53]. Another study revealed the antibacterial activity of EEO emulsion against *E. coli* and *P. aeruginosa* [54].

The calculation of MIC values from the IZD ones led to evaluating the interaction between EOs in binary combinations and EOs and conventional antibiotics in triple combinations, possibly by analyzing each FIC index (Figure 4).

Bacteria	*S. aureus*		*E. coli*		*P. aeruginosa*	
Combination	**SaFICI1**	**SaFICI2**	**EcFICI1**	**EcFICI2**	**PaFICI1**	**PaFICI2**
CEO+EEO	5.00	1.60	4.90	11.30	13.00	1.10
CEO+EEO+BAC	1.35		6.91	18.47		2.79
CEO+EEO+NEO	1.94	1.15	6.05	2.33	5.56	7.50
CEO+EEO+TET	3.52	8.52	1.97	2.46	2.93	3.54
CEO+PEO	6.00	2.20	1.00		1.80	
CEO+PEO+BAC	5.38		1.27			
CEO+PEO+NEO	9.33	2.52	1.11	2.16	1.29	
CEO+PEO+TET	3.09	0.74	0.92	1.59	1.49	2.62
EEO+PEO	0.80	1.40	3.70	7.00	9.80	4.20
EEO+PEO+BAC	1.25	3.54	8.14	10.51	18.35	3.22
EEO+PEO+NEO	1.66	0.53	3.78	3.93	2.56	3.68
EEO+PEO+TET	2.37	15.22	0.92	1.99	3.78	3.74
LEO+EEO	1.20	2.50	15.60	11.10	11.70	5.00
OEO+CEO	2.60	9.90	4.00	10.00	3.90	0.20
LEO+EEO+NEO	3.83	4.27	11.35	5.04	10.73	2.05
OEO+CEO+NEO	3.98	6.67	3.70			1.16
LEO+EEO+BAC	3.76	5.32	23.61	26.28		3.03
OEO+CEO+BAC	3.02	3.77	4.16	10.48		
LEO+EEO+TET	4.83	3.20	3.74	4.47	4.97	8.27
OEO+CEO+TET	4.87	1.80	4.42	4.96		14.23

Figure 4. Heat map of FICI values for double and triple combinations. LEO—Lavender Essential Oil; CEO—Clove Essential Oil; OEO—Oregano Essential Oil; EEO—Eucalyptus Essential Oil; PEO—Peppermint Essential Oil; NEO—Neomycin; TET—Tetracycline; and BAC—Bacitracin. FICI—fractional inhibitory concentration index. If IZD = 0 mm, it has no effect; no MIC exists. FICI $\leq$ 0.5 indicates synergism (S). 0.5 $<$ FICI $<$ 1—partial synergism (PS); FICI = 1—additive effects (Add.); 1 $<$ FICI $\leq$ 4—indifference (I); FICI $>$ 4—antagonism (Ant.) [39]; 4 $<$ FICI $\leq$ 10—low antagonism; 10 $<$ FICI $\leq$ 15—moderate antagonism; 15 $<$FICI $\leq$ 20—strong antagonism; FICI $>$ 20—very strong antagonism; FICI1—determined by DDM (disc diffusion method); and FICI2—determined by cylinder diffusion technique. Sa—*S. aureus*, Ec—*E. coli*, and Pa—*P. aeruginosa*. The FICI values $\leq$ 4 are marked with green and the color intensity decreases from dark green (synergism) to light green (indifference) the FICI values $>$ 4 are marked with brown and the color intensity increases from light brown (low antagonism) to dark brown (very strong antagonism).

Figure 4 shows that, in a binary combination, OEO and CEO synergistically act against *P. aeruginosa* (PaFICI2 = 0.2); this result was assessed using the cylinder technique. Partial synergism was revealed by another EO pair (EEO+PEO) against *S. aureus* (SaFICI1 = 0.8) using DDM. EEO+PEO in triple combinations (with NEO and TET) evidenced partial synergism against *S. aureus* (SaFICI2 = 0.53) and *E. coli* (EcFICI1 = 0.92).

Previous studies reported that a 1, 8-Cineol+Linalool combination creates more resistance in *E. coli*, compared to applying pure Linalool [55], while Carvacrol and Thymol are synergistic [56]. Synergistic effects were revealed for Eugenol, Carvacrol, Thymol, p-Cymene, and γ-Terpinene [57] on inhibiting drug resistance and the biofilm formation of oral bacteria (*S. aureus*). The Thymol+Menthol combination was reported as being antagonistic against *E. coli* and *S. aureus*, while Geraniol+Menthol was synergistic against *S. aureus* and indifferent against *E. coli* [49]. Moreover, various authors revealed synergy between Eugenol with Chloramphenicol and Norfloxacin with Oxacillin against *E. coli* and *P. aeruginosa*, Menthol with Oxytetracycline against *E coli*, Thymol with Norfloxacin and Bacitracin against *S. aureus*, and Novobiocin with Penicillin against *E. coli* [26].

On the other hand, PEO was previously tested on *S. aureus* in association with Ciprofloxacin; against *E. coli*, PEO formed synergistic combinations with Ampicilin, Erytromicin, Oxytetracycline, and Gentamycin [44,45]. In our study, a binary EO combination (CEO+PEO) recorded additive effects against *E. coli* (EcFICI1 = 1), and another triple combination (CEO+PEO+TET) showed partial synergism against *S. aureus* and *E. coli* (SaFICI2 = 0.74 and EcFICI1 = 0.92). Previous studies revealed synergistic effects of CEO combined with Ampicillin and Gentamycin against another *Staphylococcus* sp., *S. epidermidis* [46].

Moreover, CEO+PEO+TET and EEO+PEO+NEO did not show antagonism (FICI < 4). Most combinations displayed indifference and/or low antagonism. There were no recorded binary or triple combinations with an exclusive antagonism against all bacteria tested (Figure 3). However, LEO+EEO+BAC revealed powerful antagonism against *E. coli* (FICI > 20) and LEO+EEO+NEO moderate/low effects; the binary EOs combination (LEO+EEO) showed antagonism on both Gram-negative bacteria (strong/moderate against *E. coli* and moderate/low against *P. aeruginosa*). In contrast, the third triple combination (LEO+EEO+TET) recorded low antagonism only against *P. aeruginosa*. Other triple combinations with BAC, CEO+EEO+BAC, EEO+PEO+BAC, and OEO+CEO+BAC, showed antagonistic interactions against *E coli* (low/strong or low/moderate, Figure 4), while CEO+EEO and OEO+CEO acted similarly, and EEO+PEO and CEO+EEO+NEO registered low agonism against *P. aeruginosa*. CEO+EEO+TET and EEO+PEO+TET did not record an agonist behavior against both Gram-negative bacteria, while CEO+PEO+TET and EEO+PEO+NEO showed only partial synergism and indifference against all bacteria by both methods (Figure 4).

2.3. Data Analysis

An extensive data analysis was performed to support our results.

First, the correlations between chemical composition and antibacterial activity (expressed as IZD) were analyzed through Principal Component Analysis (PCA) and are illustrated in Figure 5A–C. All correlation biplots in Figure 5 show the place of each EO associated with its phytochemical profile and antibacterial activity.

Figure 5A is designed for the first group of three EOs (CEO, PEO, and EEO) and is consequently associated with three binary combinations. Figure 5B shows the same correlations in the second EO group (CEO, EEO, OEO, and LEO) associated with the other two binary combinations. Figure 5C presents an overview of the correlation of all five EOs.

The correlation biplot from Figure 5A has two principal components which explain the total data variances (PC1 = 62.52% and PC2 = 37.48%). All antibacterial activities and constituents are linked to PC1, while only four compounds are associated with PC2 (Linalool, Eugenol acetate, Eugenol, and Caryophyllene). The correlation matrix from the Supplementary Materials and Figure 5A indicates that Eucalyptol, Limonene, o-Cymene, γ-Terpinene, and Camphor are highly correlated with PaIZD and EcIZD, as evaluated by the cylinder technique (r = 0.936–0.976, $p > 0.05$), and strongly correlated with PaIZD1 (r = 0.864–0.918, $p > 0.05$). Limonene and α-Pinen display a good correlation with PaIZD2 and EcIZD2 (r = 0.807–0.873, $p > 0.05$) and a moderate one with PaIZD1 (r = 0.698, r = 0.766, $p > 0.05$). α-Pinen shows a significant negative correlation with SaIZD1 (r = −0.999, $p < 0.05$), and all the others, as previously mentioned, display a high one (r = −[0.961–0.994], $p < 0.05$). EEO contains these phytochemicals, and the PCA analysis confirmed its significant inhibitory activity against all three bacterial species.

Contrariwise, Caryophyllene, Eugenol, and Eugenol-acetate are highly and moderately correlated with SaIZD1 (r = 0.858, r = 0.721, $p < 0.05$). The CEO position in Figure 5A confirms the results from Tables 2–4. Neoisomenthol, Isomenthol, Menthofuran, and p-Thymol substantially negatively correlate with SaIZD2 (r = −0.999, $p < 0.05$), while Linalool shows a highly negative correlation (r = 0.990, $p > 0.05$). All exhibit a considerable negative correlation with EcIZD1 (r = −[0.968, 0.994], $p > 0.05$) and a good to moderate one with EcIZD2 (r = −[0.803, 0.711], $p > 0.05$). Therefore, the PEO moderate antibacterial activity on *S. aureus* and the low activity on *E. coli*, as assessed by CT, were confirmed. PEO's place in the correlation biplot supports its highly inhibitory effect against *S. aureus* evaluated by DDM and confirms its absence against *P. aeruginosa* (Figure 5A).

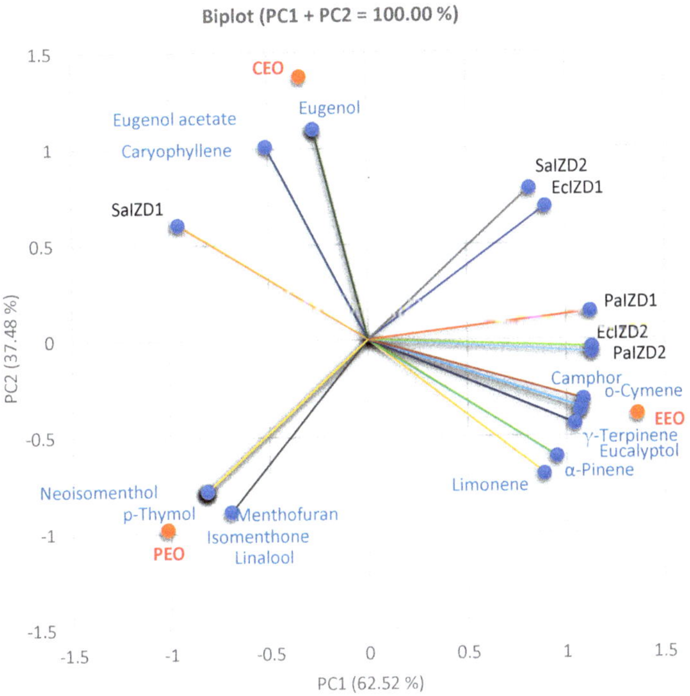

(A)

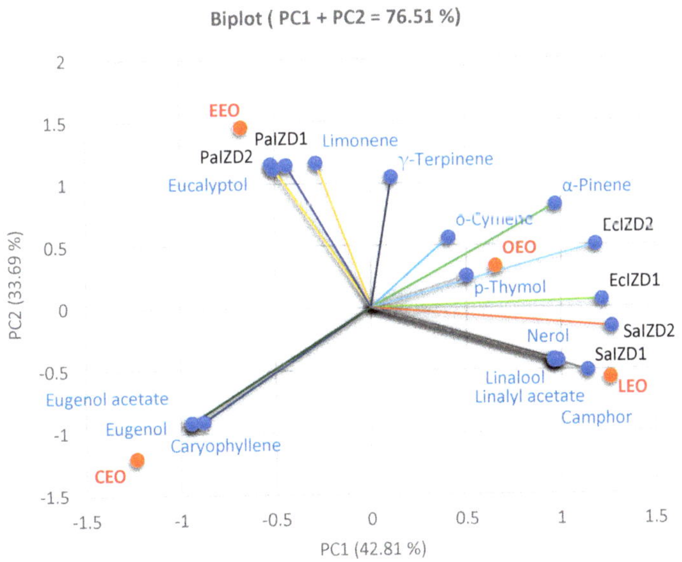

(B)

Figure 5. *Cont.*

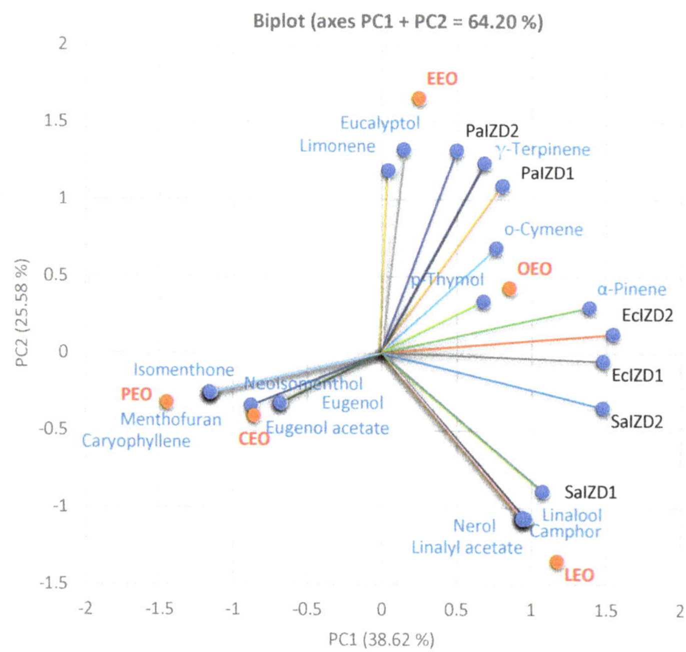

(**C**)

Figure 5. Correlations between EOs' antibacterial activity (expressed as IZD) and the main phytoconstituents: (**A**) EEO-PEO-CEO group; (**B**) CEO-OEO-EEO-LEO group; and (**C**) all EOs. IZD—inhibition zone diameter (mm), Sa—*S. aureus*, Ec—*E. coli*, Pa—*P. aeruginosa*, LEO—Lavender Essential Oil; CEO—Clove Essential Oil; OEO—Oregano Essential Oil; EEO—Eucalyptus Essential Oil; PEO—Peppermint Essential Oil; IZD1—determined by DDM (disc diffusion method); and IZD2—determined by cylinder technique.

Figure 5B displays the correlations between the antibacterial activity and phytochemicals quantified in CEO, OEO, LEO, and EEO. In the correlation biplot, the two principal components explain 76.58% of the total data variance (PC1 = 42.81% and PC2 = 33.69%). The antibacterial activities against *S. aureus* and *E. coli* and most phytochemicals are linked with PC1. In contrast, the inhibitory activities on *P. aeruginosa* and Eucalyptol, Limonene, Caryophyllene, and γ-Terpinene are associated with PC2. Eucalyptol and Limonene are significantly correlated with PaIZD (1 and 2). α-Pinen, o-Cymene, and p-Thymol show a good to moderate correlation with EcIZD (1 and 2), r = 0.821–0.589, $p < 0.05$. Finally, the antibacterial activity against *S. aureus* is moderately correlated with p-Thymol, Camphor, Linalool, Linalyl acetate, and Nerol. The EOs' places confirm that CEO is significantly different regarding its phytochemical profile and antibacterial activity compared to EEO, OEO, and LEO (Figure 5B).

In the correlation biplot from Figure 5C, the two principal components explain 64.20% of the total data variance, with 38.62% attributed to the first (PC1) and 25.58% to the second (PC2). PC1 is associated with SaIZD, EcIZD, and some EOs' constituents (α-Pinene, Neoisomenthol, Isomenthol, and Menthofuran). At the same time, PC2 is linked with PaIZD and other phytochemicals (Eucalyptol, Lymonene, γ-Terpinene, Camphor, Linalool, Linalyl-acetate, and Nerol). Figure 5C also shows the place of all EOs reported with their phytochemical profiles and antibacterial activities on Gram-positive and Gram-negative bacteria evaluated through both methods.

All interactions expressed as FICI values through DDM and CT are illustrated in Figure 6. Regarding the conventional antibiotics' influence on the EO binary combinations, TET and NEO diminished EcFICI and PaFICI. Contrariwise, BAC substantially decreased PaFICI and increased EcFICI (Figure 6).

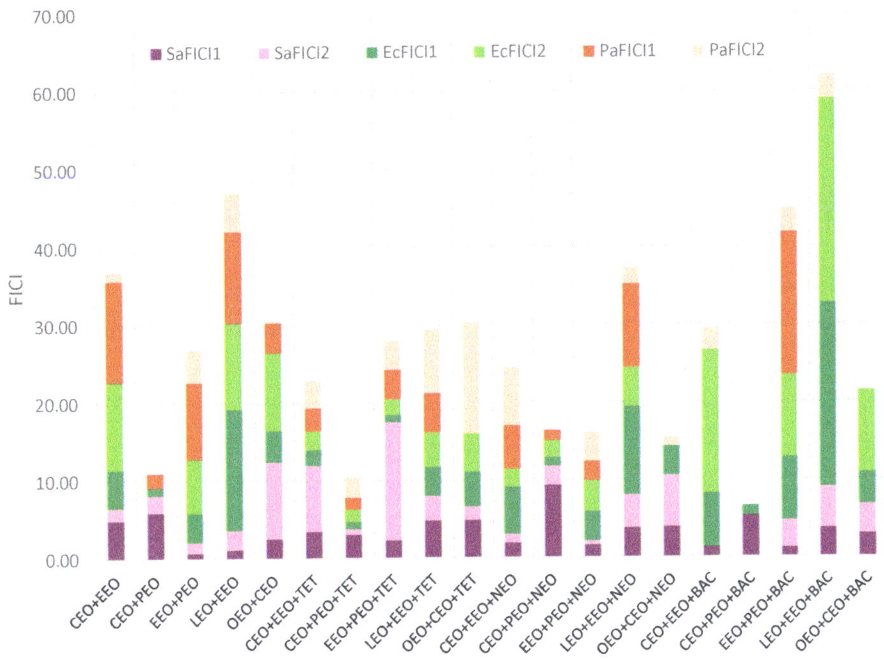

Figure 6. An overview of interactions between EOs and conventional antibiotics against Gram-positive and Gram-negative bacteria, assessed by both techniques of diffusimetric antibiogram and expressed as FICI. LEO—Lavender Essential Oil; CEO—Clove Essential Oil; OEO—Oregano Essential Oil; EEO—Eucalyptus Essential Oil; PEO—Peppermint Essential Oil; NEO—Neomycin; TET—Tetracycline; and BAC—Bacitracin. FICI—fractional inhibitory concentration index. If IZD = 0 mm, it has no effect; no MIC exists. FICI $\leq$ 0.5 indicates synergism (S). 0.5 < FICI < 1—partial synergism (PS); FICI = 1—additive effects (Add.); 1 < FICI $\leq$ 4—indifference (I); FICI > 4—antagonism (Ant.) [39]; 4 < FICI $\leq$ 10—low antagonism; 10 < FICI $\leq$ 15—moderate antagonism; 15 < FICI $\leq$ 20—strong antagonism; FICI > 20—very strong antagonism; FICI1—determined by DDM (disc diffusion method); and FICI2—determined by cylinder diffusion technique. Sa—*S. aureus*, Ec—*E. coli*, and Pa—*P. aeruginosa*.

The influence of the EOs' and conventional antibiotics' inhibitory activity (expressed as MIC (μg/mL)) on the double and triple combinations and then on interactions (expressed as FICI value) was analyzed by using PCA. All results are shown in Figure 7 and support the data in Tables 2–4 and Figure 6.

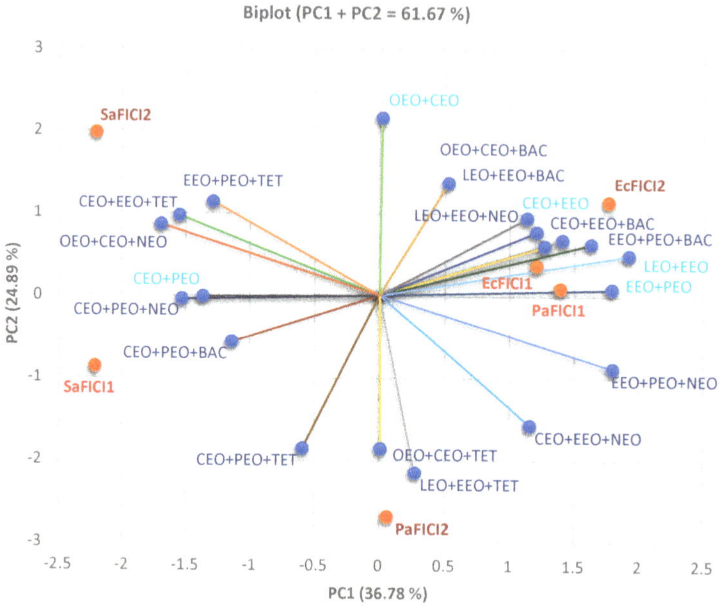

(**A**)

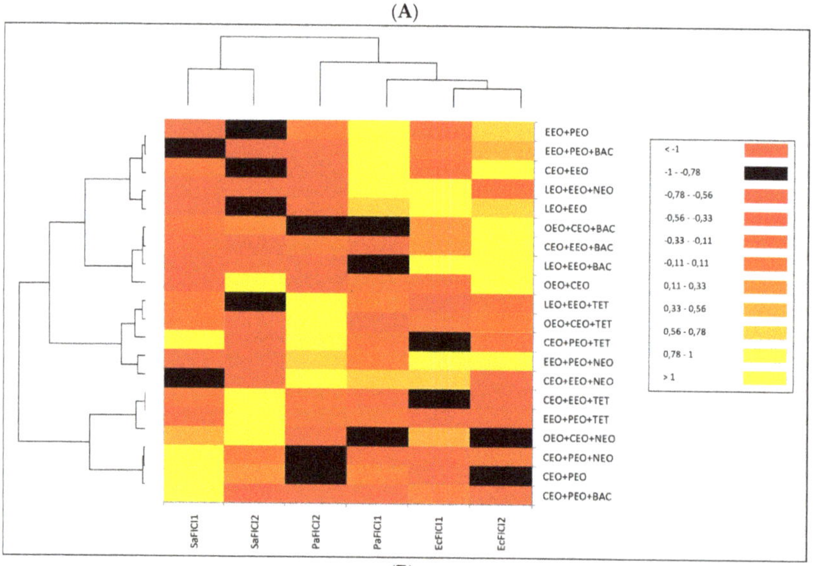

(**B**)

Figure 7. (**A**) Correlations between all tested double and triple combinations and FICI values. (**B**) AHC Dendrogram, with 5 clusters: C1 = LEO+EEO+NEO; C2 = CEO+EEO+NEO; C3 = CEO+EEO+TET; C4 = CEO+PEO; and C5 = LEO+EEO+BAC and heat map. LEO—Lavender Essential Oil; CEO—Clove Essential Oil; OEO—Oregano Essential Oil; EEO—Eucalyptus Essential Oil; PEO—Peppermint Essential Oil; NEO—Neomycin; TET—Tetracycline; and BAC—Bacitracin. FICI—Fractional inhibitory concentration index. FICI1 is determined by DDM (disc diffusion method), and the cylinder diffusion technique determines FICI2. Sa—*S. aureus*, Ec—*E. coli*, and Pa—*P. aeruginosa*.

Figure 8A–C displays the influence of each conventional antibiotic on the first three EO binary combinations; Figure 8D offers an overview of the second group of four EOs and all conventional antibiotics in binary and triple combinations.

When TET is the selected antibiotic (Figure 8A), EcMIC (1 and 2) is moderately correlated with EcFICIs (r = 0.628, r = 0.567, $p > 0.05$). The same observation is available for PaMIC1–PaFICI1 (r = 0.585, $p < 0.05$), while PaMIC2 shows a very low correlation with PaFICI2 (r = 0.075, $p > 0.05$). SaMICs display a minimal negative correlation with SaFICI (r = −0.034, r = −0.192, $p < 0.05$). TET belongs to three triple combinations with partially synergistic action against *S. aureus* (CT) and *E. coli* (DDM).

Figure 8B shows the NEO influence; thus, SaMICs (1 and 2) have a significant moderate to good correlation with SaFICI (1 and 2): r = 0.759, r = 0.892, $p < 0.05$, while PaMIC1 moderately correlates with PAFICI1 (r = 0.670, $p < 0.05$). EcMICs (1 and 2) show a low to moderate correlation with EcFICI (1 and 2), r = 0.469–0.589, $p > 0.05$ (Figure 8B).

Figure 8C, with triple combinations containing BAC, indicates a significant powerful correlation between EcMICs and EcFICIs (r = 0.913, r = 0.850, $p < 0.05$) and PaMIC1 and PaFICI1 (r = 0.865, $p < 0.05$). Moreover, EcFICI1 significantly intercorrelates with EcFICI2, and PaFICI1 reports a moderate intercorrelation with PaFICI2 (r = 0.915, r = 0.654, $p < 0.05$).

In the case of CEO+OEO+EEO+LEO combined with all three antibiotics, Figure 8D shows that EcMIC2 is significantly correlated with EcFICI2 (r = 0.905, $p < 0.05$), while EcMIC1 has an appreciable correlation with EcFICI1 (r = 0.754, $p < 0.05$). PaMIC1 and SaMIC2 are moderately correlated with PaFICI1 and SaFICI2 (r = 0.607, r = 0.579, $p < 0.05$). Moreover, EcFICI1 is highly correlated with EcFICI2 (r = 0.890, $p < 0.05$) and SaFICI1 moderately correlates with SaFICI2 (r = 0.669, $p < 0.05$).

All combinations and FICI values' places are illustrated in the correlation biplot (Figure 7A) and the dendrogram with heat map (Figure 7B), supporting our study results.

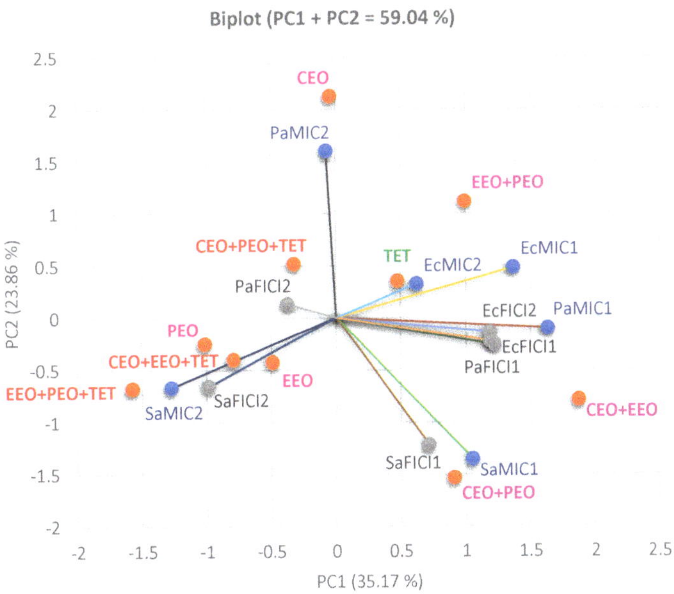

(A)

Figure 8. *Cont.*

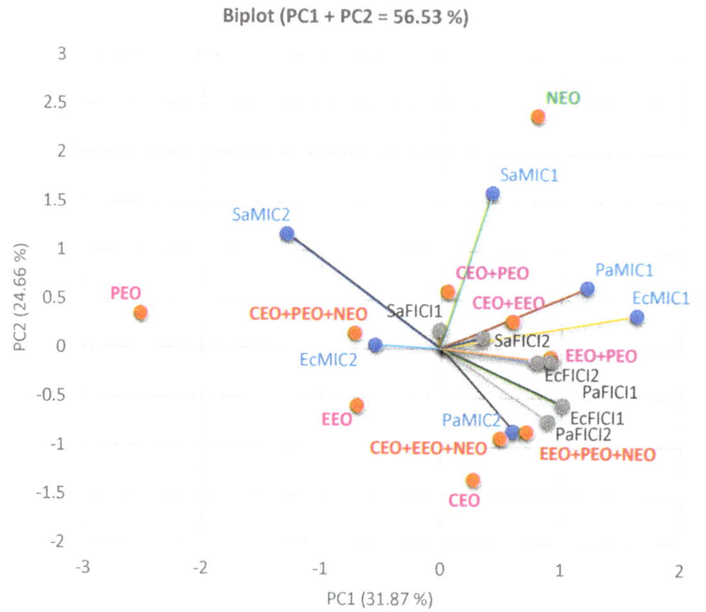

(B)

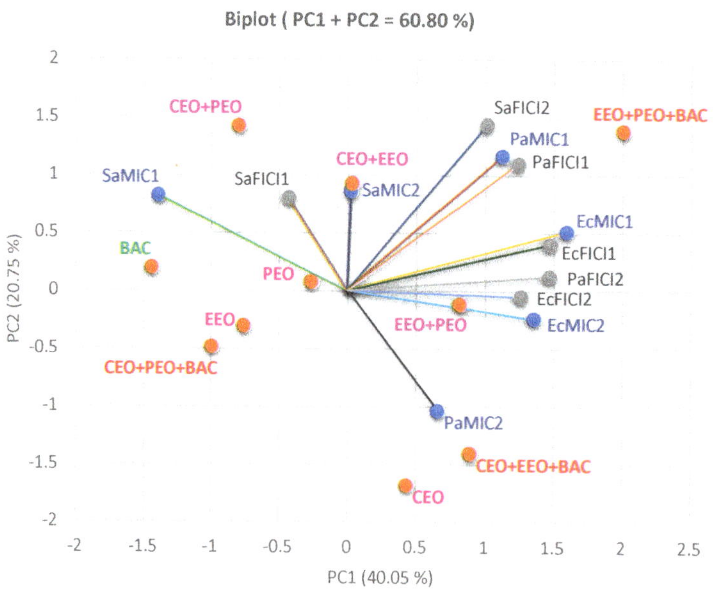

(C)

Figure 8. *Cont.*

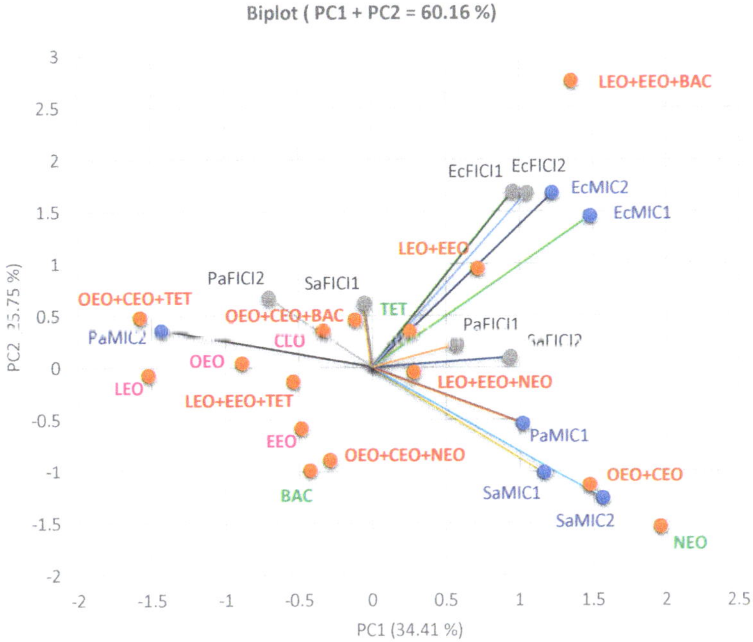

(**D**)

Figure 8. Correlation of MIC values of EOs and those of conventional antibiotics in double and triple combinations and FICI value. (**A–C**) CEO+PEO+EEO group in triple combinations with various antibiotics: (**A**) TET, (**B**) NEO, (**C**) BAC, and (**D**) CEO+OEO+EEO+LEO in double and triple combinations, with 3 antibiotics. LEO—Lavender Essential Oil; CEO—Clove Essential Oil; OEO—Oregano Essential Oil; EEO—Eucalyptus Essential Oil; PEO—Peppermint Essential Oil; NEO—Neomycin; TET—Tetracycline; BAC—Bacitracin; MIC—minimum inhibitory concentration (μg/mL), and FICI—fractional inhibitory concentration index. FICI1 was determined by DDM (disc diffusion method), and the cylinder diffusion technique provided FICI2. Sa—*S. aureus*, Ec—*E. coli*, and Pa—*P. aeruginosa*.

3. Materials and Methods

3.1. Materials and Equipment

Commercially available EOs (OEO, CEO, EEO, PEO, and LEO) were provided by "Laboratoarele Fares Biovital SRL" (Orastie, Romania) [27].

EOs used as GC-MS standards were purchased from established manufacturers: Oregano essential oil was provided by Carl Roth GmbH & Co. KG (Karlsruhe, Germany), Clove and Peppermint essential oils were supplied by Sigma Aldrich Corporation (St. Louis, MO, USA), and Lavender and Eucalyptus essential oils were provided by HWI Group (Ruelzheim, Germany). Certificates of Analysis for proving their purity are available in the Supplementary Materials.

All chemicals and reagents were of analytical grade. Conventional antibiotics (Neomycin sulfate, Bacitracin, and Tetracycline hydrochloride) and Poly (ethylene glycol)-block-poly (propylene glycol)-block-poly (ethylene glycol) (Poloxamer 407) were supplied from Sigma-Aldrich Chemie GmbH (Schnelldorf, Germany). Bacterial strains were purchased from authorized suppliers for the sale of ATCC products: *Staphylococcus aureus* (ATCC 25923), *Escherichia coli* (ATCC 25922), and *Pseudomonas aeruginosa* (ATCC 27853). The bacterial strains were cultured in Mueller–Hinton agar (Thermo Fisher, Dreieich, Germany).

The laboratory equipment consisted of an EnSight multimodal plate reader (PerkinElmer Inc., Waltham, MA, USA), adjustable incubator (Memmert GmbH + Co.KG. Büchenbach. Germany), microplate shaking incubator (Heidolph Instruments GmbH & Co. KG. Schwabach, Germany), laminar flow class II microbiological hood (Jouan SA, Saint-Herblain, Pays de la Loire, France), Evoqua double water still (Evoqua Water Technologies GmbH. Barsbüttel, Germany), an electronic balance (Ohaus Corporation, Parsippany, NJ, USA), and NUNC™ MaxiSorp™ 96-well plates (Electron Microscopy Sciences, Hatfield, PA, USA).

3.2. GC-MS Analysis

A GC/MS analysis was performed to identify and quantify the constituents of the essential oils of Cloves, Eucalyptus, Lavender, Peppermint, and Oregano by perceptible elucidation, comparing their mass spectra with reference spectra (NIST Library 2020). We used a Thermo Scientific Focus GC (Norristown, PA, USA) with an AI/AS 3000 autosampler coupled with a DSQ II mass detector and equipped with a TraceGOLD TG-624 column $60 \times 0.25 \times 1.4$ (mm). The injection volume was 1 µL, at a 1.3 mL/min flow and a split ratio 1:50 using helium as a carrier gas. The initial oven temperature was set to 110 °C and maintained for 5 min. After, it was increased to 220 °C at the rate of 2 °C/min and maintained for 15 min. The MS transfer line temperature was kept at 240 °C. The ion source temperature was 230 °C, and the electron impact ionization (EI) was set at 70 eV. The spectra were scrutinized in full scan mode over the 50 to 450 mass range, and all constituents' retention times (RT) were recorded [58].

3.3. In Vitro Evaluation of Antibacterial Activity

An in vitro evaluation of the antibacterial activity of the tested solutions was performed using a semi-quantitative method (diffusimetric antibiogram adapted from Kirby–Bauer [59]). The results—expressed as IZD (mm)—were measured after 24 h of contact between the tested solutions and pathogenic bacteria.

3.3.1. Bacterial Inoculum

The bacteria inoculum was obtained using the direct colony suspension method (CLSI) [60]. Hence, a saline suspension (0.9%) of bacterial colonies from a 24 h agar plate was prepared, adjusting it to the 0.5 McFarland standard, with around 10^8 CFU/mL (CFU = colony-forming unit) [60].

3.3.2. Sample Solutions

The O/W emulsions were prepared with an EO concentration of 30% *w*/*w*; the emulsifier was Poloxamer 407 5% in water, as previously mentioned [27,61].

Each O/W emulsion was diluted with double-distilled water to assess each EO stock solution's final concentration (25 mg/mL) [27].

The binary EO combinations were prepared in a 1:1 ratio.

The antibiotic drugs (Neomycin, Bacitracin, and Tetracycline) were dissolved in double-distilled water with a stock solution's final concentration of 5120 µg/mL.

The triple combinations contained equal parts of each constituent (1:1:1).

3.3.3. Technique

The minimum and maximum diameters of the inhibition zones (IZD) were measured in mm, and the arithmetic mean was calculated using one decimal place. Sterile cellulose discs with Ø 6 mm were impregnated with 10 µL of the sample solutions and placed on the Petri plates. Sterile borosilicate glass cylinders with Ø 6 mm were placed on the Petri plates with culture media and loaded with 100 µL of sample solutions.

Then, the Petri plates were incubated at 37 °C and read after 24 h. The minimum onhibitory concentration (MIC) was calculated from the arithmetic mean of IZD.

The negative controls were non-impregnated sterile cellulose discs or non-loaded glass cylinders.

The indicative method to evaluate the Quantity/Effect correlation (Q/Ef index) specific for antimicrobial agents screening consists of MIC value calculation:

$$\text{Effect (MIC)} = \frac{Q(\mu g)}{V(\mu L)}$$

Q—the quantity (μg) of sample solution applied.
V—the volume of the environment in which the tested sample diffused and inhibited microbial multiplication.

V was calculated from the average of the IZD, to which the classic cylinder volume formula was applied:

$$V\ (\mu L) = \pi\, R^2 \times G$$

$\pi = 3.14$ (the circle constant);
R^2—the square of the radius from the mean IZD (mm);
G—the cylinder generator, respectively, the thickness of the culture medium layer (mm).

$$V\ (\mu L) = (\phi\ (mm) \times 2^{-1})^2 \times 3.14 \times 3\ (mm)$$

The fractional inhibitory concentration index (FICI) was calculated for the double and triple combinations as follows:

$$FICI_{AB} = (MIC_{A\ in\ combination}/MIC_{A\ alone}) + (MIC_{B\ in\ combination}/MIC_{B\ alone})$$

$$FICI_{ABC} = (MIC_{A\ in\ combination}/MIC_{A\ alone}) + (MIC_{B\ in\ combination}/MIC_{B\ alone}) + (MICc_{\ in\ combination}/MICc_{\ alone})$$

The FICI values were interpreted as follows: FICI $\leq$ 0.5—synergism; 0.5 < FICI < 1—partial synergism; FICI = 1—additive effects; 1 < FICI $\leq$ 4—indifference; and FICI > 4—antagonism [39].

3.4. Data Analysis

The analyses were performed in triplicate, and the results are expressed by mean value ± SD. The differences between the combination groups were determined using Box and Whisker Plots from Microsoft 365 Excel® v.2023 (Microsoft Corporation, Redmond, WA, USA) [27]. The statistically significant values are marked in Figure 3 with superscripts [60]. The Kruskal–Wallis samples comparison, Blend–Altman comparisons between two different methods of the same determination, and correlations between variable parameters examined through Principal Component Analysis [62] were performed with XLSTAT 2024.1.0. 1418 by Lumivero (Denver, CO, USA) [63]. Statistical significance was established at $p < 0.05$ [64].

4. Conclusions

It is not easy to assess the interactions between essential oils and conventional antibiotics in various combinations and appreciate their influence on antibacterial activity. First, the tested Gram-positive and Gram-negative bacteria are crucial regarding their incidence and multidrug resistance. Then, the provenance and storage of essential oils significantly influence the type and amount of bioactive constituents. Next, the differences regarding the phytochemical profile of the combined EOs, the intrinsic mechanisms of the antibacterial actions of the main constituents, and their behavior in combinations are significant. The antibacterial activity evaluation method is essential to diminish bioactive volatile constituents' potential losses or degradation. When EOs are combined with conventional antibiotics, their structural properties and mechanisms of action are of substantial importance. The present study could enrich scientific databases, offering complex data about the antibacterial activity of five commercially available Essential Oils (Lavender, Clove, Oregano, Eucalyptus, and Peppermint) alone in 5 binary EO associations and 15 triple combinations with three conventional antibiotics (Tetracycline, Neomycin, and Bacitracin). Synergistic and additive effects were obtained between the EOs with different bioactive phytochemicals and moderate inhibitory activity on the tested bacteria. Our study reported one synergic binary combination (OEO+CEO) against *P. aeruginosa*, another partial synergy (EEO+PEO) against *S. aureus*, and another additive one (CEO+PEO) against *E. coli*. Similar observations regarding these antibacterial effects are available for the conventional antibiotics combined with EOs. Previously mentioned binary combinations led to four partially synergistic triple combinations with antibiotic drugs (TET and NEO) that were revealed against *S. aureus* and *E. coli*, three containing TET. The present study could not establish which diffusimetric technique was better, because these results were recorded through antibiotic antibiogram techniques, and all previously mentioned factors were involved. Further research could investigate the inhibitory effects of these combinations against tested bacteria through other methods. Moreover, the antibacterial effects of other potential double and triple combinations of these five essential oils and conventional antibiotics could be explored to find the most effective ones.

Supplementary Materials: The following supporting information can be downloaded at: https://www.mdpi.com/article/10.3390/antibiotics13060478/s1. Figure S1. GC-MS Chromatograms of Essential Oils References; Table S1. The Phytochemicals identified and quantified in Essential Oils by GC-MS analysis; Figure S2. Comparative presentation of antibacterial activities evaluated by both diffusimetric methods (disc diffusion—1 and cylinder technique—2) for all EOs, conventional antibiotics, Double EOs combinations (EOs binary), Triple combinations (EOs+TET, EOs+NEO, EOs+BAC). Figure S3. Comparison between DDM and CT.

Author Contributions: Conceptualization: R.N., V.P. and L.-E.I.; methodology: L.-E.I., A.B., D.M.P. and E.A.O.; software: R.N., V.P. and A.B.; validation: L.-E.I., V.O., A.B. and E.A.O.; formal analysis: V.P. and C.E.G.; investigation: R.N., V.P., V.O., D.M.P. and E.A.O.; resources: R.N. and V.P.; data curation: L.-E.I., A.B. and D.M.P.; writing—original draft preparation: R.N., V.P. and A.B.; writing—review and editing: V.P., V.O. and C.E.G.; visualization: R.N., V.P., L.-E.I., V.O., A.B., D.M.P., E.A.O. and C.E.G.; supervision: C.E.G.; project administration: L.-E.I.; funding acquisition: R.N. and V.P. All authors have read and agreed to the published version of the manuscript.

Funding: This research was funded by the Ministry of National Defense through the Internal Project "Determining the synergism of antibacterial and antibiofilm action for new phytopharmaceutical combinations coupled with antibiotics".

Institutional Review Board Statement: Not applicable.

Informed Consent Statement: Not applicable.

Data Availability Statement: Data are available in the present manuscript.

Conflicts of Interest: The authors declare no conflicts of interest.

References

1. Federal Office of Public Health. Which Are the Main Diseases Due to Bacteria? Available online: https://www.bag.admin.ch/bag/en/home/krankheiten/infektionskrankheiten-bekaempfen/antibiotikaresistenzen/welche-bakteriellen-krankheiten-gibt-es-.html (accessed on 16 March 2024).
2. Dhifi, W.; Bellili, S.; Jazi, S.; Bahloul, N.; Mnif, W. Essential Oils' Chemical Characterization and Investigation of Some Biological Activities: A Critical Review. *Medicines* **2016**, *3*, 25. [CrossRef]
3. Masyita, A.; Mustika Sari, R.; Dwi Astuti, A.; Yasir, B.; Rahma Rumata, N.; Emran, T.B.; Nainu, F.; Simal-Gandara, J. Terpenes and Terpenoids as Main Bioactive Compounds of Essential Oils, Their Roles in Human Health and Potential Application as Natural Food Preservatives. *Food Chem. X* **2022**, *13*, 100217. [CrossRef]
4. Manion, C.R.; Widder, R.M. Essentials of Essential Oils. *Am. J. Health Syst. Pharm.* **2017**, *74*, e153–e162. [CrossRef]
5. Aziz, Z.A.A.; Ahmad, A.; Setapar, S.H.M.; Karakucuk, A.; Azim, M.M.; Lokhat, D.; Rafatullah, M.; Ganash, M.; Kamal, M.A.; Ashraf, G.M. Essential Oils: Extraction Techniques, Pharmaceutical And Therapeutic Potential—A Review. *Curr. Drug Metab.* **2018**, *19*, 1100–1110. [CrossRef] [PubMed]

6. Yeddes, W.; Ouerghemmi, I.; Hammami, M.; Gadhoumi, H.; Affes, T.G.; Mohamed, S.N.; Aidi-Wannes, W.; Witrowa-Rajchert, D.; Saidani-Tounsi, M.; Nowacka, M. Optimizing the Method of Rosemary Essential Oils Extraction by Using Response Surface Methodology (RSM)-Characterization and Toxicological Assessment. *Sustainability* **2022**, *14*, 3927. [CrossRef]
7. Khodaei, N.; Nguyen, M.M.; Mdimagh, A.; Bayen, S.; Karboune, S. Compositional Diversity and Antioxidant Properties of Essential Oils: Predictive Models. *LWT* **2021**, *138*, 110684. [CrossRef]
8. Piatkowska, E.; Rusiecka-Ziółkowska, J. Influence of Essential Oils on Infectious Agents. *Adv. Clin. Exp. Med.* **2016**, *25*, 989–995. [CrossRef] [PubMed]
9. Rhayour, K.; Bouchikhi, T.; Tantaoui-Elaraki, A.; Sendide, K.; Remmal, A. The Mechanism of Bactericidal Action of Oregano and Clove Essential Oils and of Their Phenolic Major Components on *Escherichia coli* and *Bacillus subtilis*. *J. Essent. Oil Res.* **2003**, *15*, 286–292. [CrossRef]
10. Guo, F.; Chen, Q.; Liang, Q.; Zhang, M.; Chen, W.; Chen, H.; Yun, Y.; Zhong, Q.; Chen, W. Antimicrobial Activity and Proposed Action Mechanism of Linalool against *Pseudomonas fluorescens*. *Front. Microbiol.* **2021**, *12*, 562094. [CrossRef] [PubMed]
11. Merghni, A.; Belmamoun, A.R.; Urcan, A.C.; Bobiş, O.; Lassoued, M.A. 1,8-Cineol (Eucalyptol) Disrupts Membrane Integrity and Induces Oxidative Stress in Methicillin-Resistant *Staphylococcus aureus*. *Antioxidants* **2023**, *12*, 1388. [CrossRef]
12. Zhang, L.; Xu, S.; Liang, W.; Mei, J.; Di, Y.; Lan, H.; Yang, Y.; Wang, W.; Luo, Y.; Wang, H. Antibacterial Activity and Mode of Action of *Mentha Arvensis* Ethanol Extract against Multidrug-Resistant *Acinetobacter Baumannii*. *Trop. J. Pharm. Res.* **2015**, *14*, 2099. [CrossRef]
13. Tian, L.; Wang, X.; Liu, R.; Zhang, D.; Wang, X.; Sun, R.; Guo, W.; Yang, S.; Li, H.; Gong, G. Antibacterial Mechanism of Thymol against *Enterobacter sakazakii*. *Food Control* **2021**, *123*, 107716. [CrossRef]
14. Nazzaro, F.; Fratianni, F.; De Martino, L.; Coppola, R.; De Feo, V. Effect of Essential Oils on Pathogenic Bacteria. *Pharmaceuticals* **2013**, *6*, 1451–1474. [CrossRef]
15. Mutlu-Ingok, A.; Devecioglu, D.; Dikmetas, D.N.; Karbancioglu-Guler, F.; Capanoglu, E. Antibacterial, Antifungal, Antimycotoxigenic, and Antioxidant Activities of Essential Oils: An Updated Review. *Molecules* **2020**, *25*, 4711. [CrossRef]
16. Reichling, J.; Schnitzler, P.; Suschke, U.; Saller, R. Essential Oils of Aromatic Plants with Antibacterial, Antifungal, Antiviral, and Cytotoxic Properties—An Overview. *Forsch Komplementarmed* **2009**, *16*, 79–90. [CrossRef]
17. Guimarães, A.C.; Meireles, L.M.; Lemos, M.F.; Guimarães, M.C.C.; Endringer, D.C.; Fronza, M.; Scherer, R. Antibacterial Activity of Terpenes and Terpenoids Present in Essential Oils. *Molecules* **2019**, *24*, 2471. [CrossRef]
18. LaLonde, T.; Bowser, T.; Jadeja, R. Essential Oils as Antimicrobials. *Madridge J. Food Technol.* **2019**, *4*, 163–169. [CrossRef]
19. Chouhan, S.; Sharma, K.; Guleria, S. Antimicrobial Activity of Some Essential Oils—Present Status and Future Perspectives. *Medicines* **2017**, *4*, 58. [CrossRef] [PubMed]
20. Yap, P.S.X.; Yiap, B.C.; Ping, H.C.; Lim, S.H.E. Essential Oils, A New Horizon in Combating Bacterial Antibiotic Resistance. *Open Microbiol. J.* **2014**, *8*, 6–14. [CrossRef]
21. Upadhyay, R. Essential Oils: Antimicrobial, Antihelminthic, Antiviral, Anticancer and Antiinsect Properties. *J. Appl. Biosci.* **2010**, *36*, 1–22.
22. Smith-Palmer, A.; Stewart, J.; Fyfe, L. Influence of Subinhibitory Concentrations of Plant Essential Oils on the Production of Enterotoxins A and B and α-Toxin by *Staphylococcus aureus*. *J. Med. Microbiol.* **2004**, *53*, 1023–1027. [CrossRef] [PubMed]
23. Smith-palmer, A.; Stewart, J.; Fyfe, L. Inhibition of Listeriolysin O and Phosphatidylcholine-Specific Production in Listeria Monocytogenes by Subinhibitory Concentrations of Plant Essential Oils. *J. Med. Microbiol.* **2002**, *51*, 567–608. [CrossRef] [PubMed]
24. Sateriale, D.; Forgione, G.; De Cristofaro, G.A.; Facchiano, S.; Boscaino, F.; Pagliuca, C.; Colicchio, R.; Salvatore, P.; Paolucci, M.; Pagliarulo, C. Towards Green Strategies of Food Security: Antibacterial Synergy of Essential Oils from Thymus Vulgaris and *Syzygium aromaticum* to Inhibit *Escherichia coli* and *Staphylococcus aureus* Pathogenic Food Isolates. *Microorganisms* **2022**, *10*, 2446. [CrossRef] [PubMed]
25. Böhme, K.; Barros-Velázquez, J.; Calo-Mata, P.; Aubourg, S.P. Antibacterial, Antiviral and Antifungal Activity of Essential Oils: Mechanisms and Applications. In *Antimicrobial Compounds: Current Strategies and New Alternatives*; Springer: Berlin/Heidelberg, Germany, 2014; ISBN 978-3-642-40444-3.
26. Langeveld, W.T.; Veldhuizen, E.J.A.; Burt, S.A. Synergy between Essential Oil Components and Antibiotics: A Review. *Crit. Rev. Microbiol.* **2014**, *40*, 76–94. [CrossRef] [PubMed]
27. Neagu, R.; Popovici, V.; Ionescu, L.E.; Ordeanu, V.; Popescu, D.M.; Ozon, E.A.; Gîrd, C.E. Antibacterial and Antibiofilm Effects of Different Samples of Five Commercially Available Essential Oils. *Antibiotics* **2023**, *12*, 1191. [CrossRef] [PubMed]
28. Nguyen, F.; Starosta, A.L.; Arenz, S.; Sohmen, D.; Dönhöfer, A.; Wilson, D.N. Tetracycline Antibiotics and Resistance Mechanisms. *Biol. Chem.* **2014**, *395*, 559–575. [CrossRef] [PubMed]
29. Jana, S.; Deb, J.K. Molecular Understanding of Aminoglycoside Action and Resistance. *Appl. Microbiol. Biotechnol.* **2006**, *70*, 140–150. [CrossRef] [PubMed]
30. Suzuki, M.; Yamada, K.; Nagao, M.; Aoki, E.; Matsumoto, M.; Hirayama, T.; Yamamoto, H.; Hiramatsu, R.; Ichiyama, S.; Iinuma, Y. Antimicrobial Ointments and Methicillin-Resistant *Staphylococcus Aureus* USA300. *Emerg. Infect. Dis.* **2011**, *17*, 1917–1920. [CrossRef] [PubMed]
31. Vorobets, N.M.; Yavorska, H.V. Modifications of Agar Diffusion Method to DeterMination of the AntiMicrobial Effect of the Herbal Medicinal Products. *Ukraïns'kij Bìofarmacevtičnij Žurnal* **2016**, *2*, 80–84. [CrossRef]

32. Dickert, H.; Machka, K.; Braveny, I. The Uses and Limitations of Disc Diffusion in the Antibiotic Sensitivity Testing of Bacteria. *Infection* **1981**, *9*, 18–24. [CrossRef]
33. Boz, I.; Gille, E.; Necula, R.; Dunca, S.; Zamfirache, M.M. Chemical Composition and Antibacterial Activity of Essential Oils from Five Populations of *Thymus pulegioides* L. *Cellul. Chem. Technol.* **2015**, *49*, 169–174.
34. Semeniuc, C.A.; Pop, C.R.; Rotar, A.M. Antibacterial Activity and Interactions of Plant Essential Oil Combinations against Gram-Positive and Gram-Negative Bacteria. *J. Food Drug Anal.* **2017**, *25*, 403–408. [CrossRef] [PubMed]
35. Tahric, A.; Kolic, H.; Lavic, A.; Latinovic, D.; Pramenkovic, E. Antibacterial Activity of Oregano Essential Oil and Its Effect on Biofilm Formation. *J. Pure Appl. Microbiol.* **2023**, *17*, 1205–1213. [CrossRef]
36. Truong, S.; Mudgil, P. The Antibacterial Effectiveness of Lavender Essential Oil against Methicillin-Resistant *Staphylococcus aureus*: A Systematic Review. *Front. Pharmacol.* **2023**, *14*, 1306003. [CrossRef] [PubMed]
37. Horváth, P.; Koščová, J. In Vitro Antibacterial Activity of Mentha Essential Oils Against *Staphylococcus aureus*. *Folia Vet.* **2017**, *61*, 71–77. [CrossRef]
38. Damjanović-Vratnica, B.; Đakov, T.; Šuković, D.; Damjanović, J. Antimicrobial Effect of Essential Oil Isolated from *Eucalyptus globulus* Labill. from Montenegro. *Czech J. Food Sci.* **2011**, *29*, 277–284. [CrossRef]
39. Pereira, V.; Dias, C.; Vasconcelos, M.C.; Rosa, E.; Saavedra, M.J. Antibacterial Activity and Synergistic Effects between *Eucalyptus globulus* Leaf Residues (Essential Oils and Extracts) and Antibiotics against Several Isolates of Respiratory Tract Infections (*Pseudomonas aeruginosa*). *Ind. Crops Prod.* **2014**, *52*, 1–7. [CrossRef]
40. Damborg, P.; Pirolo, M.; Schøn Poulsen, L.; Frimodt-Møller, N.; Guardabassi, L. Dogs Can Be Reservoirs of *Escherichia coli* Strains Causing Urinary Tract Infection in Human Household Contacts. *Antibiotics* **2023**, *12*, 1269. [CrossRef]
41. Batinović, F.; Kujundžić, S.L.; Golec, N.; Sunara, D. Malignant Otitis Externa in an Adult Diabetic Patient Caused by Escherichia Coli: A Case Report. *SN Compr. Clin. Med.* **2023**, *5*, 190. [CrossRef]
42. Ishida, K.; Noborio, M.; Nakamura, M.; Ieki, Y.; Sogabe, T.; Sadamitsu, D. Spontaneous *Escherichia Coli* Bacterial Meningitis Mimicking Heatstroke in an Adult. *Clin. Case Rep.* **2016**, *4*, 323–326. [CrossRef]
43. Angeletti, S.; Cella, E.; Prosperi, M.; Spoto, S.; Fogolari, M.; De Florio, L.; Antonelli, F.; Dedej, E.; De Flora, C.; Ferraro, E.; et al. Multidrug Resistant *Pseudomonas aeruginosa* Nosocomial Strains: Molecular Epidemiology and Evolution. *Microb. Pathog.* **2018**, *123*, 233–241. [CrossRef] [PubMed]
44. Schelz, Z.; Molnar, J.; Hohmann, J. Antimicrobial and Antiplasmid Activities of Essential Oils. *Fitoterapia* **2006**, *77*, 279–285. [CrossRef] [PubMed]
45. van Vuuren, S.; Viljoen, A. Plant-Based Antimicrobial Studies—Methods and Approaches to Study the Interaction between Natural Products. *Planta Med.* **2011**, *77*, 1168–1182. [CrossRef] [PubMed]
46. Moon, S.-E.; Kim, H.-Y.; Cha, J.-D. Synergistic Effect between Clove Oil and Its Major Compounds and Antibiotics against Oral Bacteria. *Arch. Oral Biol.* **2011**, *56*, 907–916. [CrossRef] [PubMed]
47. Sousa, F.L.; Santos, M.; Rocha, S.M.; Trindade, T. Encapsulation of Essential Oils in SiO2 Microcapsules and Release Behavior of Volatile Compounds. *J. Microencapsul.* **2014**, *31*, 627–635. [CrossRef] [PubMed]
48. Inouye, S.; Takizawa, T.; Yamaguchi, H. Antibacterial Activity of Essential Oils and Their Major Constituents against Respiratory Tract Pathogens by Gaseous Contact. *J. Antimicrob. Chemother.* **2001**, *47*, 565–573. [CrossRef] [PubMed]
49. Gurtler, J.B.; Garner, C.M. A Review of Essential Oils as Antimicrobials in Foods with Special Emphasis on Fresh Produce. *J. Food Prot.* **2022**, *85*, 1300–1319. [CrossRef] [PubMed]
50. Hyldgaard, M.; Mygind, T.; Meyer, R.L. Essential Oils in Food Preservation: Mode of Action, Synergies, and Interactions with Food Matrix Components. *Front. Microbiol.* **2012**, *3*, 12. [CrossRef]
51. Zhao, H.; Ren, S.; Yang, H.; Tang, S.; Guo, C.; Liu, M.; Tao, Q.; Ming, T.; Xu, H. Peppermint Essential Oil: Its Phytochemistry, Biological Activity, Pharmacological Effect and Application. *Biomed. Pharmacother.* **2022**, *154*, 113559. [CrossRef]
52. Puvača, N.; Milenković, J.; Galonja Coghill, T.; Bursić, V.; Petrović, A.; Tanasković, S.; Pelić, M.; Ljubojević Pelić, D.; Miljković, T. Antimicrobial Activity of Selected Essential Oils against Selected Pathogenic Bacteria: In Vitro Study. *Antibiotics* **2021**, *10*, 546. [CrossRef]
53. Predoi, D.; Iconaru, S.; Buton, N.; Badea, M.; Marutescu, L. Antimicrobial Activity of New Materials Based on Lavender and Basil Essential Oils and Hydroxyapatite. *Nanomaterials* **2018**, *8*, 291. [CrossRef] [PubMed]
54. Clavijo-Romero, A.; Quintanilla-Carvajal, M.X.; Ruiz, Y. Stability and Antimicrobial Activity of Eucalyptus Essential Oil Emulsions. *Food Sci. Technol. Int.* **2019**, *25*, 24–37. [CrossRef] [PubMed]
55. Syafri, S.; Jaswir, I.; Yusof, F.; Rohman, A.; Ahda, M.; Hamidi, D. The Use of Instrumental Technique and Chemometrics for Essential Oil Authentication: A Review. *Results Chem.* **2022**, *4*, 100622. [CrossRef]
56. Nostro, A.; Roccaro, A.S.; Bisignano, G.; Marino, A.; Cannatelli, M.A.; Pizzimenti, F.C.; Cioni, P.L.; Procopio, F.; Blanco, A.R. Effects of Oregano, Carvacrol and Thymol on *Staphylococcus aureus* and *Staphylococcus epidermidis* Biofilms. *J. Med. Microbiol.* **2007**, *56*, 519–523. [CrossRef] [PubMed]
57. Miladi, H.; Zmantar, T.; Kouidhi, B.; Al Qurashi, Y.M.A.; Bakhrouf, A.; Chaabouni, Y.; Mahdouani, K.; Chaieb, K. Synergistic Effect of Eugenol, Carvacrol, Thymol, p-Cymene and γ-Terpinene on Inhibition of Drug Resistance and Biofilm Formation of Oral Bacteria. *Microb. Pathog.* **2017**, *112*, 156–163. [CrossRef] [PubMed]

58. Luță, E.A.; Biță, A.; Moroșan, A.; Mihaiescu, D.E.; Ghica, M.; Mihai, D.P.; Olaru, O.T.; Deculescu-Ioniță, T.; Duțu, L.E.; Popescu, M.L.; et al. The Influence of Phytosociological Cultivation and Fertilization on Polyphenolic Content of Menthae and Melissae Folium and Evaluation of Antioxidant Properties through In Vitro and In Silico Methods. *Plants* **2022**, *11*, 2398. [CrossRef] [PubMed]
59. Hudzicki, J. Kirby-Bauer Disk Diffusion Susceptibility Test Protocol. *Am. Soc. Microbiol.* **2012**, 1–13. Available online: https://asm.org/getattachment/2594ce26-bd44-47f6-8287-0657aa9185ad/Kirby-Bauer-Disk-DiffusionSusceptibility-Test-Protocol-pdf.pdf (accessed on 3 March 2024).
60. Popovici, V.; Bucur, L.; Calcan, S.I.; Cucolea, E.I.; Costache, T.; Rambu, D.; Schröder, V.; Gîrd, C.E.; Gherghel, D.; Vochita, G.; et al. Elemental Analysis and In Vitro Evaluation of Antibacterial and Antifungal Activities of *Usnea barbata* (L.) Weber Ex F.H. Wigg from Călimani Mountains, Romania. *Plants* **2022**, *11*, 32.
61. Popovici, V.; Matei, E.; Cozaru, G.C.; Bucur, L.; Gîrd, C.E.; Schröder, V.; Ozon, E.A.; Karampelas, O.; Musuc, A.M.; Atkinson, I.; et al. Evaluation of *Usnea barbata* (L.) Weber Ex F.H. Wigg Extract in Canola Oil Loaded in Bioadhesive Oral Films for Potential Applications in Oral Cavity Infections and Malignancy. *Antioxidants* **2022**, *11*, 1601. [CrossRef]
62. Ivan, I.M.; Popovici, V.; Chițescu, C.L.; Popescu, L.; Luță, E.A.; Ilie, E.I.; Brașoveanu, L.I.; Hotnog, C.M.; Olaru, O.T.; Nițulescu, G.M.; et al. Phytochemical Profile, Antioxidant and Cytotoxic Potential of *Capsicum annuum* (L.) Dry Hydro-Ethanolic Extract. *Pharmaceutics* **2024**, *16*, 245. [CrossRef]
63. Popovici, V.; Bucur, L.; Gîrd, C.E.; Popescu, A.; Matei, E.; Caraiane, A.; Botnarciuc, M. Phenolic Secondary Metabolites and Antiradical and Antibacterial Activities of Different Extracts of *Usnea barbata* (L.) Weber Ex F.H. Wigg from Călimani Mountains, Romania. *Pharmaceuticals* **2022**, *15*, 829. [CrossRef] [PubMed]
64. Luță, E.-A.; Biță, A.; Moroșan, A.; Mihaiescu, D.E.; Mihai, D.P.; Popescu, L.; Bejenaru, L.E.; Bejenaru, C.; Popovici, V.; Olaru, O.T.; et al. Implications of the Cultivation of Rosemary and Thyme (Lamiaceae) in Plant Communities for the Development of Antioxidant Therapies. *Int. J. Mol. Sci.* **2023**, *24*, 11670. [CrossRef] [PubMed]

antibiotics

MDPI

Article

Spent Material Extractives from Hemp Hydrodistillation as an Underexplored Source of Antimicrobial Cannabinoids

Simon Vlad Luca [1,*], Krzysztof Wojtanowski [2], Izabela Korona-Głowniak [3], Krystyna Skalicka-Woźniak [4], Mirjana Minceva [1] and Adriana Trifan [5,*]

1 Biothermodynamics, TUM School of Life Sciences, Technical University of Munich, 85354 Freising, Germany; mirjana.minceva@tum.de
2 Department of Pharmacognosy with Medicinal Plant Unit, Medical University of Lublin, 20-093 Lublin, Poland; krzysztof.wojtanowski@umlub.pl
3 Department of Pharmaceutical Microbiology, Faculty of Pharmacy, Medical University of Lublin, 20-093 Lublin, Poland; iza.glowniak@umlub.pl
4 Department of Natural Products Chemistry, Medical University of Lublin, 20-093 Lublin, Poland; kskalicka@pharmacognosy.home.pl
5 Department of Pharmacognosy-Phytotherapy, Faculty of Pharmacy, "Grigore T. Popa" University of Medicine and Pharmacy Iasi, 700115 Iasi, Romania
* Correspondence: vlad.luca@tum.de (S.V.L.); adriana.trifan@umfiasi.ro (A.T.)

Abstract: Hemp (*Cannabis sativa* L.) has been used for millennia as a rich source of food and fibers, whereas hemp flowers have only recently gained an increased market interest due to the presence of cannabinoids and volatile terpenes. Currently, the hemp flower processing industry predominantly focuses on either cannabinoid or terpene extraction. In an attempt to maximize the valorization of hemp flowers, the current study aimed to evaluate the phytochemical composition and antimicrobial properties of several extracts obtained from post-distillation by-products (e.g., spent material, residual distillation water) in comparison to the essential oil and total extract obtained from unprocessed hemp flowers. A terpene analysis of the essential oil revealed 14 monoterpenes and 35 sesquiterpenes. The cannabinoid profiling of extracts showed seven acidic precursors and 14 neutral derivatives, with cannabidiol (CBD) reaching the highest concentration (up to 16 wt.%) in the spent material extract. The antimicrobial assessment of hemp EO, cannabinoid-containing extracts, and single compounds (i.e., CBD, cannabigerol, cannabinol, and cannabichromene) against a panel of 20 microbial strains demonstrated significant inhibitory activities against Gram-positive bacteria, *Helicobacter pylori*, and *Trichophyton* species. In conclusion, this work suggests promising opportunities to use cannabinoid-rich materials from hemp flower processing in functional foods, cosmetics, and pharmaceuticals with antimicrobial properties.

Keywords: *Cannabis sativa* L.; cannabidiol; CBD; essential oil; by-products; terpenes; antimicrobial; fatty acids

1. Introduction

Hemp (*Cannabis sativa* L., Cannabaceae) has served for over 6000 years as a versatile resource for food, fibers, oils, medicines, and recreational or religious activities [1]. Throughout the Middle Ages, hemp became a vital fiber crop for the production of textiles and ropes [2]. However, following the discovery of the central nervous system (CNS)-intoxicating effects of Δ^9-tetrahydrocannabinol (Δ^9-THC), an important constituent in its flowers, hemp cultivation has decreased over the 20th century [3]. Nevertheless, at the beginning of the 2000s, controlled genotypes (containing less than 0.2–0.3% Δ^9-THC) were re-authorized to be marketed for agricultural purposes in the European Union [2]. Consequently, more than 60 cultivars have been registered since [4]. Thanks to the wide application of its derivatives, hemp cultivation has ramped up in the last few decades; for instance, fibers and hurds are used for the textile, paper, construction, and automotive industries; biomass is used for the production of biofuels (biodiesel, bioethanol, biogas); and lastly, the oils derived from seeds are regarded as food, cosmetic ingredients, or animal feed [2,3]. Nevertheless, from the global perspective of a sustainable circular economy, more attention has been recently paid to utilizing agro-industrial waste within the hemp production chain, with particular emphasis on the flowers. Hemp flowers/inflorescences are valuable sources of high-adding value compounds, in particular cannabinoids and volatile terpenes.

Cannabinoids are a class of terpeno-phenolics encompassing more than 150 members. These constituents are biosynthesized in the secretory cells of glandular trichomes, primarily concentrated in unfertilized female flowers [5]. Unlike Δ^9-THC, cannabidiol (CBD) lacks CNS-intoxicating properties and has antiemetic, anti-inflammatory, antioxidant, antimicrobial, and neuromodulatory effects. CBD is currently under clinical evaluation for the treatment of 26 medical conditions and has already been granted orphan drug status for 11 other diseases [6]. Additionally, the Food and Drug Administration and the European Medicines Agency have approved Epidiolex®/Epidyolex® as the first CBD-based medication for adjunctive therapy in treating seizures linked to Dravet and Lennox-Gastaut syndromes [7]. As compared to CBD, other cannabinoids, such as cannabigerol (CBG), cannabidivarin (CBDV), cannabinol (CBN), or cannabichromene (CBC), are regarded as minor constituents since they are present in significantly lower amounts in hemp flowers. Preliminary pharmacological investigations have shown promising therapeutic attributes for some of the minor cannabinoids, such as antimicrobial, anti-inflammatory, neuroprotective, anticonvulsant, antiemetic, anti-psoriatic, anticancer, anti-melanogenic, or antidepressant effects [6,8–11].

Besides cannabinoids, volatile terpenes are another class of constituents secreted in the trichomes of female hemp flowers [12]. More than 90 terpenes have been reported in hemp, formally grouped into monoterpenes (e.g., α-pinene, β-pinene, β-myrcene, limonene) and sesquiterpenes (e.g., β-caryophyllene, α-humulene, β-caryophyllene oxide) [13]. These compounds generally confer a mild, light floral fragrance to the hemp flowers, allowing them to be used in perfumery or cosmetics. In addition, hemp terpenes were shown to exert interesting pharmacological properties, such as insecticidal, neuromodulatory, antimicrobial, anticancer, or anti-inflammatory effects [14–16]. Thus, niche products, such as hemp essential oils (EO), usually isolated by steam distillation or hydrodistillation, are currently available on the market [15]. However, in light of the low EO yields achieved from hemp flowers, the residual (spent) plant material remaining at the end of steam or hydrodistillation is considerably high (up to 99.7 wt.%). In addition, a diverse array of other post-distillation by-products are generated, such as aqueous condensates (hydrolates, hydrosols) and distillation waters (residual waters, leachates). Since there is no information concerning the further use of these materials, it can be assumed that they are simply regarded as waste and disposed of in the environment, similar to other waste stream products from the aroma and EO industry [17].

In an attempt to find uses for the above-mentioned by-products that could arise from hemp flower hydrodistillation, this study evaluated the phytochemical profile and antimicrobial activity of several post-distillation extracts in comparison to the essential oil and total extract obtained from unprocessed hemp flowers. Besides our ongoing multi-team research efforts to find novel naturally derived chemotherapeutic agents for further use in treating human infections, our interest in screening hemp's antibacterial and antifungal properties has also been motivated by several preliminary reports that scarcely explored the inhibitory properties of hemp extracts and purified cannabinoids against Gram-positive bacteria, yeasts, and certain Gram-negative bacteria [18–24]. Thus, this work aimed to obtain a hydrodistilled hemp flower essential oil and use the remaining by-products (e.g., spent material, residual distillation water) as an underexplored source of cannabinoids for a further assessment of their antibacterial and antifungal properties.

2. Results and Discussion

In this study, the essential oil from hemp flowers obtained by hydrodistillation (HEO) was analyzed by gas chromatography coupled with mass spectrometry (GC–MS). In parallel, the residual water used for hydrodistillation was concentrated (HWE), whereas the processed hemp flowers (spent material after hydrodistillation) were extracted with hexane (HSE). For comparison purposes, the unprocessed hemp flowers were also extracted under the same conditions (HTE) (Table 1). The fatty acid composition of the solvent extracts was analyzed by GC–MS, whereas the cannabinoid profile was assessed qualitatively by liquid chromatography hyphenated with high-resolution tandem mass spectrometry (LC-HRMS/MS) and quantitatively by liquid chromatography coupled with diode array detection (LC-DAD). Subsequently, the antimicrobial activity of hemp essential oil, cannabinoid-containing extracts, and individual cannabinoids (i.e., CBD, CBG, CBN, and CBC) was tested against a panel of 20 microbial strains.

Table 1. Extraction yields of hemp essential oil and extracts.

Extract	Code	Unit	Yield
Hemp flower essential oil	HEO	mL/100 g	0.7 ± 0.1 [a]
Total (unprocessed) hemp flower extract	HTE	g/100 g	7.3 ± 0.3 [b]
Spent (processed) hemp flower extract	HSE		8.2 ± 0.4 [c]
Hydrodistillation water extract	HWE		6.4 ± 0.2 [d]

Data are reported as mean ± SD of three experiments; different letters within columns indicate significant differences ($p < 0.05$).

2.1. Terpene Profile of Hemp Essential Oil

According to the obtained terpene profile data (Table 2), sesquiterpenes in HEO accounted for $85.94 \pm 0.63\%$ of the total peaks, with almost an equal distribution between oxygenated and non-oxygenated (hydrocarbon) sesquiterpenes. (*E*)-*β*-Caryophyllene (Figure 1) was the dominant compound in HEO ($17.17 \pm 0.24\%$), followed by β-caryophyllene oxide ($8.22 \pm 0.13\%$), α-bisabolol ($8.20 \pm 0.12\%$), α-humulene ($6.12 \pm 0.53\%$), *β*-eudesmol ($5.79 \pm 0.05\%$), and *β*-bisabolene ($5.14 \pm 0.05\%$). From the group of monoterpenes, *α*-pinene and *β*-myrcene are worth mentioning, even though their relative amounts were only $2.80 \pm 0.04\%$ and $1.09 \pm 0.02\%$, respectively (Table 2).

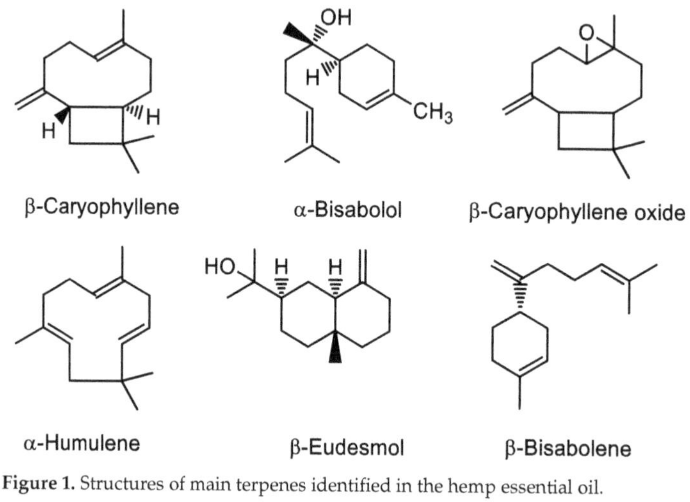

Figure 1. Structures of main terpenes identified in the hemp essential oil.

Table 2. Terpene profile of hemp essential oil.

No.	Compound	LRI [a]	LRI [b]	Relative Abundance (%) [c]
1	*α*-Pinene	936	936	2.80 ± 0.04
2	*β*-Pinene	980	978	0.73 ± 0.00
3	*β*-Myrcene *	989	989	1.09 ± 0.02
4	Limonene *	1032	1033	0.24 ± 0.00
5	Eugenol	1035	1034	0.41 ± 0.01
6	Linalool *	1099	1103	0.55 ± 0.01
7	Fenchyl alcohol	1124	1117	0.32 ± 0.01
8	*cis-p*-Menth-2-en-1-ol	1131	1121	0.24 ± 0.00
9	*trans-p*-Menth-2-en-1-ol	1140	1140	0.32 ± 0.01
10	Camphor	1150	1148	0.12 ± 0.00
11	Borneol	1178	1177	0.24 ± 0.01
12	Terpinen-4-ol	1185	1180	0.13 ± 0.01
13	*p*-Cymen-8-ol	1199	1185	0.49 ± 0.02
14	α-Terpineol	1198	1198	0.46 ± 0.05
15	Copaene	1376	1378	0.20 ± 0.01
16	(Z)-*β*-Caryophyllene	1414	1409	0.20 ± 0.01
17	*γ*-Elemene	1418	1430	0.33 ± 0.01
18	**(*E*)-*β*-Caryophyllene** *	**1431**	**1428**	**17.17 ± 0.24**
19	*α*-Bergamotene	1438	1435	1.87 ± 0.03
20	*allo*-Aromadendrene	1443	1444	0.49 ± 0.01
21	(*E*)-*β*-Farnesene	1453	1456	1.59 ± 0.03
22	**α-Humulene** *	**1467**	**1458**	**6.12 ± 0.53**
23	*γ*-Muurolene	1472	1476	0.25 ± 0.02
24	*α*-Guaiene	1482	1499	0.34 ± 0.06
25	Selina-4(14),7(11)-diene	1485	1497	0.60 ± 0.08
26	*β*-Selinene	1495	1486	0.80 ± 0.03
27	*α*-Selinene	1501	1497	2.85 ± 0.01
28	***β*-Bisabolene**	**1511**	**1509**	**5.14 ± 0.05**
29	*δ*-Cadinene	1520	1519	0.73 ± 0.02
30	*γ*-Patchoulene	1525	1522	0.47 ± 0.01
31	*β*-Sesquiphellandrene	1530	1525	1.06 ± 0.03
32	*α*-Cadinene	1544	1530	1.48 ± 0.03
33	Selina-3,7(11)-diene	1549	1542	0.76 ± 0.02
34	Selina-4(15),7(11)-diene	1554	1532	0.38 ± 0.01
35	Germancrene B	1563	1157	1.48 ± 0.03
36	(*E*)-Nerolidol	1568	1560	0.15 ± 0.01
37	Spathulenol	1575	1573	0.42 ± 0.02
38	***β*-Caryophyllene oxide** *	**1597**	**1581**	**8.22 ± 0.13**
39	Aromadendrene oxide	1606	1650	2.87 ± 0.04
40	*cis*-(*Z*)-*α*-Bisabolene epoxide	1613	1619	0.85 ± 0.01
41	Humulene epoxide	1619	1610	0.41 ± 0.02
42	Cubenol	1625	1614	2.70 ± 0.04
43	*γ*-Eudesmol	1638	1642	4.02 ± 0.06
44	Caryophylla-4(12),8(13)-dien-5*α*-ol	1648	1640	2.29 ± 0.02
45	Caryophylla-4(12),8(13)-dien-5*β*-ol	1651	1634	1.70 ± 0.08
46	*α*-Cadinol	1657	1647	0.53 ± 0.03
47	***β*-Eudesmol**	**1670**	**1651**	**5.79 ± 0.05**
48	*epi*-*α*-Bisabolol	1677	1682	3.49 ± 0.07
49	**α-Bisabolol** *	**1692**	**1685**	**8.20 ± 0.12**
50	Cannabidiol *	–	–	1.75 ± 0.03
			Total	**95.82 ± 1.35**
			Hydrocarbon monoterpenes	4.86 ± 0.02
			Oxygenated monoterpenes	3.33 ± 0.10
			Hydrocarbon sesquiterpenes	44.33 ± 0.62
			Oxygenated sesquiterpenes	41.61 ± 0.64

[a] Linear retention index on ZB-5MS column; [b] Linear retention indices on ZB-5MS column or equivalent columns according to the NIST library (webbook.nist.gov, accessed on 2 February 2024); [c] Expressed as the mean percentage area extracted from the GC–MS chromatograms of three repeated analyses; in bold are the major terpenes; * confirmed by standard.

Our data are consistent with those reported in a previous work, when β-caryophyllene, α-humulene, β-caryophyllene oxide, and β-eudesmol were found in relative amounts of 16.20 ± 0.96%, 6.51 ± 0.45%, 6.39 ± 0.53%, and 5.20 ± 0.26%, respectively [25]. In another study, Pieracci et al. [26] showed that hemp EOs obtained from 10 different genotypes of *C. sativa* were characterized by a predominance of oxygenated sesquiterpenes ranging from 30.90 ± 7.02% to 60.90 ± 2.48%, with β-caryophyllene, α-humulene, β-caryophyllene oxide, and humulene epoxide as the typical constituents. Menghini et al. [27] reported the presence of β-caryophyllene (19.3%), α-humulene (8.3%), and β-caryophyllene oxide (4.3%) in an EO obtained from *C. sativa* cv. Futura 75; in addition, the same group of authors also noticed higher relative amounts of monoterpenes, such as α-pinene (14.9%), β-myrcene (11.8%), and β-pinene (3.8%), than in our study. Furthermore, Zheljazkow et al. [28] claimed that the concentrations of the main hemp terpenes in cultivated and wild hemp ascensions varied between 11.0–29.6% for β-caryophyllene, 4.4–11.9% for α-humulene, 0.2–31.2% for β-caryophyllene oxide, and 1.2–9.5% for humulene epoxide.

2.2. *Cannabinoid Profile of Hemp Extracts*

According to the extraction yields (Table 1), the processed plant material gave a yield of extraction higher than the unprocessed one. This could be related to the fact that the water used for the hydrodistillation extracted many polar components from the initial hemp flowers. Therefore, the same dried mass of processed hemp flowers contains more hydrophobic components (easily extracted with hexane) than the same mass of the initial hemp flowers [17]. The cannabinoid profile of HWE, HSE, and HTE was assessed by LC—HRMS/MS. Due to the lack of standards, the identity of the cannabinoids was tentatively annotated according to previously developed methodologies [29–32]. For instance, the discrimination between CBD-, THC-, and CBC-type acidic cannabinoids was already presented by Piccolella et al. [32] for 20 different compounds identified in hemp pollen samples from Italy. Borille et al. [30] proposed the annotation of around 70 different cannabinoids from 68 Brazilian samples of *Cannabis* spp. (leaves, stems, and flowers). In addition, Berman et al. [29] performed the comprehensive metabolic profiling of 36 samples of *Cannabis* plants from Israel, identifying 94 cannabinoids belonging to 10 distinct subclasses. Barhdadi et al. [31] identified and quantified 17 cannabinoids in 20 CBD-based e-cigarette liquids from the Belgium market.

According to the above-mentioned literature records and standard injections, 21 compounds were putatively identified in HWE, HSE, and HTE (Table 3, Figure 2). Based on the obtained data (Table 3), there were no essential qualitative differences between the extracts, i.e., most peaks were detected in all samples. The annotated cannabinoids can be grouped into two categories: acidic and neutral cannabinoids. With few exceptions, the acidic precursors and the corresponding decarboxylated derivatives were both present in the extracts; e.g., cannabielsoic acid (CBEA, **2**) + cannabielsoin (CBE, **10**); cannabinodiolic acid (CBNDA, **3**) + cannabinodiol (CBND, **7**); cannabitriolic acid (CBTA, **4**) + cannabitriol (CBT, **6**); cannabidivarinic acid (CBDVA, **5**) + CBDV (**9**); tetrahydrocannabinolic acid (Δ^9-THCA, **11**) + Δ^9-THC (**18**); cannabicoumaronic acid (**13**) + cannabicoumarone (**19**); cannabidiolic acid (CBDA, **14**) + CBD (**15**). On the other hand, the neutral form of cannabichromanonic acid (**1**) was not present, while the acidic precursors from epoxycannabigerol (**8**), hydroxycannabinol (**12**), CBN (**16**), dehydrocannabifuran (**17**), cannabicyclol (CBL, **20**), and CBC (**21**) were not identified.

Table 3. Cannabinoid profile of hemp extracts.

No	Proposed Identity	T_R (min)	HRMS (*m*/*z*)	MF	HRMS/MS (*m*/*z*)	Sample	Ref.
1	Cannabichromanonic acid	4.1	377.1974	$C_{21}H_{28}O_6$	301.1463, 283.1369, 273.1480, 255.1386, 245.1524	HTE, HSE	[29]
2	Cannabielsoic acid	5.2	375.2153	$C_{22}H_{30}O_5$	357.2051, 339.1941, 297.1521, 275.1269, 245.1509, 233.1140, 219.0995, 207.1004	HTE, HSE, HWE	[29,30]
3	Cannabinodiolic acid	5.9	355.1892	$C_{22}H_{26}O_4$	323.1605, 313.1776, 299.1625, 273.1465, 253.0850, 239.1048, 225.0896, 211.0977, 197.0944, 187.0741	HTE, HSE, HWE	[29,30,32]
4	Cannabitriolic acid	7.0	391.2091	$C_{22}H_{30}O_6$	301.1454, 283.1310, 273.1460, 255.1353, 245.1510, 235.0933, 217.0834	HTE, HSE, HWE	[29]
5	Cannabidivarinic acid *	8.2	331.1893	$C_{20}H_{26}O_4$	313,1781, 295.0933, 273.1487, 255.1346, 215.1504, 193.1198, 173.0941, 145.0996, 141.0894	HTE, HSE, HWE	[29,30,32]
6	Cannabitriol	10.2	347.2195	$C_{21}H_{30}O_4$	291.1574, 271.1515, 245.1521, 231.1353, 217.1195, 207.1053, 201.0889, 193.1195	HTE, HSE, HWE	[29,30]
7	Cannabinodiol	11.6	311.1995	$C_{21}H_{26}O_2$	281.1500, 231.1358, 225.0907, 217, 1210, 213.0901, 199.0744, 193.1200, 173.0945, 165.0893, 145.0978	HTE, HSE, HWE	[29,30]
8	Epoxycannabigerol	12.7	333.2339	$C_{21}H_{32}O_3$	273.1846, 259.1678, 247.1653, 193.1203, 177.1216, 135.1143, 123.0422	HTE, HSE, HWE	[29,30]
9	Cannabidivarin *	13.9	287.2002	$C_{19}H_{26}O_2$	231.1375, 219.136, 203.1074, 189.0911, 179.059, 165.0907, 123.0439	HTE, HSE, HWE	[29–31]
10	Cannabielsoin	15.3	331.2249	$C_{21}H_{30}O_3$	271.1670, 231.1360, 193.1204, 135.0426, 109.1000	HTE, HSE, HWE	[29–31]
11	Δ^9/Δ^8-Tetrahydrocannabinolic acid	16.5	345.2059	$C_{21}H_{28}O_4$	327.1949, 297.1498, 285.1515, 268.1521, 229.0866, 211.0762, 197.0623	HTE, HSE, HWE	[29–32]
12	Hydroxycannabinol	18.2	327.1978	$C_{21}H_{26}O_3$	313.2111, 287.1663, 271.1692, 259.1686, 247.1362, 231.1355, 211.0768, 201.0899, 193.1208	HTE, HSE, HWE	[30]
13	Cannabicoumaronic acid	20.6	373.1988	$C_{22}H_{28}O_5$	299.1685, 273.1462, 233.1158, 193.1122, 183.0999, 147.0787, 127.0376	HTE, HSE	[29]
14	Cannabidiolic acid *	21.5	359.2201	$C_{22}H_{30}O_4$	341.2096, 299.1638, 285.1469, 273.1474, 261.1467, 233.1159, 219.1001	HTE, HSE, HWE	[29,30,32]
15	Cannabidiol *	22.6	315.2311	$C_{21}H_{30}O_2$	273.1825, 259.1677, 247.1676, 231.1373, 217.1209, 207.21363, 193.1209, 177.1189, 165.0901, 151.0744, 137.0588, 135.1155, 123.0424, 107.0844	HTE, HSE, HWE	[29–32]
16	Cannabinol *	26.1	311.2001	$C_{21}H_{26}O_2$	273.1866, 259.1670, 241.1544, 231.1374, 217.1217, 193.1213, 135.1154, 123.0433, 107.0848	HTE, HSE, HWE	[29–32]
17	Dehydrocannabifuran	29.7	309.1829	$C_{21}H_{24}O_2$	294.1607, 281.1576, 253.1205, 238.0974, 235.1106, 225.1198	HTE, HSE	[29,30]
18	Δ^9/Δ^8-Tetrahydrocannabinol *	32.2	315.2304	$C_{21}H_{30}O_2$	273.1844, 259.1680, 247.1673, 231.1377, 217.125, 207.1355, 193.1211, 135.1158, 123.0433, 107.847	HTE, HSE, HWE	[29–32]
19	Cannabicoumaronone	33.3	329.2095	$C_{21}H_{28}O_3$	287.1522, 273.1444, 259.1655, 247.1320, 229.0827, 209.1138, 153.0543	HTE, HSE	[30]
20	Cannabicyclol *	38.1	315.2308	$C_{21}H_{30}O_2$	273.1792, 259.1681, 247.1697, 233.1519, 217.1224, 207.1359, 193.1208, 177.1217, 135.1160, 123.0431	HTE, HSE, HWE	[29–31]
21	Cannabichromene *	39.1	315.2316	$C_{21}H_{30}O_2$	259.1659, 247.1670, 231.1372, 217.1351, 193.1193, 177.1175, 165.0884, 151.0731, 135.1141, 123.0420	HTE, HSE	[29–31]

* confirmed by standard.

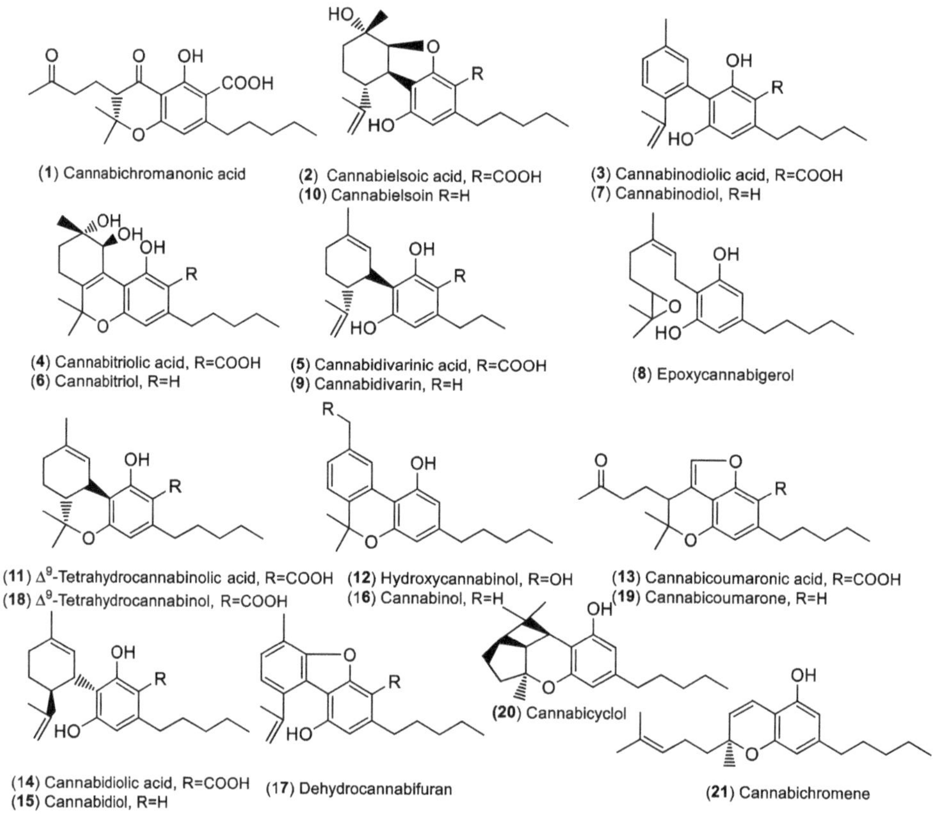

Figure 2. Chemical structures of cannabinoids identified in the hemp extracts.

2.3. Cannabinoid Composition of Hemp Extracts

The quantification of cannabinoids (also referred to as 'cannabinoid potency') is an important step in assessing the quality of a hemp flower-derived product. According to the results presented in Table 4, CBD was the major cannabinoid in all extracts (up to 15.93 ± 0.02 wt.%). Compared to HTE, HSE displayed CBD levels around two times higher, while the CBD concentration in the residual water extract HWE was ten times lower. This tendency could be correlated to the lipophilic character of the cannabinoids, conferring them a low extractability in water and high extractability in organic solvents. Interestingly, the acidic precursor CBDA was found in significant amounts in the unprocessed flower extracts HTE but in very low levels in the spent material extracts (Table 4). This could indicate a thermal conversion of CBDA to CBD during hydrodistillation due to exposing CBDA to a high temperature (~100 °C) for a long time (3 h). The remaining cannabinoids (e.g., CBDV, CBG, CBC, CBN, etc.) were found in concentrations between 0.01 to 0.55 wt.% (Table 4).

Our data are comparable with previous studies reporting on the cannabinoid concentration of various hemp extracts. For example, the concentrations of CBD, CBC, CBG, CBN, and CBL in methanol hemp flower extracts were 2.50 ± 0.10, 0.33 ± 0.12, 0.13 ± 0.01, 0.02 ± 0.00, and 0.02 ± 0.00 wt.%, respectively [33]. Muscara et al. [34] reported the presence of CBD, CBDA, cannabigerolic acid, CBN, and Δ^9-THCA in amounts of 23.51 ± 0.05, 14.65 ± 0.01, 0.29 ± 0.01, 0.36 ± 0.01, 0.23 ± 0.01, and 0.03 ± 0.01 wt.%, respectively, in a hemp flower hexane extract.

Table 4. Cannabinoid concentration of the hemp flower extracts.

Cannabinoid \ Sample	HTE	HSE	HWE
		wt.%	
Cannabidivarinic acid	–	–	–
Cannabidivarin	0.35 ± 0.01 [a]	0.21 ± 0.05 [b]	0.02 ± 0.00 [c]
Cannabidiolic acid	1.11 ± 0.01 [a]	0.06 ± 00 [b]	0.01 ± 0.00 [c]
Cannabigerolic acid	–	–	–
Cannabigerol	0.07 ± 0.00 [a]	0.38 ± 0.00 [b]	0.01 ± 0.00 [c]
Cannabidiol	7.02 ± 0.02 [a]	15.93 ± 0.02 [b]	0.76 ± 0.00 [c]
Tetrahydrocannabivarin	0.11 ± 0.00 [a]	0.02 ± 0.00 [b]	-
Cannabinol	0.54 ± 0.02 [a]	0.55 ± 0.01 [a]	0.04 ± 0.00 [b]
Δ^9-Tetrahydrocannabinol	0.25 ± 0.02 [a]	0.29 ± 0.01 [b]	0.01 ± 0.00 [c]
Δ^8-Tetrahydrocannabinol	0.23 ± 0.01 [a]	0.02 ± 0.00 [b]	-
Cannabicyclol	0.02 ± 0.00 [a]	0.06 ± 0.00 [b]	0.01 ± 0.00 [c]
Cannabichromene	0.02 ± 0.00 [a]	0.02 ± 0.00 [a]	-
Δ^9-Tetrahydrocannabinolic acid	0.03 ± 0.00 [a]	0.02 ± 0.00 [b]	-

Data are expressed as mean ± SD of three repeated analyses; sample codes as in Table 1; different letters within rows indicate significant differences ($p < 0.05$).

2.4. Fatty Acid Composition of Hemp Extracts

Even though hemp flower extracts are known to contain lipids, their fatty acid composition is usually neglected. In all samples (Table 5), linoleic acid (C18:2) and linolenic acid (C18:3) were found in the highest amounts (up to 1.75 ± 0.07 wt.%). Significant concentrations of palmitic (C16:0) and arachidic acid (C20:1) were also noticed. Concerning its high polarity, the residual water extract HWE contained 15–50 times lower quantities of fatty acids than the solvent extracts. Generally, HSE displayed slightly higher values of fatty acids than HTE. This can be related to the fact that, during hydrodistillation, lipids are neither entrained with the volatile terpenes nor transferred into the residual water.

Table 5. Quantification of fatty acid derivatives in the hemp flower extracts.

Fatty acid \ Sample	HTE	HSE	HWE
		wt.%	
Myristic acid (C14:0)	0.09 ± 0.00 [a]	0.17 ± 0.01 [b]	-
Palmitic acid (C16:0)	0.88 ± 0.03 [a]	0.86 ± 0.01 [a]	0.03 ± 0.01 [b]
Stearic acid (C18:0)	0.09 ± 0.00 [a]	0.10 ± 0.00 [b]	-
Oleic acid (C18:1)	0.29 ± 0.01 [a]	0.36 ± 0.01 [b]	-
Linoleic acid (C18:2)	1.52 ± 0.08 [a]	1.75 ± 0.07 [b]	0.03 ± 0.01 [c]
Linolenic acid (C18:3)	1.39 ± 0.08 [a]	1.26 ± 0.09 [a]	0.14 ± 0.00 [b]
Arachidic acid (C20:1)	0.58 ± 0.02 [a]	0.89 ± 0.03 [b]	-
Behenic acid (C22:0)	0.13 ± 0.01 [a]	0.20 ± 0.02 [b]	-

Data are expressed as mean ± SD of three repeated analyses; sample codes as in Table 1; different letters within rows indicate significant differences ($p < 0.05$).

To our knowledge, there are no previous data concerning the fatty acid composition of hemp flower extracts. However, there are numerous similarities with the composition of hemp seeds. Kriese et al. [35] showed that linoleic acid (C18:2) and linolenic acid (C18:3) are the predominant fatty acids in the oils extracted from the seeds of 51 *C. sativa* genotypes. Their levels varied from 15.00–19.89 wt.% for linoleic acid and 5.15–8.24 wt.% for linolenic acid. Furthermore, numerous studies confirmed that linoleic acid (C18:2) is the major constituent of hempseed oils (up to 60% of the total fatty acids) [36–39].

2.5. Antimicrobial Activity of Hemp Essential Oil and Extracts

All obtained extracts were tested against a panel of human pathogens comprising nine Gram-positive bacteria, six Gram-negative bacteria, and five fungi (Table 6). Except for

H. pylori (minimum inhibitory concentration, MIC ≤ 62.5 mg/L), the samples were inactive against the other Gram-negative bacteria. Nevertheless, most Gram-positive strains were sensitive to the treatment with the hemp extracts. The inactivity of hemp against most Gram-negative bacteria has already been proven by other authors [34]. According to the criteria proposed by Kuete and Efferth [40], a sample is considered to have a significant antimicrobial activity if the MIC value is below 100 mg/L. Thus, HEO displayed the most potent antibacterial activity against *S. aureus* (MIC = 62.5 mg/L) and *M. luteus* (MIC = 15.6 mg/L); for the other strains, the MIC values were ≥125 mg/L. In general, the previous literature data also showed a modest activity of hemp EO. For instance, the MIC values ranged from 1200–4700 mg/L for a panel of bacteria comprising *S. aureus*, *B. subtilis*, *M. luteus*, *E. coli*, *P. aeruginosa*, and *K. pneumoniae* [41]. Similar values (MIC = 8000 mg/L) were also reported for hemp EO against different strains of *S. aureus* [42].

Table 6. Antimicrobial activity of hemp essential oil and extracts.

Microorganism \ Extract	HEO	HSE	HTE	HWE	Antibiotics
	MIC [mg/L]				
Gram-positive bacteria					
Staphylococcus aureus	62.5 [a]	0.98 [b]	0.98 [b]	31.3 [c]	0.98 [#]
Staphylococcus aureus *	1000 [a]	3.9 [b]	3.9 [b]	125 [c]	0.98 [#]
Staphylococcus epidermidis	1000 [a]	3.9 [b]	3.9 [b]	125 [c]	0.12 [#]
Micrococcus luteus	15.6 [a]	1.95 [b]	3.9 [c]	62.5 [d]	1.95 [#]
Enterococcus faecalis	2000 [a]	7.8 [b]	7.8 [b]	250 [c]	0.24 [#]
Bacillus cereus	500 [a]	1.95 [b]	3.9 [c]	62.5 [d]	0.98 [#]
Streptococcus pneumoniae	125 [a]	31.3 [b]	62.5 [c]	62.5 [c]	0.24 [#]
Streptococcus pyogenes	250 [a]	31.3 [b]	62.5 [c]	1000 [d]	0.24 [#]
Streptococcus mutans	250 [a]	125 [b]	125 [b]	2000 [c]	0.98 [#]
Gram-negative bacteria					
Helicobacter pylori	15.6 [a]	7.8 [b]	7.8 [b]	62.5 [c]	0.98 [$]
Salmonella Typhimurium	>2000	>2000	>2000	>2000	0.06 [§]
Escherichia coli	>2000	>2000	>2000	>2000	0.02 [§]
Proteus mirabilis	>2000	>2000	>2000	>2000	0.03 [§]
Klebsiella pneumoniae	>2000	>2000	>2000	>2000	0.12 [§]
Pseudomonas aeruginosa	>2000	>2000	>2000	>2000	0.49 [§]
Yeasts					
Candida albicans	1000 [a]	2000 [b]	250 [c]	>2000	0.49 [ß]
Candida parapsilosis	500 [a]	2000 [b]	250 [c]	2000 [b]	0.24 [ß]
Candida glabrata	500 [a]	>2000	250 [b]	>2000	0.24 [ß]
Trichophyton rubrum	125 [a]	125 [a]	31.3 [b]	1000 [c]	0.0049 [&]
Trichophyton mentagrophytes	125 [a]	31.3 [b]	31.3 [b]	250 [c]	0.0012 [&]

[#] vancomycin; [$] ofloxacin; [§] ciprofloxacin; [ß] nystatin; [&] terbinafine; sample codes as in Table 1; * methicillin-resistant strain; different letters within rows indicate significant differences ($p < 0.05$); MIC, minimum inhibitory concentration.

HTE showed the highest activity against *S. aureus* (MIC = 0.98 mg/L). For *S. aureus* MRSA, *S. epidermidis*, *M. luteus*, and *B. cereus*, the MIC values displayed by HTE were identical (3.9 mg/L). *Candida* species were not significantly inhibited; however, a strong anti-*Trichophyton* activity (MIC = 31.3 mg/L) was noticed for HTE. Similarly, HSE acted as a potent inhibitor against the growth of *S. aureus* (MIC = 0.98 mg/L), *M. luteus* (MIC = 1.95 mg/L), *E. faecalis* (MIC = 7.8 mg/L), and *B. cereus* (MIC = 1.95 mg/L), and *T. mentagrophytes* (MIC = 31.3 mg/L). The antimicrobial potential of HWE was better than HEO but considerably lower than that exhibited by HTE and HSE (Table 6). This behavior could be linked to the low cannabinoid concentration noticed in HWE compared to HTE and HSE (Table 4).

Our findings are in line with those observed by other authors. For instance, various hemp flower extracts displayed MIC values between 10 and 66 mg/L against *E. coli*, *P. aeruginosa*, *B. subtilis*, and *S. aureus* [43]. Muscara et al. [34] reported MIC values of 39 mg/L against *S. aureus*. On the other hand, Serventi et al. [4] documented for various hemp extracts a potent inhibitory activity against *B. subtilis* (MIC = 1.5–25 mg/L), *S. aureus* (MIC = 12.5–100 mg/L), *P. aeruginosa* (MIC = 25–100 mg/L), and *E. coli* (MIC = 6.2–12.5 mg/L); however, the inhibition of *B. cereus* and *S. typhy* was negligible (MIC > 200 mg/L). Furthermore, the same group of authors also presented the activity of hemp extracts against several dermatophyte strains, with MIC values ranging from 25 to 100 mg/L against *T. mentagrophytes*, *T. tonsurans*, and *T. rubrum* [4].

In an attempt to link the antimicrobial activity of hemp flower extracts to the presence of cannabinoids, the effects of four previously isolated cannabinoids (i.e., CBD, CBG, CBN, and CBC) [44] were subsequently evaluated against the same panel of microbial strains (Table 7). All tested cannabinoids were inactive against Gram-negative bacteria except for *H. pylori* (MIC ≤ 0.98 mg/L for CBG, CBD, and CBN). Furthermore, the four compounds showed no significant inhibitory activity against *Candida* spp. and *Trichophyton* spp. However, the MIC values against the Gram-positive strains indicated promising antibacterial activity, in particular against *S. aureus*, methicillin-resistant *S. aureus* (MRSA), *S. epidermidis*, *M. luteus*, *E. faecalis*, and *B. cereus* (MIC between 0.49 and 15.6 mg/mL). Overall, the following decreasing order of the inhibitory activity against Gram-positive bacteria could be proposed: CBD > CBN > CBG > CBC.

Table 7. Antimicrobial activity of selected cannabinoids.

Compound / Microorganism	CBG	CBD	CBN	CBC	Antibiotics
	MIC [mg/L]				
Gram-positive bacteria					
Staphylococcus aureus	1.95 ᵃ	0.49 ᵇ	1.95 ᵃ	3.95 ᶜ	0.98 #
Staphylococcus aureus *	1.95 ᵃ	1.95 ᵃ	1.95 ᵃ	15.6 ᵇ	0.98 #
Staphylococcus epidermidis	1.95 ᵃ	1.95 ᵃ	1.95 ᵃ	1.95 ᵃ	0.12 #
Micrococcus luteus	1.95 ᵃ	0.98 ᵇ	0.98 ᵇ	0.98 ᵇ	1.95 #
Enterococcus faecalis	3.9 ᵃ	1.95 ᵇ	1.95 ᵇ	0.98 ᶜ	0.24 #
Bacillus cereus	1.95 ᵃ	0.98 ᵇ	1.95 ᵃ	7.8 ᶜ	0.98 #
Streptococcus pneumoniae	31.3 ᵃ	31.3 ᵃ	62.5 ᵇ	62.5 ᵇ	0.24 #
Streptococcus pyogenes	31.3 ᵃ	31.3 ᵃ	62.5 ᵇ	62.5 ᵇ	0.24 #
Streptococcus mutans	31.3 ᵃ	31.3 ᵃ	62.5 ᵇ	62.5 ᵇ	0.98 #
Gram-negative bacteria					
Helicobacter pylori	0.49 ᵃ	0.98 ᵇ	0.98 ᵇ	15.6 ᶜ	0.98 $
Salmonella Typhimurium	>2000	>2000	>2000	>2000	0.06 §
Escherichia coli	>2000	>2000	>2000	>2000	0.02 §
Proteus mirabilis	>2000	>2000	>2000	>2000	0.03 §
Klebsiella pneumoniae	>2000	>2000	>2000	>2000	0.12 §
Pseudomonas aeruginosa	>2000	>2000	>2000	>2000	0.49 §
Yeasts					
Candida albicans	500 ᵃ	500 ᵃ	1000 ᵇ	500 ᵃ	0.49 ß
Candida parapsilosis	1000 ᵃ	1000 ᵃ	1000 ᵃ	500 ᵇ	0.24 ß
Candida glabrata	250 ᵃ	1000 ᵇ	1000 ᵇ	1000 ᵇ	0.24 ß
Trichophyton rubrum	1000 ᵃ	1000 ᵃ	1000 ᵃ	1000 ᵃ	0.0049 &
Trichophyton mentagrophytes	1000 ᵃ	1000 ᵃ	1000 ᵃ	1000 ᵃ	0.0012 &

vancomycin; $ ofloxacin; § ciprofloxacin; ß nystatin; & terbinafine; * methicillin-resistant strain; different letters within rows indicate significant differences ($p < 0.05$); CBG, cannabigerol; CBD, cannabidiol; CBN, cannabinol; CBC, cannabichromene; MIC, minimum inhibitory concentration.

The evidence of antimicrobial activity for CBD has already been proven. For instance, MIC values of 1–4 mg/L were noticed against a diverse array of Gram-positive bacteria, including MRSA, MDR *S. pneumoniae*, *E. faecalis*, and anaerobic *Clostridium difficile* and *Cutibacterium acnes* [45]. CBD also showed different levels of antibacterial activity against Gram-positive bacteria, including susceptible and MDR strains (MIC = 2–4 mg/L for *E. faecium*, *Enterococcus* spp., *Staphylococcus* spp., *M. luteus*, and *Rhodococcus equi*. Furthermore, CBD displayed MIC values of 4–8 mg/L against MRSA, *E. faecalis*, and *L. monocytogenes* [46].

The antimicrobial activity of other minor cannabinoids has been scarcely investigated. For instance, Appendino et al. [5] showed that CBC, Δ^9-THC, CBN, and CBG could exert antibacterial activity against various MDR *S. aureus* strains, with MIC values between 0.5 and 2 mg/L. In addition, the same cannabinoids also inhibited MRSA, *Streptococcus mutans*, *Streptococcus sanguis*, *Streptococcus sobrinus*, and *Streptococcus salivarius* with MIC below 5 mg/L [47]. Furthermore, *S. aureus*, *S. epidermidis*, and *S. pyogenes* were also shown to be impacted by CBG treatment (MIC = 10–75 µM) [48,49]. Nevertheless, the inhibitory effects of CBG, CBN, and CBC against other human pathogens are reported herein for the first time.

Concerning the possible mechanisms of antimicrobial activity of cannabinoids, it was previously shown that CBD, in combination with bacitracin, caused defects in cell division and irregularities in the cell envelope [46]. The treatment with CBG led to intracellular accumulation of membrane structures, induced membrane hyperpolarization, and decreased membrane fluidity of various bacterial strains [48]. The antibacterial activity of cannabinoids against MRSA was shown to be mediated through the inhibition of biofilm formation, the eradication of pre-formed biofilms and stationary phase cells persistent to antibiotics [47]. Furthermore, CBD proved a potent inhibitor of membrane vesicle release from *E. coli*, altering cell communication [50]. Nevertheless, the antimicrobial mechanisms of CBD and minor cannabinoids remain elusive, requiring subsequent investigations.

3. Materials and Methods

3.1. Materials

Alkane standard solution (C8–C20, ~40 mg/L each, in hexane), fatty acid methyl ester (FAME) mix (C8–C24), Mueller–Hinton (MH) broth, bisabolol (≥93%), caryophyllene (≥98%), myrcene (≥98%), linalool (≥99%), tridecanoic acid (≥98%), hydrochloric acid (≥37%), and glucose (≥99%) were acquired from Merck KGaA (Darmstadt, Germany). Hexane (≥95%), ethanol (≥99%), methanol (≥99%), acetonitrile (≥99%), and water (≥99%) were from VWR Chemicals (Ismaning, Germany). Caryophyllene oxide (≥95%) was bought from Thermo Scientific (Olching, Germany), whereas humulene (≥97%) was purchased from Biomol (Hamburg, Germany). The certified reference materials of 13 cannabinoids containing cannabidivarinic acid (CBDVA), cannabidivarin (CBDV), cannabigerolic acid (CBGA), cannabigerol (CBG), cannabidiol (CBD), tetrahydrocannabivarin (THCV), cannabinol (CBN), Δ^9-tetrahydrocannabinol (Δ^9-tTHC), Δ^8-tetrahydrocannabinol (Δ^8-THC), cannabicyclol (CBL), cannabichromene (CBC), and Δ^9-tetrahydrocannabinolic acid (Δ^9-THCA) was supplied from LGC Standards (Kielpin, Poland). The dried *Cannabis sativa* (hemp) flowers (cv. Futura 75) were provided by Hempartis GmbH (Malsch, Germany).

3.2. Preparation of Essential Oil and Solvent Extracts

The powdered hemp flowers (50 g) were subjected to hydrodistillation on a Clevenger-type apparatus with 500 mL water for 3 h. At the end of the hydrodistillation process, the amount of essential oil (HEO) was measured using the apparatus scale (in mL), collected, and dried over anhydrous sulfate. The water in the flask was filtered, and 250 mL were freeze-dried to afford the residual water extract (HWE). The solid plant material residue (spent hemp flowers) was dried in an oven at 40 °C for 48 h, and 7.5 g were extracted with hexane (75 mL) in an ultrasound bath for 30 min, for 3 repeated cycles, each time with the same volume of fresh solvent. After filtration, the solvent was evaporated under reduced pressure, yielding the spent extract (HSE). For comparison purposes, the unprocessed powdered hemp flowers were extracted with the same solvent under the same conditions, affording the total (unprocessed material) extract (HTE). For HEO, the yield was calculated with the following formula:

$$\%\text{yield} = \frac{\text{volume of the obtained oil(mL)}}{\text{mass of the hemp flowers(g)}} \times 100$$

For the spent extract (HSE), the yield was determined as follows:

$$\%\text{yield} = \frac{\text{mass of the obtained extract after drying(g)}}{\text{mass of the spent hemp flowers(g)}} \times 100$$

For the total extract (HTE), the yield was calculated with the formula:

$$\%\text{yield} = \frac{\text{mass of the obtained extract after drying(g)}}{\text{mass of the hemp flowers(g)}} \times 100$$

For HWE, the yield was determined using the formula:

$$\%\text{yield} = \frac{m_1(g)}{m_2(g)} \times \frac{V_1(mL)}{V_2(mL)} \times 100$$

where m_1 is the mass of the obtained water extract after freeze-drying, m_2 is the mass of the hemp flowers, V_1 is the volume of the water introduced in the hydrodistillation, and V_2 is the volume of the freeze-dried water.

All extractions were performed in triplicate, with the extraction yields provided in Table 1.

3.3. *Analytical Methods*

3.3.1. Terpene Profile of Hemp Essential Oil (GC–MS)

The terpene profile of HEO was assessed by GC–MS performed on a TRACE GC Ultra instrument (Thermo Fischer, Waltham, MA, USA). The chromatographic separations were conducted on a Zebron ZB-5MS (30 m × 0.25 mm, 0.25 μm) column (Phenomenex, Torrance, CA, USA), with helium as the carrier gas at a flow rate of 1.43 mL/min. The injection temperature was 250 °C; 1 μL was injected with a split ratio of 50:1. The column temperature was initially held at 60 °C for 4 min, then increased to 280 °C at a rate of 10 °C/min, and maintained at 280 °C for 5 min. The MS parameters included: transfer line temperature, 320 °C; source temperature, 230 °C; ionization energy, 70 eV. The linear retention indices (LRI) were determined for all compounds in the chromatograms using a standard mixture of alkanes ranging from 8 to 20 carbon atoms. Peak identification was achieved by referencing the NIST 11 Mass Spectra Library and comparing the calculated LRI with those found in the relevant literature. All analyses were performed in triplicate.

3.3.2. Cannabinoid Profile and Composition

The cannabinoid profile of solvent extracts (HTE, HSE, and HWE) was assessed by LC–HRMS/MS. An Agilent 1200 HPLC (Agilent Technologies, PaloAlto, CA, USA), comprising an auto-sampler (G1329B), a binary pump (G1312C), and a column thermostat (G1316A), was used. The instrument was connected to an accurate-mass quadrupole time-of-flight (QTOF) MS/MS system from the Agilent 6530B series via a dual electrospray ionization (ESI) interface. The chromatographic separations were conducted on a Gemini C18 (100 mm × 2 mm, 3 μm) column (Phenomenex, Torrance, CA, USA) operated at 25 °C. The mobile phase comprised (A) water and (B) acetonitrile, both containing 0.1% formic acid. The phases were delivered at a flow rate of 0.2 mL/min, in the following gradient: 50–55% B from 0 to 12 min, 55–58% from 12 to 25 min, 58–75% B from 25 to 30 min, 75–90% B from 30 to 45 min, 90% B from 45 to 50 min. The sample injection volume was 10 μL.

The detection was carried out in positive electrospray ionization mode, with the spectra recorded in the m/z 100–1000 Da range. The ion source parameters were as follows: drying gas (nitrogen) flow rate, 12 L/min; heated capillary temperature, 300 °C; nebulizer pressure, 35 psi; sheath gas temperature, 275 °C; sheath gas flow rate, 12 L/min; capillary voltage, 4000 V; nozzle voltage, 2000 V; fragmentor, 110 V; skimmer, 65 V; octupole radiofrequency peak voltage, 750 V. The MS/MS spectra were generated by automated fragmentation at a fixed collision energy of 30 V.

The cannabinoid concentration of solvent extracts (HTE, HSE, and HWE) was assessed by LC–DAD on a Shimadzu HPLC (Tokyo, Japan) containing a binary pump (LC-20AD), autosampler (SIL-20A), degasser (DGU-20A), and UV/VIS detector (SPD-M20A). The chromatographic separations were conducted on a Zorbax XDB-C18 (150 mm × 4.6 mm, 3.5 μm) column (Agilent Technologies, Palo Alto, CA, USA) at 30 °C. The mobile phase comprised (A) water and (B) acetonitrile, both containing 0.1% formic acid. The phases were delivered at a flow rate of 1.5 mL/min in the following gradient: 72% B for 4 min, 75–80% B from 4.01 to 11 min, and 90% B from 11.01 to 12 min. The sample injection volume was 10 μL, with the chromatograms recorded at 228 nm. The concentrations of cannabinoids were assessed using the calibration curves of the corresponding standards. All analyses were performed in triplicate.

3.3.3. Fatty Acid Composition of Hemp Extracts (GC–MS)

The fatty acid composition of solvent extracts (HTE, HSE, and HWE) was assessed by GC–MS on an Agilent 6890 GC (Agilent Technologies, PaloAlto, CA, USA) coupled to a mass selective detector (MSD). The chromatographic separations were conducted on a Zebron Rtx-Wax (30 m × 0.25 mm, 0.25 μm) column (Restek, Centre County, PA, USA), with helium as the carrier gas at a flow rate of 1 mL/min. The injection temperature was 200 °C; 1 μL was injected with a split ratio of 7.5:1. The column temperature was initially held at 80 °C for 2 min, then increased to 180 °C at a rate of 7 °C/min, maintained at 180 °C for 10 min; then ramped up to 230 °C at a rate of 1 °C/min, and finally held at 230 °C for 10 min. The MS parameters included: transfer line temperature, 230 °C; source temperature, 230 °C; ionization energy, 70 eV. Before analyses, the samples (~20 mg) were dissolved in 1 mL hexane. Then, 0.2 mL of the obtained solutions were incubated at 90 °C for 1 h in the presence of 6 mL of methanol/hydrochloric acid (11:1, v/v) and 0.1 mL tridecanoic acid (1 mg/mL in hexane) as internal standard. Next, 1.7 mL hexane and 2 mL water were added; after shaking and phase separation, 1 mL of the upper phase was analyzed by GC–MS. The fatty acid content in the samples was expressed based on a calibration curve performed with the FAME mixture. All analyses were performed in triplicate.

3.4. Antimicrobial Assays

The antimicrobial assays were performed using the microdilution method, according to the European Committee on Antimicrobial Susceptibility Testing [51,52]. The antimicrobial activity was evaluated against Gram-positive bacteria (*Staphylococcus aureus* ATCC 25923, *Staphylococcus aureus* ATCC BAA-1707, *Staphylococcus epidermidis* ATCC 12228, *Micrococcus luteus* ATCC 10240, *Enterococcus faecalis* ATCC 29212, *Bacillus cereus* ATCC 10876, *Streptococcus pneumoniae* ATCC 49619, *Streptococcus pyogenes* ATCC 19615, *Streptococcus mutans* ATCC 25175), Gram-negative bacteria (*Helicobacter pylori* ATCC 43504, *Salmonella* Typhimurium ATCC 14028, *Escherichia coli* ATCC 25922, *Proteus mirabilis* ATCC 12453, *Klebsiella pneumoniae* ATCC 13883, *Pseudomonas aeruginosa* ATCC 9027), and fungi (*Candida albicans* ATCC 102231, *Candida parapsilosis* ATCC 22019, *Candida glabrata* ATCC 2091, *Trichophyton rubrum* ATCC 28188, *Trichophyton mentagrophytes* ATCC 9533). Serial double dilutions of samples were prepared in MH broth or RPMI 1640 medium 2% glucose buffered with 0.165 M MOPS and supplemented with chloramphenicol 50 mg/L and cycloheximide 300 mg/L, for non-fastidious bacteria and fungi, respectively. The sterile 96-well flat-bottom polystyrene microtitrate plates (Nunc, Denmark) were prepared by dispensing 100 μL of the appropriate dilution of the tested extracts in broth medium per well by serial two-fold dilutions, to obtain the final concentrations of the tested extracts ranging from 2000 to 0.25 mg/L for bacteria and yeasts. The inocula were prepared with fresh microbial cultures in sterile 0.85% NaCl to match the turbidity of 0.5 McFarland standard, and were added to wells to obtain the final density of 5×10^5 colony forming units (CFU)/mL for bacteria, 5×10^4 CFU/mL for yeasts and 5×10^5 CFU/mL for dermatophytes. After incubation (non-fastidious bacteria and yeasts—35 °C for 24 h and dermatophytes—28 °C for 5 days), the growth of microorganisms was measured spectrophotometrically at 600 nm (BioTEK ELx808, BioTek Instruments, Inc., Winooski, VT, USA). The minimum inhibitory concentration (MIC) for *H. pylori* ATCC 43504 was determined using a two-fold microdilution method in MH broth with 7% of lysed horse blood at an extract concentration ranging from 2000 to 0.25 mg/L with a *H. pylori* suspension of 3 McFarland standard diluted 100 times (9×10^6 CFU/mL). After incubation at 35 °C for 72 h under microaerophilic conditions (5% O_2, 15% CO_2, and 80% N_2), the growth of *H. pylori* was visualized with the addition of 10 µL of 0.04% resazurin. The MIC endpoint was recorded after 4 h incubation as the lowest concentration of extract that completely inhibits bacterial growth. An appropriate DMSO control (at a final concentration of 10%), a positive control (containing inoculum without the tested extracts), and a negative control (containing the tested extracts without inoculum) were included on each microplate. The MIC was determined and reported for each sample and strain. Vancomycin, ciprofloxacin, ofloxacin, and nystatin/terbinafine were used as the standard reference drugs. All experiments were performed in triplicate.

3.5. Statistical Analysis

Data are provided as mean ± standard deviation of three repeated experiments; ANOVA with Tukey's post hoc test was conducted; $p < 0.05$ was considered statistically significant.

4. Conclusions

The seminal findings provided by the phytochemical analysis are as follows: (i) 15 monoterpenes and 36 sesquiterpenes were identified in the EO, with sesquiterpenes accounting for ~85% of total peak area; (ii) a total of 7 acidic cannabinoids and 14 neutral derivatives were annotated in the post-distillation by-products, with CBD as the dominant compound (up to 16 wt.%); (iii) linoleic and linolenic acid were the representative fatty acids in the solvent extracts (up to 2 wt.%); (iv) the spent extracts displayed cannabinoid levels around 2–3 times higher than the unprocessed flower solvent extracts. Concerning the biological study, the hemp EO and extracts demonstrated potent antimicrobial activity (MIC < 62.5 mg/L) against Gram-positive bacteria (e.g., *S. aureus*, *S. epidermidis*, *M. luteus*, *E. faecalis*, *B. cereus*, *S. pneumoniae*), *H. pylori*, and *Trichophyton* spp. In addition, when CBD, CBG, CBN, and CBC were individually tested against the same panel of microorganisms, MIC values ranging from 0.49 and 15.6 mg/mL against Gram-positive bacteria were retrieved. The inhibitory activity generally decreased in the following order: CBD > CBN > CBG > CBC. For some cannabinoids, the antimicrobial properties against certain microbial strains were proven for the first time in the current study.

Considering that the hemp flower essential oil industry generates significant amounts of unused biomass rich in cannabinoids, the strategy implemented in the current work could afford high-added-value by-products within the hemp production chain, contributing to the principles of the circular economy and sustainability. Altogether, this work can open promising avenues for utilizing cannabinoid-rich materials obtained during hemp flower processing in functional foods or cosmeceutical and pharmaceutical products with antimicrobial properties.

Author Contributions: Conceptualization, S.V.L.; methodology, S.V.L., K.W., I.K.-G. and A.T.; software, S.V.L., K.W. and I.K.-G.; validation, S.V.L. and A.T.; formal analysis, S.V.L.; investigation, S.V.L., K.W. and I.K.-G.; resources, S.V.L., K.S.-W. and M.M.; data curation, S.V.L.; writing—original draft preparation, S.V.L.; writing—review and editing, K.W., I.K.-G., K.S.-W., M.M. and A.T.; visualization, S.V.L.; supervision, S.V.L.; project administration, S.V.L.; funding acquisition, K.S.-W. and M.M. All authors have read and agreed to the published version of the manuscript.

Funding: This research received no external funding.

Institutional Review Board Statement: Not applicable.

Informed Consent Statement: Not applicable.

Data Availability Statement: Data are contained within the article.

Acknowledgments: The authors would like to thank Hempartis GmbH (Malsch, Germany) for providing the hemp materials.

Conflicts of Interest: The authors declare no conflicts of interest.

References

1. Luca, S.V.; Roehrer, S.; Kleigrewe, K.; Minceva, M. Approach for simultaneous cannabidiol isolation and pesticide removal from hemp extracts with liquid-liquid chromatography. *Ind. Crop. Prod.* **2020**, *155*, 112726. [CrossRef]
2. Moscariello, C.; Matassa, S.; Esposito, G.; Papirio, S. From residue to resource: The multifaceted environmental and bioeconomy potential of industrial hemp (*Cannabis sativa* L.). *Resour. Conserv. Recycl.* **2021**, *175*, 105864. [CrossRef]
3. Benelli, G.; Pavela, R.; Lupidi, G.; Nabissi, M.; Petrelli, R.; Ngahang Kamte, S.L.; Cappellacci, L.; Fiorini, D.; Sut, S.; Dall'Acqua, S. The crop-residue of fiber hemp cv. Futura 75: From a waste product to a source of botanical insecticides. *Environ. Sci. Pollut. Res.* **2018**, *25*, 10515–10525. [CrossRef] [PubMed]
4. Serventi, L.; Flores, G.A.; Cusumano, G.; Barbaro, D.; Tirillini, B.; Venanzoni, R.; Angelini, P.; Acquaviva, A.; Di Simone, S.C.; Orlando, G. Comparative investigation of antimicrobial and antioxidant effects of the extracts from the inflorescences and leaves of the *Cannabis sativa* L. cv. strawberry. *Antioxidants* **2023**, *12*, 219. [CrossRef] [PubMed]
5. Appendino, G.; Gibbons, S.; Giana, A.; Pagani, A.; Grassi, G.; Stavri, M.; Smith, E.; Rahman, M.M. Antibacterial cannabinoids from Cannabis sativa: A structure– activity study. *J. Nat. Prod.* **2008**, *71*, 1427–1430. [CrossRef] [PubMed]
6. Russo, E.B.; Marcu, J. Cannabis pharmacology: The usual suspects and a few promising leads. *Adv. Pharmacol.* **2017**, *80*, 67–134. [PubMed]
7. Alhadid, A.; Luca, S.V.; Nasrallah, S.; Minceva, M. Experimental investigation and thermodynamic modeling of cannabidiol solubility in plant oils and hydrophobic eutectic systems. *J. Mol. Liq.* **2023**, *372*, 121172. [CrossRef]
8. Dos Santos, R.G.; Hallak, J.E.; Crippa, J.A.S. Neuropharmacological effects of the main phytocannabinoids: A narrative review. In *Cannabinoids and Neuropsychiatric Disorders*; Springer: Berlin/Heidelberg, Germany, 2021; pp. 29–45.
9. Russo, C.; Lavorgna, M.; Nugnes, R.; Orlo, E.; Isidori, M. Comparative assessment of antimicrobial, antiradical and cytotoxic activities of cannabidiol and its propyl analogue cannabidivarin. *Sci. Rep.* **2021**, *11*, 22494. [CrossRef] [PubMed]
10. Agrawal, A.; Gupta, A. A review on exploring the antibacterial activity of cannabinoids with respect to biofilm formation. *J. Integr. Health Sci.* **2020**, *8*, 101–104.
11. Gaweł-Bęben, K.; Czech, K.; Luca, S.V. Cannabidiol and minor phytocannabinoids: A preliminary study to assess their anti-melanoma, anti-melanogenic, and anti-tyrosinase properties. *Pharmaceuticals* **2023**, *16*, 648. [CrossRef]
12. Iseppi, R.; Brighenti, V.; Licata, M.; Lambertini, A.; Sabia, C.; Messi, P.; Pellati, F.; Benvenuti, S. Chemical characterization and evaluation of the antibacterial activity of essential oils from fibre-type *Cannabis sativa* L. (Hemp). *Molecules* **2019**, *24*, 2302. [CrossRef] [PubMed]
13. Fiorini, D.; Scortichini, S.; Bonacucina, G.; Greco, N.G.; Mazzara, E.; Petrelli, R.; Torresi, J.; Maggi, F.; Cespi, M. Cannabidiol-enriched hemp essential oil obtained by an optimized microwave-assisted extraction using a central composite design. *Ind. Crop. Prod.* **2020**, *154*, 112688. [CrossRef]
14. Micalizzi, G.; Vento, F.; Alibrando, F.; Donnarumma, D.; Dugo, P.; Mondello, L. *Cannabis Sativa* L.: A comprehensive review on the analytical methodologies for cannabinoids and terpenes characterization. *J. Chromatogr. A* **2020**, *1637*, 461864. [CrossRef] [PubMed]
15. Eschlwech, F.; Ruedenauer, F.; Leonhardt, S.D.; Minceva, M.; Luca, S.V. Liquid-liquid chromatography separation of hemp terpenes with repellent properties against *Hyalomma marginatum*: A multi-methodological approach. *Ind. Crop. Prod.* **2023**, *197*, 116562. [CrossRef]
16. Benelli, G.; Pavela, R.; Petrelli, R.; Cappellacci, L.; Santini, G.; Fiorini, D.; Sut, S.; Dall'Acqua, S.; Canale, A.; Maggi, F. The essential oil from industrial hemp (*Cannabis sativa* L.) by-products as an effective tool for insect pest management in organic crops. *Ind. Crop. Prod.* **2018**, *122*, 308–315. [CrossRef]
17. Luca, S.V.; Zengin, G.; Sinan, K.I.; Skalicka-Woźniak, K.; Trifan, A. Post-distillation by-products of aromatic plants from Lamiaceae family as rich sources of antioxidants and enzyme inhibitors. *Antioxidants* **2023**, *12*, 210. [CrossRef] [PubMed]

18. Schofs, L.; Sparo, M.D.; Sanchez Bruni, S.F. The antimicrobial effect behind *Cannabis sativa*. *Pharmacol. Res. Perspect.* **2021**, *9*, e00761. [CrossRef]
19. Hong, H.; Sloan, L.; Saxena, D.; Scott, D.A. The antimicrobial properties of cannabis and cannabis-derived compounds and relevance to CB2-targeted neurodegenerative therapeutics. *Biomedicines* **2022**, *10*, 1959. [CrossRef] [PubMed]
20. Scott, C.; Neira Agonh, D.; Lehmann, C. Antibacterial effects of phytocannabinoids. *Life* **2022**, *12*, 1394. [CrossRef] [PubMed]
21. Fathordoobady, F.; Singh, A.; Kitts, D.D.; Pratap Singh, A. Hemp (*Cannabis sativa* L.) extract: Anti-microbial properties, methods of extraction, and potential oral delivery. *Food Rev. Int.* **2019**, *35*, 664–684. [CrossRef]
22. Gabarin, A.; Yarmolinsky, L.; Budovsky, A.; Khalfin, B.; Ben-Shabat, S. Cannabis as a source of approved drugs: A new look at an old problem. *Molecules* **2023**, *28*, 7686. [CrossRef]
23. Karas, J.A.; Wong, L.J.; Paulin, O.K.; Mazeh, A.C.; Hussein, M.H.; Li, J.; Velkov, T. The antimicrobial activity of cannabinoids. *Antibiotics* **2020**, *9*, 406. [CrossRef] [PubMed]
24. Kumar, P.; Mahato, D.K.; Kamle, M.; Borah, R.; Sharma, B.; Pandhi, S.; Tripathi, V.; Yadav, H.S.; Devi, S.; Patil, U. Pharmacological properties, therapeutic potential, and legal status of *Cannabis sativa* L.: An overview. *Phytother. Res.* **2021**, *35*, 6010–6029. [CrossRef] [PubMed]
25. Luca, S.V.; Kittl, T.; Minceva, M. Supercritical CO_2 extraction of hemp flowers: A systematic study to produce terpene-rich and terpene-depleted cannabidiol fractions. *Ind. Crop. Prod.* **2022**, *187*, 115395. [CrossRef]
26. Pieracci, Y.; Ascrizzi, R.; Terreni, V.; Pistelli, L.; Flamini, G.; Bassolino, L.; Fulvio, F.; Montanari, M.; Paris, R. Essential oil of *Cannabis sativa* L: Comparison of yield and chemical composition of 11 hemp genotypes. *Molecules* **2021**, *26*, 4080. [CrossRef] [PubMed]
27. Menghini, L.; Ferrante, C.; Carradori, S.; D'Antonio, M.; Orlando, G.; Cairone, F.; Cesa, S.; Filippi, A.; Fraschetti, C.; Zengin, G. Chemical and bioinformatics analyses of the anti-leishmanial and anti-oxidant activities of hemp essential oil. *Biomolecules* **2021**, *11*, 272. [CrossRef] [PubMed]
28. Zheljazkov, V.D.; Sikora, V.; Dincheva, I.; Kačániová, M.; Astatkie, T.; Semerdjieva, I.B.; Latkovic, D. Industrial, CBD, and wild hemp: How different are their essential oil profile and antimicrobial activity? *Molecules* **2020**, *25*, 4631. [CrossRef] [PubMed]
29. Berman, P.; Futoran, K.; Lewitus, G.M.; Mukha, D.; Benami, M.; Shlomi, T.; Meiri, D. A new ESI-LC/MS approach for comprehensive metabolic profiling of phytocannabinoids in *Cannabis*. *Sci. Rep.* **2018**, *8*, 14280. [CrossRef]
30. Borille, B.T.; Ortiz, R.S.; Mariotti, K.C.; Vanini, G.; Tose, L.V.; Filgueiras, P.R.; Marcelo, M.C.; Ferrão, M.F.; Anzanello, M.J.; Limberger, R.P. Chemical profiling and classification of cannabis through electrospray ionization coupled to Fourier transform ion cyclotron resonance mass spectrometry and chemometrics. *Anal. Methods* **2017**, *9*, 4070–4081. [CrossRef]
31. Barhdadi, S.; Courselle, P.; Deconinck, E.; Vanhee, C. The analysis of cannabinoids in e-cigarette liquids using LC-HRAM-MS and LC-UV. *J. Pharm. Biomed. Anal.* **2023**, *230*, 115394. [CrossRef]
32. Piccolella, S.; Formato, M.; Pecoraro, M.T.; Crescente, G.; Pacifico, S. Discrimination of CBD-, THC-and CBC-type acid cannabinoids through diagnostic ions by UHPLC-HR-MS/MS in negative ion mode. *J. Pharm. Biomed. Anal.* **2021**, *201*, 114125. [CrossRef] [PubMed]
33. Motiejauskaitė, D.; Ullah, S.; Kundrotaitė, A.; Žvirdauskienė, R.; Bakšinskaitė, A.; Barčauskaitė, K. Isolation of biologically active compounds from *Cannabis sativa* L. inflorescences by using different extraction solvents and evaluation of antimicrobial activity. *Antioxidants* **2023**, *12*, 998. [CrossRef] [PubMed]
34. Muscarà, C.; Smeriglio, A.; Trombetta, D.; Mandalari, G.; La Camera, E.; Grassi, G.; Circosta, C. Phytochemical characterization and biological properties of two standardized extracts from a non-psychotropic *Cannabis sativa* L. cannabidiol (CBD)-chemotype. *Phytother. Res.* **2021**, *35*, 5269–5281. [CrossRef] [PubMed]
35. Kriese, U.; Schumann, E.; Weber, W.; Beyer, M.; Brühl, L.; Matthäus. Oil content, tocopherol composition and fatty acid patterns of the seeds of 51 *Cannabis sativa* L. genotypes. *Euphytica* **2004**, *137*, 339–351. [CrossRef]
36. Aiello, A.; Pizzolongo, F.; Scognamiglio, G.; Romano, A.; Masi, P.; Romano, R. Effects of supercritical and liquid carbon dioxide extraction on hemp (*Cannabis sativa* L.) seed oil. *Int. J. Food Sci. Technol.* **2020**, *55*, 2472–2480. [CrossRef]
37. Petrović, M.; Debeljak, Ž.; Kezić, N.; Džidara, P. Relationship between cannabinoids content and composition of fatty acids in hempseed oils. *Food Chem.* **2015**, *170*, 218–225. [CrossRef] [PubMed]
38. Da Porto, C.; Decorti, D.; Tubaro, F. Fatty acid composition and oxidation stability of hemp (*Cannabis sativa* L.) seed oil extracted by supercritical carbon dioxide. *Ind. Crop. Prod.* **2012**, *36*, 401–404. [CrossRef]
39. Da Porto, C.; Voinovich, D.; Decorti, D.; Natolino, A. Response surface optimization of hemp seed (*Cannabis sativa* L.) oil yield and oxidation stability by supercritical carbon dioxide extraction. *J. Supercrit. Fluids* **2012**, *68*, 45–51. [CrossRef]
40. Kuete, V.; Efferth, T. Cameroonian medicinal plants: Pharmacology and derived natural products. *Front. Pharmacol.* **2010**, *1*, 123. [CrossRef]
41. Nafis, A.; Kasrati, A.; Jamali, C.A.; Mezrioui, N.; Setzer, W.; Abbad, A.; Hassani, L. Antioxidant activity and evidence for synergism of *Cannabis sativa* (L.) essential oil with antimicrobial standards. *Ind. Crop. Prod.* **2019**, *137*, 396–400. [CrossRef]
42. Zengin, G.; Menghini, L.; Di Sotto, A.; Mancinelli, R.; Sisto, F.; Carradori, S.; Cesa, S.; Fraschetti, C.; Filippi, A.; Angiolella, L. Chromatographic analyses, in vitro biological activities, and cytotoxicity of *Cannabis sativa* L. essential oil: A multidisciplinary study. *Molecules* **2018**, *23*, 3266. [CrossRef] [PubMed]

43. Jokić, S.; Jerković, I.; Pavić, V.; Aladić, K.; Molnar, M.; Kovač, M.J.; Vladimir-Knežević, S. Terpenes and cannabinoids in supercritical CO_2 extracts of industrial hemp inflorescences: Optimization of extraction, antiradical and antibacterial activity. *Pharmaceuticals* **2022**, *15*, 1117. [CrossRef] [PubMed]
44. Luca, S.V.; Braumann, L.; Gerigk, M.; Frank, O.; Minceva, M. Separation of minor cannabinoids from hemp extract with trapping multiple dual mode liquid-liquid chromatography. *J. Chromatogr. A* **2021**, *1658*, 462608. [CrossRef] [PubMed]
45. Blaskovich, M.A.; Kavanagh, A.M.; Elliott, A.G.; Zhang, B.; Ramu, S.; Amado, M.; Lowe, G.J.; Hinton, A.O.; Pham, D.M.T.; Zuegg, J. The antimicrobial potential of cannabidiol. *Commun. Biol.* **2021**, *4*, 7. [CrossRef] [PubMed]
46. Wassmann, C.S.; Højrup, P.; Klitgaard, J.K. Cannabidiol is an effective helper compound in combination with bacitracin to kill Gram-positive bacteria. *Sci. Rep.* **2020**, *10*, 4112. [CrossRef]
47. Farha, M.A.; El-Halfawy, O.M.; Gale, R.T.; MacNair, C.R.; Carfrae, L.A.; Zhang, X.; Jentsch, N.G.; Magolan, J.; Brown, E.D. Uncovering the hidden antibiotic potential of cannabis. *ACS Infect. Dis.* **2020**, *6*, 338–346. [CrossRef] [PubMed]
48. Aqawi, M.; Sionov, R.V.; Gallily, R.; Friedman, M.; Steinberg, D. Anti-bacterial properties of cannabigerol toward *Streptococcus mutans*. *Front. Microbiol.* **2021**, *12*, 656471. [CrossRef] [PubMed]
49. Luz-Veiga, M.; Amorim, M.; Pinto-Ribeiro, I.; Oliveira, A.L.; Silva, S.; Pimentel, L.L.; Rodríguez-Alcalá, L.M.; Madureira, R.; Pintado, M.; Azevedo-Silva, J. Cannabidiol and cannabigerol exert antimicrobial activity without compromising skin microbiota. *Int. J. Mol. Sci.* **2023**, *24*, 2389. [CrossRef] [PubMed]
50. Kosgodage, U.S.; Matewele, P.; Awamaria, B.; Kraev, I.; Warde, P.; Mastroianni, G.; Nunn, A.V.; Guy, G.W.; Bell, J.D.; Inal, J.M. Cannabidiol is a novel modulator of bacterial membrane vesicles. *Front. Cell. Infect. Microbiol.* **2019**, *9*, 324. [CrossRef]
51. Arendrup, M.C.; Kahlmeter, G.; Guinea, J.; Meletiadis, J.; the Subcommittee on Antifungal Susceptibility Testing (AFST) of the ESCMID European Committee for Antimicrobial Susceptibility Testing (EUCAST). How to: Perform antifungal susceptibility testing of microconidia-forming dermatophytes following the new reference EUCAST method E. Def 11.0, exemplified by Trichophyton. *Clin. Microbiol. Infect.* **2021**, *27*, 55–60. [CrossRef]
52. European Committee for Antimicrobial Susceptibility Testing (EUCAST) of the European Society of Clinical Microbiology and Infectious Diseases (ESCMID). Determination of minimum inhibitory concentrations (MICs) of antibacterial agents by broth dilution. *Clin. Microbiol. Infect.* **2003**, *9*, ix–xv.

antibiotics

MDPI

Article

Anti-Infective Activity of *Momordica charantia* Extract with Molecular Docking of Its Triterpenoid Glycosides

Mohammed Aldholmi [1,*], Rizwan Ahmad [1], Mohammad Habeeb Shaikh [2], Ayad Mohammed Salem [2], Maher Alqurashi [3] and Mansour Alturki [4]

1 Department of Natural Products, College of Clinical Pharmacy, Imam Abdulrahman Bin Faisal University, Dammam 31441, Saudi Arabia
2 Department of Physiology, College of Medicine, Imam Abdulrahman Bin Faisal University, Dammam 34224, Saudi Arabia
3 Department of Microbiology, College of Medicine, Imam Abdulrahman Bin Faisal University, Dammam 31441, Saudi Arabia
4 Department of Pharmaceutical Chemistry, College of Clinical Pharmacy, Imam Abdulrahman Bin Faisal University, Dammam 31441, Saudi Arabia
* Correspondence: mjaldholami@iau.edu.sa

Abstract: *Momordica charantia*, commonly known as bitter melon, is a fruiting plant that has been used for several diseases including infectious diseases. In this study, we report the antibacterial, antifungal, and antiviral activity of different bitter melon fruit parts originating from India and Saudi Arabia. The in vitro experiments are supported by the molecular docking of karavilosides to verify their role in the bioactivity. The antimicrobial assays revealed activity against *Candida albicans*, *Escherichia coli*, and *Staphylococcus aureus*. The extracts exhibited the potent inhibition of HIV-I reverse transcriptase, with an IC50 of 0.125 mg/mL observed for the pith extract originating from Saudi Arabia and the standard drug doxorubicin. The molecular docking of karavilosides exhibited a significant affinity to reverse transcriptase comparable to Rilpivirine and higher than that of doxorubicin. These outcomes encourage the precious bioactive components of the seed and pith of the Saudi bitter melon fruits to be further studied for isolation and structure elucidation.

Keywords: karavilosides; antiviral; antibacterial; antifungal; bitter melon

1. Introduction

Infectious diseases caused by pathogens, including viruses, bacteria, fungi, and parasites, are still a substantial risk to public health, despite the availability of numerous classes of anti-infective agents. Some reports estimated that by 2050, ten million people will die every year from antimicrobial resistance, unless a universal response is projected to tackle this problem [1,2]. Many of the approved anti-infective agents are becoming less effective as the pathogens develop resistance mechanisms to avoid the effect of the utilized drugs, while other medications have crucial issues related to their pharmacokinetic and pharmacodynamic properties, as well as leading to unpleasant side effects [3,4]. Therefore, exploring novel anti-infective agents with unique mechanisms of action and fewer side effects is a research propriety supported by the governments to combat infectious diseases. Natural compounds from plants, microorganisms, and marine organisms represent an essential source of anti-infective agents. It has been estimated that 66% of all small-molecule anti-infective drugs approved by the FDA between January 1981 and September 2019 are either natural products or based on natural products pharmacophore [5].

Traditional medicinal plants have been used for over fifty centuries as a source of antimicrobial, anticancer, analgesic, cardioprotective, antihypertensive, antidiabetic, and other medications [6]. Plants produce bioactive compounds known as secondary metabolites with a chemical structural diversity. Examples of the major classes of bioactive secondary metabolites include alkaloids, phenolics, and terpenoids. Tamiflu (oseltamivir) is an antiviral phenolic-based medication for influenza A and B produced from the shikimic acid obtained from star anise [7]. The traditional Chinese *Artemisia annua*, commonly known as sweet wormwood, is the source of the antimalarial terpenoid artemisinin, which has saved the lives of millions of people around the world [8]. Terpenoids have been reported to demonstrate potent antiviral activity and reduce viral infectivity [9,10]. For instance, artemisinin and its derivatives have been reported to show antiviral activity against various viruses, including the human cytomegalovirus (HCMV), Hepatitis B virus (HBV), and Hepatitis C virus (HCV) [11]. A synergetic effect was noticed when these terpenoids were combined with the antiviral lamivudine, which might be an effective strategy to minimize the serious adverse effects of the latter.

Momordica charantia, known as bitter melon or bitter gourd, is a fruiting vine belonging to the family Cucubitaceae, probably native to tropical and subtropical Africa and Asia [12]. The plant is wildly cultivated in India, China, and Southeast Asia. It is used as a vegetable crop, as well as a medicine against several diseases, including diabetes, cancer, dementia, hyperlipemia, hypertension, digestive problems, and infectious diseases [12]. The ethanol extract of *M. charantia* leaves was found to have an antibacterial activity against *Escherichia coli*, *Staphylococcus aureus*, *Bacillus subtilis*, *Pseudomonas aeruginosa*, *Salmonella typhi*, and *Klebsiella pneumonia* [13,14]. The antifungal activity of the leaves extract was also reported to work against *Candida albicans* [15]. The leaves extract has also shown activity against *Aeromonas hydrophila*, the most common pathogenic bacteria of freshwater fish [16]. The ground seeds of bitter melon were shown to present an antibacterial activity against *Pasteurella multocida* and *Staphylococcus aureus* [17]. The antimicrobial activity was evaluated using disk diffusion; hence, the minimum inhibitory concentration and minimum bactericidal concentration were not reported. Several proteins, e.g., MAP 30, alpha and beta-momorcharin, and lectin, have been isolated from the Chinese bitter melon seeds and are reported to have an antiviral activity against coxsackievirus B3, human immunodeficiency virus (HIV), herpes, Epstein–Barr, and polio viruses [18–20]. The antiviral activity of MAP30, alpha-momorcharin, and beta-momorcharin was attributed to the inhibition of HIV-1 integrase [18]. The bitter melon lectin was reported to show its antiviral bioactivity via the inhibition of the reverse transcriptase enzyme [21]. Another protein known as MRK29 was isolated from the Thai bitter melon seed and reported to act through the same mechanism [22]. Nevertheless, there are limited studies on the antiviral, antibacterial, and antifungal activity of bitter melon fruits and the role of karavilosides in this bioactivity, particularly for reverse transcriptase inhibition. Moreover, most previous studies have been limited to the Chinese and Thai bitter melon, while the bitter melon commonly used in other Asian countries originates from India. There are also local varieties of bitter melon cultivated in most countries and consumed, along with the imported ones.

In this context, we evaluated the antiviral, antibacterial, and antifungal activity of different parts of bitter melon fruits to determine the specific part responsible for bioactivity. Our study included one local origin (Saudi Arabia) and one global origin (India) of BM fruits to compare the bioactivity of the fruits produced by plants grown under different climatic and environmental conditions. Finally, a detailed molecular docking study of the major phytochemicals of the BM fruits (karavilosides; KVs) was performed to evaluate the role of KVs in the antiviral activity via the inhibition of reverse transcriptase enzyme.

2. Results

2.1. Extract Yield and KV Quantity

The highest extract yield was observed for K1 (5.41%) and I1 (2.78%), followed by I3 (1.18%), K3 (0.56%), K2 (0.18%), and I2 (0.09%). The extracts with the highest total amount of KVs (KV-VIII, -X, and -XI)/BM fruit part were I3 (1217.81 μg/L), K3 (970.67 μg/L), and I2 (734.17 μg/L). On an individual basis, KV-VIII (592.1 μg/L) and KV-XI (405.32 μg/L) were the highest in I3, while the highest amount of KV-X was found in K3 (269.55 μg/L).

The details regarding extract yields and KV quantities for all the samples are provided in Table S1.

2.2. Antiviral Activity

The non-radioactive HIV-RT colorimetric ELISA kit revealed an antiviral activity for the BM extracts, with a mean inhibition (±SD) of 76.35 (±3.46%) against HIV-I reverse transcriptase. The inhibition (%) for HIV-RT was observed in the range of 72.40–81.60, where K1 exhibited the greatest inhibition of 81.6%, followed by K3 (79.2%). The standard drug doxorubicin showed an inhibition of 89.6% in the study. The inhibition data for all the samples (I1-I3 and K1-K3) against HIV-I RT are shown in Table S1. The active extract of K1 was further evaluated for the IC_{50} calculation, where a similar IC_{50} of 0.125 mg/mL was observed for K3 and the standard drug doxorubicin. The IC_{50} data, along with the negative and positive controls for HIV-RT, are shown in Table S2. The descending order for the antiviral activity of the extracts may be summarized as K1 (81.6%) > K3 (79.2%) > I2 (76%) > I3 (75.2%) > K2 (73.7%) > I1 (72.4%).

2.3. Antimicrobial Activity (MIC and MBC)

The antimicrobial assays revealed minimum inhibitory concentrations (MICs) of the extracts in the range of 0.08–1.0 mg/mL against *Candida albicans* (CA), 0.3–3.0 mg/mL against *Escherichia coli* (EC), and 3.0–3.0 mg/mL against *Staphylococcus aureus* (SA). The minimum bactericidal concentrations (MBCs) ranged between 0.08 and 1.0 mg/mL against *C. albicans*, 1.0 and 3.0 mg/mL against *E. coli*, and 3.0 and >3 mg/mL for *S. aureus*. The detailed antimicrobial activity for all the samples (I1-I3 and K1-K3) against bacteria and yeast is shown in Table S1.

On an individual basis, the lowest MIC (0.08 mg/mL) was observed for the I1 and I3 extracts against *C. albicans*, followed by 0.3 mg/mL for I2 and K3 against *C. albicans*. The lowest MIC against *E. coli* (0.3 mg/mL) was observed for I1, I3, and K3, whereas all the I and K samples presented a similar MIC of 3 mg/mL against *S. aureus*. For MBC against *C. albicans*, the lowest concentration of 0.08 mg/mL was observed for I3, followed by 0.3 mg/mL for I1 and I2. The lowest MBC against *E. coli* was observed to be 1 mg/mL for I1, I3, and K3. All the samples exhibited MBCs of 3 or >3 mg/mL against *S. aureus*. The details regarding the MIC and MBC for all the samples against *C. albicans*, *E. coli*, and *S. aureus* are provided in Table S1. The ascending order for the MICs and MBCs is summarized as *C. albicans*_MIC (I1 and I3) = *C. albicans*_MBC (I3) > *C. albicans*_MIC (I2 and K3) = *C. albicans*_MBC (I1, I2, and K3) > *E. coli*_MIC (I1, I3, and K3) > (MIC/MBC for the remaining samples against *E. coli* and *S. aureus*).

2.4. Molecular Docking Studies

The ligands were docked using the extra precision mode (XP) without any constraints and a 0.80 van der Waals (vdw) radius scaling factor and a 0.15 partial charge cut-off were used. The GlideScore, implemented in Glide, was used to estimate the binding affinity and to rank the ligands, whereas the XP-Pose rank was used to select the best-docked pose for each ligand. The compounds were then thoroughly analyzed based on the binding scores along with a comprehensive examination of all the binding interactions. The results revealed good binding modes of superimposition, with an RMSD of 0.3571, which initially reflects the accuracy of pose prediction for the Glide (Figure 1).

The karavilosides ligands were then docked in the RT active site using the extra precision mode on Glide [9]. The results showed that all karavilosides ligands adapted a relatively similar docking pattern in the active site that is close to the reference drug Rilpivirine, and doxorubicin poses, which suggests the potential affinity of these molecules to the RT active site (Figure 2). The calculated RMSD for the best binding poses in reference to Rilpivirine and doxorubicin ranged from 0.0.5 to 0.2 for karavilosides VIII, X, and XI. The relative binding affinities (Glide score) of the docked compounds on the enzyme were analyzed, and the results showed that karaviloside XI has the highest glide score for the inhibition of HIV-1 reverse transcriptase (Table 1).

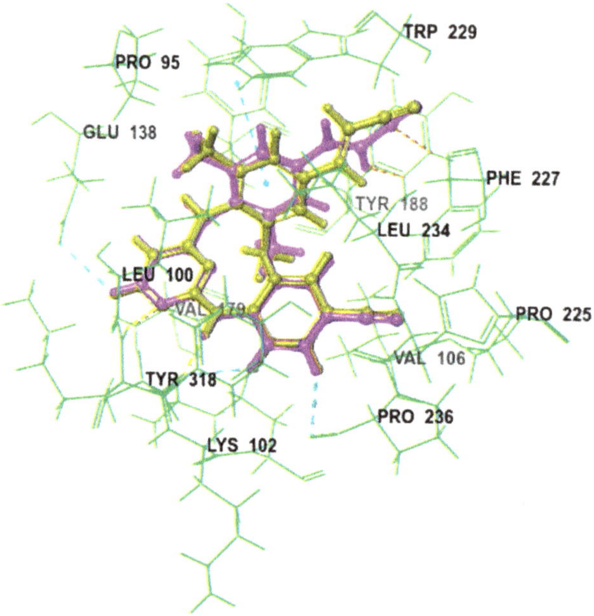

Figure 1. Comparative binding positions of cognate inhibitor (light green) and redocked inhibitor (purple) bound to RT active site, RMSD of 0.3571.

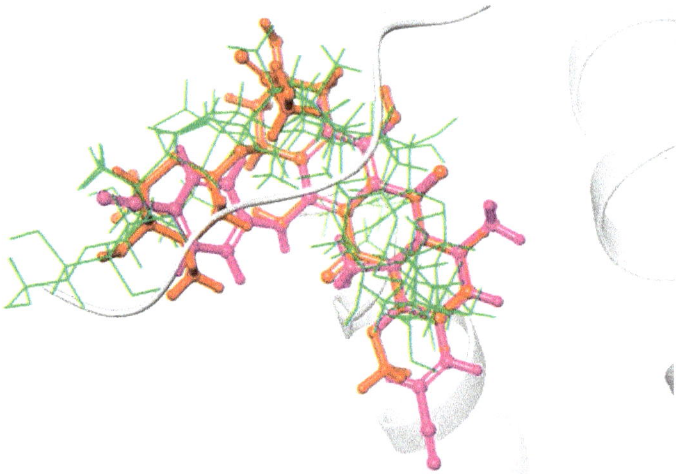

Figure 2. Comparative binding positions of Rilpivirine (magenta); doxorubicin (orange); and karavilosides VIII, X, and XI (green), bound to RT active site.

Table 1. Docking energies of investigated karavilosides (KV-VIII, -X, and -XI) and co-crystallized reference controls at the binding site of HIV-1 reverse transcriptase (PDB entry: 4G1Q), throughout the preliminary docking analysis. A higher negative value indicates higher binding interactions within the binding pocket.

Compounds	Glide Score Docking Energy (Kcal/mol)	Free Binding Energy (Kcal/mol)	H-Bond Interactions	Pi-Interactions	Van Der Waals Interactions
KV-VIII	−9.213	−60.74	LEU100 and GLU138	PHE227, TYR181, TYR188 and LYS103	GLU138, PHE227, LYS103, PHE227 and LEU234
KV-X	−11.829	−76.91	TYR181, TYR188 and MET230	TYR181 and PHE227	GLU138, PHE227, LYS103, VAL106 VAL179, LEU100 and LEU234
KV-XI	−12.732	−93.23	TYR181, TYR188, TRP229 and LYS101	TYR181 and PHE227	GLU138, PHE227, LYS103, VAL106, VAL179, LEU100 and LEU234
Rilpivirine	−13.350	−116.07	LYS101 and GLU138	TYR181	PRO236 and LYS101
Doxorubicin	−8.352	−40.34	LEU100, LYS103, GLU138, GLY99 and ILE180	PHE227	TYR188, LEU234 and VAL179

2.5. Statistical Models

2.5.1. Correlation Analysis

Pearson's bivariate model was applied to establish the significant correlations for the samples' origin vs. dependent variables (DVs) in the dataset. For sample origin vs. DV, a positive correlation ($p = 0.05$) was seen for the pairs of I2 and DV (0.670; 0.02), I3 and DV (0.638; 0.03), and K3 and DV (0.620; 0.04). The correlation within the samples showed a positive correlation ($p = 0.01$) for I1 and K1 (0.997; 0.00), whereas I2, I3, K2, and K3 revealed a lack of any correlation with I1 and K1. The details for the pairwise correlation for the sample origin and DV is given in Table S3.

2.5.2. Principal Component Analysis (PCA)

The principal component analysis (PCA) exhibited a cumulative variance of 90.52%, where PC1 computed the major variance of 58.82%, followed by PC2 (31.70%). Interestingly, PC1 was loaded with DV, I2, I3, K2, and K3, whereas I1 and K1 were loaded in PC2. The outcome confirms Pearson's test data, i.e., a greater correlation was observed for DV vs. the I2, I3, K2, and K3 samples, which is confirmed herein by the greater variance in the data shown by the mentioned samples vs. DV. The test was significant at $p = 0.05$, with a high X2 value of 141.33. The % variance for the components with detailed data regarding the KMO and Bartlett's test is shown in Figure 3 and Table S4.

2.5.3. Paired Samples *t*-Test

The dataset was evaluated for pair differences using the sample origins vs. individual biological activities for the six extracts. The data revealed only three pairs with significant pair differences—antiviral activity t(−60.32) at $p = 0.00$ (CI = −75.95–69.75), MIC for *C. albicans* t(4.40) at $p = 0.01$ (CI = 1.27–4.82), and MBC for *C. albicans* t(4.21) at $p = 0.01$ (CI = 1.17–4.84). The remaining data were inconsiderable, as shown in Table S5.

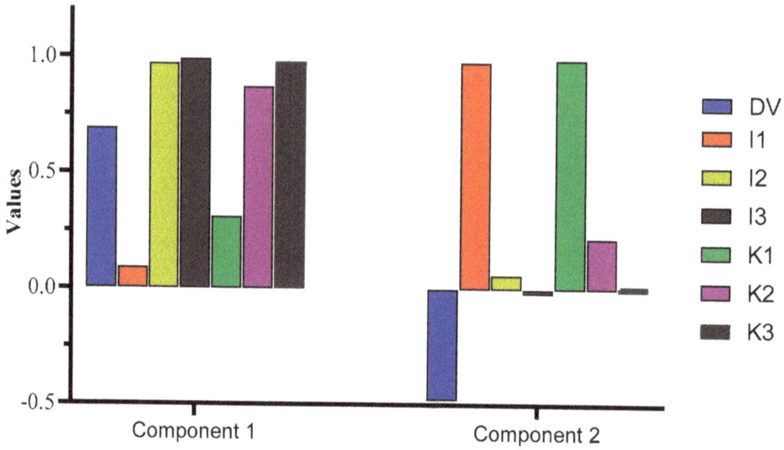

Figure 3. PCA representation of the components loaded with DVs in component 1 and 2.

3. Discussion

This study investigated the antiviral and antimicrobial potential of three different parts (skin, seeds, and pith) of the BM fruits from two distinct origins of India (I) and the Kingdom of Saudi Arabia (K). Following the green extraction and LCMSMS quantitative analysis for the presence of the three different cucurbitane triterpenoid glycosides (karavilosides; KVs), i.e., KV-VIII, -X, and -XI in these BM fruit parts, molecular docking was executed to determine the affinity and binding properties for the KVs. The ligands for the KV compounds exhibited a significant affinity to the reverse transcriptase (RT) protein. To broaden the findings, the RT inhibition profile for the KVs was studied through docking in the active site of the named target. The binding pattern, target interactions, and binding affinities for the KVs were investigated and compared to the reference drugs Rilpivirine and doxorubicin. For the validation of the docking phenomenon, the process was repeated via the redocking of the co-crystalline ligand in their respective target (HIV-1 reverse transcriptase) while utilizing the alike protocols, as applied for the KV ligands. Subsequently, rigid-body superposition was performed for the predicted lowest energy conformation of the target with its corresponding crystalline ligand using the structure superposition tool on Maestro. The classical root mean square deviation (RMSD) for the predicted binding poses from the co-crystalline pose was calculated at an RMSD < 2Å cut-off value [23]. The detailed analysis for the calculated binding affinities of the KVs exhibited a comparable inhibition to Rilpivirine and higher than that of doxorubicin. A horseshoe conformation was seen for the KVs during RT binding (Figure 4), which is similar to the NNRTIs in the DAPY family [24,25].

The KVs exhibited an impressive structural complementarity to the non-nucleoside inhibitor-binding pocket (NNIBP). Moreover, KV-X and -XI (left-wing structures of these compounds) interact with GLU138, LYS103, VAL106, VAL179, and LEU100 through van der Waals (VDW) interactions. The presence of the steroidal ring (center) enables the establishment of Pi and non-polar interactions with Phe227 and LEU234. Likewise, the right wing for these compounds interacts with TYR181, TYR188, and MET230 (Figure 5A,B). For KV-VIII, less interaction was observed with the non-nucleoside inhibitor-binding pocket (Figure 5C). The variation in affinity and bindings may be due to the absence of the third functional region compared to KV-X and -XI. A superior binding affinity for the protein binding based on the Pi and non-polar interactions with GLU138, LYS103, VAL106, VAL179, and LEU100 have been reported for KV-XI [26].

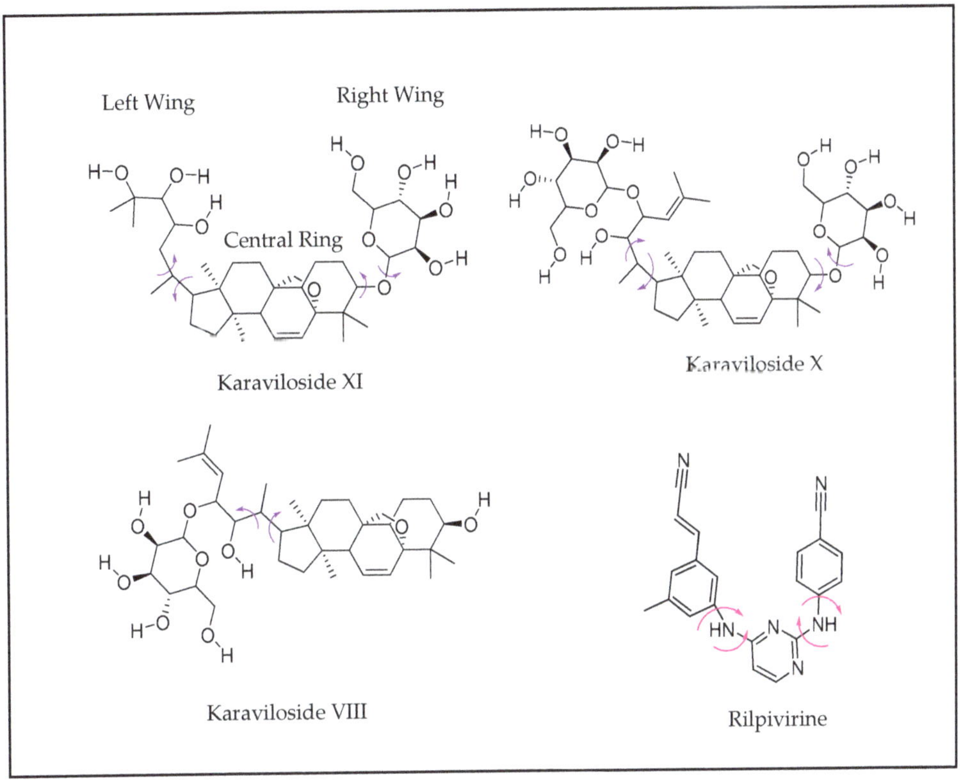

Figure 4. Chemical structures of NNRTI Rilpivirine (RPV). The torsion angles defining the rotatable bonds are labeled using purple-colored arrows in Karaviloside VIII, X, and XI. The equivalent torsion angle RPV is labeled using magenta-colored arrows. The structures of karavilosides, apart from karaviloside VIII, can be divided into three functional regions—a steroidal central ring, a hexose sugar/ethylene glycol left wing, and a hexose sugar right wing.

The findings from the molecular docking suggested valuable insights into the probable role of KVs in combating the ongoing challenges in HIV treatment. The six extracts were evaluated against HIV-I RT, where K1 and K3 revealed the highest inhibition of the HIV-I RT. The extract with the highest inhibition (K1) was compared to the standard drug doxorubicin to determine the IC_{50} value. Both tested samples were seen with relatable IC_{50} values, suggesting the potential inhibitory role for K1 against HIV-I RT. With regard to the BM fruit, these extracts represent the seeds (K1) and pith (K3) parts of the fruit. The correlation with an in-depth analysis of the extract yield and phytochemical profile established a paramount link between the KV amount in these samples. Although the K1 sample had the lowest amount of KVs, a high extract yield among the samples was reported for K1. This favors the exploration of the K1 sample to investigate the presence of other phytochemical classes and to establish the biological activity reported herein. Though antimicrobial and anti-HIV activities have been reported for bitter melon metabolites, several triterpenoid glycosides are present in the fruit and might contribute to the overall bioactivity [27]. Therefore, a detailed study of the mentioned phytochemicals is required to link them to the anti-HIV activity of the BM fruit.

On the contrary, the K3 sample was quantified with the highest amount of KV-X on an individual basis, as well as the highest amounts of the total KVs (KV-VIII, -X, and -XI) in an individual fruit part, following I3. Previous studies reported the antiviral role of the BM against herpes, Epstein–Barr, and HIV [28,29], where the juice from BM fruit was effective in suppressing HIV via an enhanced T-cell count and a stabilization of the CD4/CD8 cells in the HIV-positive patients. This anti-HIV role was demonstrated by the α- and β-momorcharin proteins present in the BM seed. In addition, the presence of anti-retroviral protein (MAP30) was studied for the probable anti-HIV, anti-HBV, and anti-herpes activities of the BM [28,30]. Yet again, there is a lack of research to establish the expected role of KV-VII, -X, and -XI in HIV inhibition. The isolation with appropriate in-depth molecular-level studies may be helpful in uncovering the novel role of these KVs. The outcome of the antiviral study herein supports the traditional use of BM in treating viral infections [27,31].

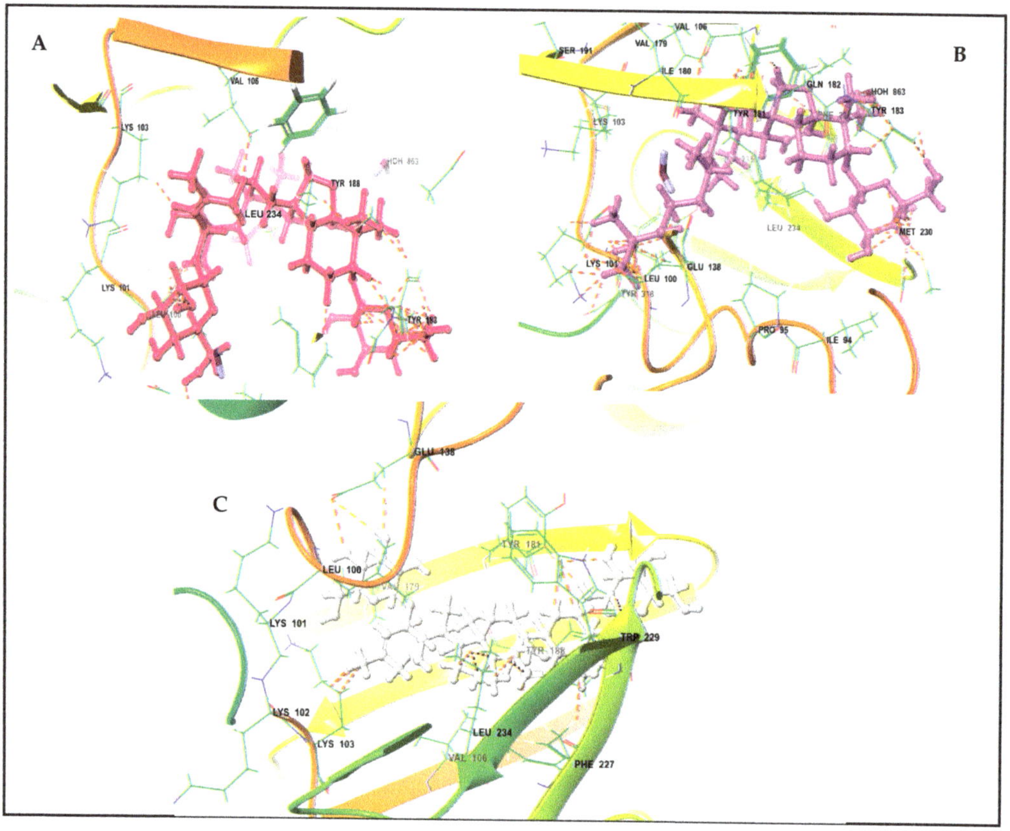

Figure 5. Karavilosides bound to RT active site. (**A**) Karaviloside X. (**B**) Karaviloside XI. (**C**) Karaviloside VIII.

In order to explore the antimicrobial potential, the extracts were screened against the fungal strain of *C. albicans*, as well as the bacterial strains of *E. coli* and *S. aureus*. For the antimicrobial demarcation, the results for MIC and MBC manifested a substantial activity against *C. albicans*, followed by a significant MIC against *E. coli*. However, the MIC and MBC values for *S. aureus* exhibited insignificant results for the extracts. In contrast to the antiviral activity, the lowest MIC value against *C. albicans* was noted for I1 and I3, followed by K3, whereas the lowest MBC value against *C. albicans* was recorded for I3 only. Likewise, the lowest MIC values against *E. coli* were seen for I1, I3, and K3 among the six extracts. The comprehensive analysis of these extracts reveals a high extract yield for I1, following K1. The I3 extract contains the highest amount of both, the total KVs (KV-III, -X, -XI), along with the amount of KV-VIII and -XI in an individual sample among the six extracts. The

K3 extract has been mentioned previously to contain the highest amount of KV-X on an individual basis. Herein, the seeds and pith parts of the BM fruit also presented antifungal and antibacterial activities, which underscore the significance of these parts of the BM fruits in anti-infective activities.

The samples with significant antimicrobial activity observed were I1 and I3. For I3, though an enormous amount of KVs was observed during quantitative analysis, no studies have reported the KVs' correlation with antimicrobial activities, which entails the study for an individual antimicrobial model where the KVs' potential is explored against these species with appropriate mechanisms. For I1, there was a high extract yield with the least detected amount of KVs, suggesting the presence of other phytochemicals in the extract, imparting an antifungal activity to the extract. A study reported the presence of flavonoids in the BM leaves with significant antimicrobial activity [32]; however, the phenolics and flavonoids content in the fruit parts remain unexplored. Likewise, the presence of trans-nerolidol in the seeds' essential oil has been reported for the antimicrobial potential of the BM fruit [33]. The results for the antimicrobial activity in this study are in line with the reported literature. However, a detailed characterization of the I1 and I3 extracts may foster the presence of novel chemical entities in the BM fruit parts. The outcomes in the shape of KVs' correlation with the biological activity are a call for the important prospect of the KVs in having antiviral and antifungal activities to be able to establish the essential role of these KVs in the emerging and resistant diseases of microbial and HIV infections.

In terms of geographical origin, a significant antiviral activity was witnessed for the KSA sample, whereas the Indian sample revealed a substantial activity against *C. albicans* and *E. coli*. Both the fruits and their parts were noted with sizeable differences in the extract yield and KV amounts. The disparity may be due to the change in climate, air, soil, salinity, altitude, and various geographical factors, as previously reported [34,35]. A detailed investigation of the effect of these factors may be helpful in categorizing the optimum conditions for the growth and marketing of the BM fruit with the maximum extract yield, KV amount, and bioactivity. Nevertheless, the exciting commonality observed in the study was that the seeds and pith parts from the fruits of both origins exhibited an intriguing antiviral and antimicrobial activity. The authors are of the opinion that the KVs were the critical contributors to these activities, irrespective of the nature of the KVs, i.e., KV-VIII, -X, or -XI present in these samples.

To depict the biological activities with the KVs profile of the BM fruit, statistical tests were computed for the dependent variables (DVs) to highlight the significant correlations and differences among the different origin extracts. The bivariate Pearson analysis displayed a significant pairwise correlation for I3 and K3 vs. DVs at $p = 0.05$, whereas I1 and K1 exhibited a positive correlation at $p = 0.01$. I1 and K1 showed the highest extract yields, whereas I3 and K3 had the highest KV amounts, in addition to their antiviral and antimicrobial activities. The PCA further confirmed the loading with more % variability for DVs with I3 and K3 in PC1. The extracts for I1 and K1 were loaded in PC2 with less % variability. This suggests the seed parts with higher extract yield and the pith parts enriched with the KVs are responsible for the respective activities observed in the study. Though the seed part revealed the lowest KVs among the fruit parts, an in-depth exploration is certainly recommended to investigate the chemical compounds present in these extracts, which play a pivotal role in the tested activities. The exploration may be a source of promising novel drug discovery from natural products in order to combat the challenges faced during HIV treatment and to alleviate the burden of microbial infections.

Moreover, the sample origin was tested against the individual biological activity using the two-tailed paired *t*-test. The pairs with significant *p*-values observed were sample origin vs. antiviral and antifungal activity against *C. albicans*. The statistical models finalized the seeds and pith parts from the BM fruit with significant antiviral activity against HIV and antifungal activity against *C. albicans*, suggested due to the presence of KVs.

4. Methods

4.1. Sample Collection and Preparation

Green and fresh BM fruits (Saudi and Indian origins) were purchased from the local farm house and markets in the Eastern province of Saudi Arabia and were prepared as previously reported in our LCMS/MS quantitative study [36]. Briefly, the fruits were sliced to remove the seeds and pith parts, while the skin part of the fruit was carefully abrased to separate it from the pith part using a sharp blade. The fruit parts were then placed on an aluminum sheet and dried in an oven at 60 °C.

4.2. Extraction and Analysis of KVs (Karavilosides)

The green ultrasound-assisted extraction with liquid chromatography–tandem mass spectrometry (LCMSMS) quantification for the three KVs (KV-VIII, -X, and -XI) using different parts (skin, seeds, and pith) of the bitter melon fruit (BM) from two different origins (India; I, and Saudi Arabia; K) have been reported in our previous study [36]. Concisely, one gram of each fruit part was subjected to ultrasonic extraction with a 1:1 mixture of acetone and ethanol. The extracted samples were dried and used for final % extract yield calculation and LCMSMS analysis. The details regarding extract yields (%) and KVs amount (μg/L) for the six extracts of the BM fruit—I1 (Indian seeds extract), I2 (Indian skin extract), I3 (Indian pith extract), K1 (Saudi Arabia seeds extract), K2 (Saudi Arabia skin extract), and K3 (Saudi Arabia pith extract)—are given in detail in Table S1.

4.3. Antiviral Activity

The antiviral activity was investigated against the specific drug target known as reverse transcriptase enzyme, a key enzyme in the retroviral life cycle, particularly for human immunodeficiency virus (HIV). The antiviral activity of *Momordica charantia* fruit part extracts was evaluated using a non-radioactive HIV-RT colorimetric ELISA kit (Roche Diagnostics, Mannheim, Germany). The protocol outlined in the kit was followed, using 2 ng of the enzyme in a well and incubating the reaction for 2 h at 37 °C [37]. The extracts were diluted with assay buffer and were tested at 2 mg/mL. The assay was carried out in duplicate, and the absorbance measurement of samples was carried out utilizing a microplate reader at 405 nm wavelength (ELISA; BioTek, Winooski, VT, USA). The negative control (water instead of sample) and positive control (Doxorubicin—Zydus Celexa, Ahmedabad, India) [38,39] were tested under the same conditions. The emerging color intensity is directly related to the reverse transcriptase action. The percentage of reverse transcriptase inhibition was calculated by comparing the absorbance of the samples to the negative control using the equation below [40]:

$$\text{HIV} - 1\ \text{RT inhibition}\ (\%) = \left\{1 - \left(\frac{\text{sample absorbance}}{\text{negative control absorbance}}\right)\right\} \times 100 \quad (1)$$

The extract showing the highest inhibition was investigated further, along with doxorubicin at concentrations ranging from zero to 2 mg/mL, to obtain a dose–response curve. The IC_{50} value for the extract and positive control were calculated by plotting the inhibition % against different concentrations using GraphPad Prism software (version 8.0) [40].

4.4. Antimicrobial Activity

4.4.1. Microbial Strains and Media

The bacterial strains consisted of *Staphylococcus aureus* (ATCC-25923), *Escherichia coli* (ATCC-25922), and *Candida albicans* (ATCC-14053). The culture media used for the agar-well diffusion method were blood (SPLM; cat-1009), MacConkey (Criterion; cat-C16131), and Sabouraud dextrose agar (SAB; criterion cat C6811), whereas Muller–Hinton broth (MHB, Oxoid, CM0405) was used for the broth dilution method to determine both MIC (minimum inhibitory concentration) and MBC (minimum bactericidal concentration).

4.4.2. Standard Inoculum

The selected colonies from the blood-, MacConkey-, and SAB-grown microorganisms (37 °C; 24 h) were inoculated in Muller–Hinton broth to form a homogenous microbial suspension and were standardized up to 0.5 McFarland turbidity (DensiCHEK plus, bioMérieux, Taguig, Philippines).

4.4.3. Determination of MIC and MBC

The six selected extracts (I1, I2, I3, K1, K2, and K3) were serially diluted with Muller–Hinton broth in a 96-well microtiter plate and were mixed to provide eight different concentrations for the assays in the range of 0.0005–3 mg/mL. The 0.5 McFarland standard (50 μL) for the bacterial strains in MHB was poured into each well of the six selected extracts, as well as into the control. The plates were incubated overnight at 37 °C, and the MIC and MBC were calculated as per the guidelines of the Clinical and Laboratory Standards Institute [41]. All the antimicrobial results were analyzed using Microsoft Excel 2021 and GraphPad Prism 8.0 software.

4.5. Molecular Docking Studies

4.5.1. Molecular Modeling Optimization

The computational simulations were performed using the Maestro 13.6; Schrödinger 2023-2, LLC, New York, NY, USA. The crystal structure of the studied enzyme, i.e., wild-type HIV-1 reverse transcriptase (RT), was retrieved from the Research Collaboratory for Structural Bioinformatics (RCSB) Protein Data Bank (PDB), with a corresponding PDB ID of 4G1Q and a resolution of 1.51 Å [42]. The crystal structure was prepared using the Schrödinger Maestro's Protein preparation wizard tool at a pH of 7.4, with the correction of ionization states, the addition of polar hydrogens, and the removal of the non-essential water molecules. The entire structure was minimized and optimized with the OPLS3 force field to optimize protein energies and remove any steric hindrance. The default RMSD value of 0.30 Å was used for non-hydrogen atoms. In order to prepare the ligand, Maestro Ligprep was used, where the SDF format structures were downloaded and converted to Maestro's 3D format. Epik, with several treatments, was utilized to determine the optimal chirality and ionization states at a physiological pH of 7.4 ± 2.0. Finally, the geometries were optimized using the OPLS3 force field. These conformations were used as the initial input structures for docking [43].

4.5.2. Binding Pocket Determination and Validation of Molecular Docking

The binding pocket was identified by the workspace co-crystallized inhibitor, non-nucleoside reverse transcriptase inhibitor (NNRTI), and Rilpivirine using Schrödinger's Maestro. A docking grid was created using the Glide software; Schrödinger 2023-2, LLC, New York, NY, USA, which was identified via a ligand-binding pocket from the crystal structure that was co-crystallized with the inhibitor. To ensure accuracy, the co-crystallized inhibitor was redocked using the alike protocol order. The docking poses and interactions were verified using maestro structure superimposition and root mean square deviation calculations (RMSD) of the alignment [23,44]. The receptor grids were generated using a 1.0 van der Waals (vdw) radius scaling factor, a 0.25 partial charge cut-off, and they were centered on the bound inhibitor. The binding site was enclosed within the grid box using default parameters and without constraints. Finally, the docking process was repeated and verified using three screening settings.

4.6. Statistical Analysis

SPSS V 22.0 and GraphPad Prism 8 were used to analyze the data and create the graphs. The results are presented as the mean ± standard deviation (SD) from three experiments. All data were analyzed with a *p*-value of <0.05 considered as significant in order to find the statistical significance between treated groups and controls.

5. Conclusions

The BM fruit samples from two different origins were investigated for their antiviral and antimicrobial role based on the KVs amount quantified using LCMSMS. The extracts exhibited >70% inhibition of HIV-RT, with K1 and K3 showing the highest inhibition (≈80%) among the samples. For the antimicrobial activity, the lowest MIC and MBC were observed for I3, I1, and K3 against *C. albicans* and *E. coli*. The correlation studies for the KVs and biological activities in these extracts established a prospective impact for the KVs in K3 and I3. For K1 and I1, the presence of other chemical compounds seems to impart biological activities to the extracts. The bioactive compounds might be other karaviloside derivatives or entirely different chemical entities. These outcomes unfold the crucial role of the seeds and pith parts of the BM fruits to be further studied for the isolation and identification of the novel molecules from these fruit parts. This may result in the discovery of novel medications to alleviate the disease burden of HIV and resistant microbial infections and to improve the patients' quality of life in the community.

Supplementary Materials: The following supporting information can be downloaded at: https://www.mdpi.com/article/10.3390/antibiotics13060544/s1, Table S1. Antiviral and antibacterial results for BM fruit parts with extract yields and KV amounts (* I1: Indian seeds; I2: Indian skin; I3: Indian pith; K1: KSA seeds; K2: KSA skin; K3: KSA pith). Table S2. IC50 determination for the active extract based on the antiviral inhibition results. Table S3. Bivariate correlation for the extracts vs. DVs at $p = 0.05$. * Correlation is significant at the 0.05 level (2-tailed). ** Correlation is significant at the 0.01 level (2-tailed). Table S4. Principial component analysis for the DVs with component loadings and KMO and Bartlett's test at $p = 0.05$. Table S5. Paired differences for the sample origin vs. individual activity at $p = 0.05$.

Author Contributions: Conceptualization, M.A. (Mohammed Aldholmi) and R.A.; Methodology, M.A. (Mohammed Aldholmi) and R.A.; Software, M.A. (Mansour Alturki) and R.A.; Validation, M.A. (Mohammed Aldholmi) and R.A.; Formal Analysis, M.A. (Mansour Alturki) and R.A.; Investigation, M.A. (Maher Alqurashi), M.H.S., and A.M.S.; Resources, M.A. (Mohammed Aldholmi), M.A. (Maher Alqurashi), and A.M.S.; Data Curation, M.A. (Mohammed Aldholmi) and R.A.; Writing—Original Draft Preparation, M.A. (Mohammed Aldholmi), R.A., and M.A. (Mansour Alturki); Writing—Review and Editing, M.A. (Mohammed Aldholmi), R.A., M.A. (Maher Alqurashi), M.H.S., and A.M.S.; Visualization, M.A. (Mohammed Aldholmi) and R.A.; Supervision, M.A. (Mohammed Aldholmi) and R.A.; Project Administration, M.A. (Mohammed Aldholmi). All authors have read and agreed to the published version of the manuscript.

Funding: This research received no external funding.

Institutional Review Board Statement: This study does not involve the use of any animal or human data or tissue.

Informed Consent Statement: Not applicable.

Data Availability Statement: All data generated or analyzed during this study are included in this published article.

Conflicts of Interest: The authors declare no conflict of interest.

Abbreviations

HCMV	human cytomegalovirus
HBV	Hepatitis B virus
HCV	Hepatitis C virus
BM	bitter melon
HIV	human immunodeficiency virus
LCMSMS	liquid chromatography–tandem mass spectrometry
KVs	karavilosides
I	India
K	Kingdom of Saudi Arabia
RT	reverse transcriptase

ELISA	enzyme-linked immunosorbent assay
SAB	Sabouraud dextrose agar
MHB	Muller–Hinton broth
MIC	minimum inhibitory concentration
MBC	minimum bactericidal concentration
CA	*Candida albicans*
EC	*Escherichia coli*
SA	*Staphylococcus aureus*
RCSB	Research Collaboratory for Structural Bioinformatics
PDB	Protein Data Bank
NNRTI	non-nucleoside reverse transcriptase inhibitor
RMSD	root mean square deviation
vdw	van der Waals
SD	standard deviation
PCA	principal component analysis
GLU	glutamic acid
LYS	lysine
VAL	valine
LEU	leucine
Phe	phenylalanine
TYR	tyrosine
MET	methionine

References

1. De Kraker, M.E.A.; Stewardson, A.J.; Harbarth, S. Will 10 Million People Die a Year due to Antimicrobial Resistance by 2050? *PLoS Med.* **2016**, *13*, e1002184. [CrossRef] [PubMed]
2. O'Neill, J. *Antimicrobial Resistance: Tackling a Crisis for the Health and Wealth of Nations*; Review on Antimicrobial Resistance: London, UK, 2014.
3. Okeke, I.N.; Laxminarayan, R.; Bhutta, Z.A.; Duse, A.G.; Jenkins, P.; O'Brien, T.F.; Pablos-Mendez, A.; Klugman, K.P. Antimicrobial resistance in developing countries. Part I: Recent trends and current status. *Lancet Infect. Dis.* **2005**, *5*, 481–493. [CrossRef] [PubMed]
4. Aldholmi, M.; Marchand, P.; Ourliac-Garnier, I.; Le Pape, P.; Ganesan, A. A Decade of Antifungal Leads from Natural Products: 2010–2019. *Pharmaceuticals* **2019**, *12*, 182. [CrossRef] [PubMed]
5. Newman, D.J.; Cragg, G.M. Natural Products as Sources of New Drugs over the Nearly Four Decades from 01/1981 to 09/2019. *J. Nat. Prod.* **2020**, *83*, 770–803. [CrossRef] [PubMed]
6. Anand, U.; Jacobo-Herrera, N.; Altemimi, A.; Lakhssassi, N. A Comprehensive Review on Medicinal Plants as Antimicrobial Therapeutics: Potential Avenues of Biocompatible Drug Discovery. *Metabolites* **2019**, *9*, 258. [CrossRef] [PubMed]
7. Patra, J.K.; Das, G.; Bose, S.; Banerjee, S.; Vishnuprasad, C.N.; del Pilar Rodriguez-Torres, M.; Shin, H. Star anise (*Illicium verum*): Chemical compounds, antiviral properties, and clinical relevance. *Phyther. Res.* **2020**, *34*, 1248–1267. [CrossRef] [PubMed]
8. Su, X.-Z.; Miller, L.H. The discovery of artemisinin and the Nobel Prize in Physiology or Medicine. *Sci. China Life Sci.* **2015**, *58*, 1175–1179. [CrossRef] [PubMed]
9. Astani, A.; Reichling, J.; Schnitzler, P. Comparative study on the antiviral activity of selected monoterpenes derived from essential oils. *Phyther. Res.* **2010**, *24*, 673–679. [CrossRef] [PubMed]
10. Guo, Y.; Ma, A.; Wang, X.; Yang, C.; Chen, X.; Li, G.; Qiu, F. Research progress on the antiviral activities of natural products and their derivatives: Structure–activity relationships. *Front. Chem.* **2022**, *10*, 1005360. [CrossRef]
11. Wohlfarth, C.; Efferth, T. Natural products as promising drug candidates for the treatment of hepatitis B and C. *Acta Pharmacol. Sin.* **2009**, *30*, 25–30. [CrossRef]
12. Gayathry, K.S.; John, J.A. A comprehensive review on bitter gourd (*Momordica charantia* L.) as a gold mine of functional bioactive components for therapeutic foods. *Food Prod. Process. Nutr.* **2022**, *4*, 10. [CrossRef]
13. Ingle, A.; Kapgatte, R. Phytochemical screening and antimicrobial activity of *Momordica charantia* Linn. *Int. J. Pharmacol. Res.* **2018**, *8*, 63–65.
14. Leelaprakash, G.; Rose, J.C.; Gowtham, B.M.; Javvaji, P.K.; Prasad, S.A. In vitro antimicrobial and antioxidant activity of *Momordica charantia* leaves. *Pharmacophore* **2011**, *2*, 244–252.
15. Jagessar, R.C.; Mohamed, A.; Gomes, G. An evaluation of the antibacterial and antifungal activity of leaf extracts of *Momordica charantia* against *Candida albicans*, *Staphylococcus aureus* and *Escherichia coli*. *Nat. Sci.* **2008**, *6*, 1–14.
16. Masithoh, D.A.; Kusdarwati, R.; Handijatno, D. Antibacterial activity of bitter gourd (*Momordica charantia* L.) leaf extract against Aeromonas hydrophila. *IOP Conf. Ser. Earth Environ. Sci.* **2019**, *236*, 012096. [CrossRef]
17. Mahmood, A.; Raja, G.K.; Mahmood, T.; Gulfraz, M.; Khanum, A. Isolation and characterization of antimicrobial activity conferring component (s) from seeds of bitter gourd (*Momordica charantia*). *J. Med. Plants Res.* **2012**, *6*, 566–573. [CrossRef]

18. Grover, J.; Yadav, S. Pharmacological actions and potential uses of *Momordica charantia*: A review. *J. Ethnopharmacol.* **2004**, *93*, 123–132. [CrossRef] [PubMed]
19. NG, T.B.; WONG, C.M.; LI, W.W.; YEUNG, H.W. Isolation and characterization of a galactose binding lectin with insulinomimetic activities. *Int. J. Pept. Protein Res.* **1986**, *28*, 163–172. [CrossRef]
20. Yao, X.; Li, J.; Deng, N.; Wang, S.; Meng, Y.; Shen, F. Immunoaffinity purification of α-momorcharin from bitter melon seeds (*Momordica charantia*). *J. Sep. Sci.* **2011**, *34*, 3092–3098. [CrossRef] [PubMed]
21. Wang, H.X.; Ng, T.B. Examination of Lectins, Polysaccharopeptide, Polysaccharide, Alkaloid, Coumarin and Trypsin Inhibitors for Inhibitory Activity Against Human Immunodeficiency Virus Reverse Transcriptase and Glycohydrolases. *Planta Med.* **2001**, *67*, 669–672. [CrossRef]
22. Jiratchariyakul, W.; Wiwat, C.; Vongsakul, M.; Somanabandhu, A.; Leelamanit, W.; Fujii, I.; Suwannaroj, N.; Ebizuka, Y. HIV Inhibitor from Thai Bitter Gourd. *Planta Med.* **2001**, *67*, 350–353. [CrossRef]
23. Alves, M.; Froufe, H.; Costa, A.; Santos, A.; Oliveira, L.; Osório, S.; Abreu, R.; Pintado, M.; Ferreira, I. Docking Studies in Target Proteins Involved in Antibacterial Action Mechanisms: Extending the Knowledge on Standard Antibiotics to Antimicrobial Mushroom Compounds. *Molecules* **2014**, *19*, 1672–1684. [CrossRef]
24. Das, K.; Bauman, J.D.; Clark, A.D.; Frenkel, Y.V.; Lewi, P.J.; Shatkin, A.J.; Hughes, S.H.; Arnold, E. High-resolution structures of HIV-1 reverse transcriptase/TMC278 complexes: Strategic flexibility explains potency against resistance mutations. *Proc. Natl. Acad. Sci. USA* **2008**, *105*, 1466–1471. [CrossRef]
25. Das, K.; Clark, A.D.; Lewi, P.J.; Heeres, J.; de Jonge, M.R.; Koymans, L.M.H.; Vinkers, H.M.; Daeyaert, F.; Ludovici, D.W.; Kukla, M.J.; et al. Roles of Conformational and Positional Adaptability in Structure-Based Design of TMC125-R165335 (Etravirine) and Related Non-nucleoside Reverse Transcriptase Inhibitors That Are Highly Potent and Effective against Wild-Type and Drug-Resistant HIV-1 Varian. *J. Med. Chem.* **2004**, *47*, 2550–2560. [CrossRef] [PubMed]
26. Shivanagoudra, S. Differential Inhibition of Carbolytic Enzymes and Suppression of LPS-Induced Inflammation by Isolated Compounds from Bitter Melon (*Momordica charantia*). Ph.D. Thesis, Texas A&M University, College Station, TX, USA, 2019.
27. Li, Z.; Xia, A.; Li, S.; Yang, G.; Jin, W.; Zhang, M.; Wang, S. The Pharmacological Properties and Therapeutic Use of Bitter Melon (*Momordica charantia* L.). *Curr. Pharmacol. Rep.* **2020**, *6*, 103–109. [CrossRef]
28. Bourinbaiar, A.S.; Lee-Huang, S. The Activity of Plant-Derived Antiretroviral Proteins MAP30 and GAP31 against Herpes Simplex Virus Infectionin Vitro. *Biochem. Biophys. Res. Commun.* **1996**, *219*, 923–929. [CrossRef] [PubMed]
29. Lee-Huang, S.; Huang, P.L.; Chen, H.-C.; Huang, P.L.; Bourinbaiar, A.; Huang, H.I.; Kung, H. Anti-HIV and anti-tumor activities of recombinant MAP30 from bitter melon. *Gene* **1995**, *161*, 151–156. [CrossRef] [PubMed]
30. Huang, W.-H.; Su, W.-M.; Wang, C.-W.; Fang, Y.-H.; Jian, Y.-W.; Hsu, H.-J.; Peng, C.-W. *Momordica* anti-HIV protein MAP30 abrogates the Epstein-Barr virus nuclear antigen 1 dependent functions in host cells. *Heliyon* **2023**, *9*, e21486. [CrossRef]
31. Eryılmaz Pehlivan, F. Bitter Melon: A Multifunctional Medicinal Plant with Powerful Bioactive Compounds. In *Functional Foods—Phytochemicals and Health Promoting Potential*; IntechOpen: London, UK, 2021.
32. Guarniz, W.A.S.; Canuto, K.M.; Ribeiro, P.R.V.; Dodou, H.V.; Magalhaes, K.N.; Miranda Sa, K.; do Nascimento, P.G.G.; Silva, K.L.; Passos Sales, G.W.; Monteiro, M.P.; et al. *Momordica charantia* L. Variety from Northeastern Brazil: Analysis of Antimicrobial Activity and Phytochemical Components. *Pharmacogn. J.* **2019**, *11*, 1312–1324. [CrossRef]
33. Braca, A.; Siciliano, T.; D'Arrigo, M.; Germanò, M.P. Chemical composition and antimicrobial activity of *Momordica charantia* seed essential oil. *Fitoterapia* **2008**, *79*, 123–125. [CrossRef]
34. Ahmad, R.; Aldholmi, M.; Mostafa, A.; Alqathama, A.; Aldarwish, A.; Abuhassan, A.; Alateeq, L.; Bubshait, S.; Aljaber, M.; Aldossary, S. A novel green extraction and analysis technique for the comprehensive characterization of mangiferin in different parts of the fresh mango fruit (*Mangifera indica*). *LWT* **2022**, *159*, 113176. [CrossRef]
35. Ahmad, R.; Aldholmi, M.; Alqathama, A.; Aldossary, S.; Bubshait, S.; Aljaber, M.; Abuhassan, A.; Aldarwish, A.; Alateeq, L. Green and novel ultrasonic extraction with UHPLC-MSMS analysis of natural sweetener (Glycyrrhizic acid) from *Glycyrrhiza glabra*; a multifactorial mechanistic evaluation based on statistical analysis. *Ultrason. Sonochem.* **2021**, *77*, 105696. [CrossRef] [PubMed]
36. Ahmad, R.; Aldholmi, M.; Alqathama, A.; Al Nahab, H.Z.; Almutawah, A.I. A comprehensive LCMS/MS characterization for the green extracted cucurbitane-triterpenoid glycosides from bitter melon (*Momordica charantia*) fruit. *Food Chem.* **2024**, *445*, 138479. [CrossRef]
37. Harnett, S.M.; Oosthuizen, V.; van de Venter, M. Anti-HIV activities of organic and aqueous extracts of *Sutherlandia frutescens* and *Lobostemon trigonus*. *J. Ethnopharmacol.* **2005**, *96*, 113–119. [CrossRef]
38. Kapewangolo, P.; Hussein, A.A.; Meyer, D. Inhibition of HIV-1 enzymes, antioxidant and anti-inflammatory activities of *Plectranthus barbatus*. *J. Ethnopharmacol.* **2013**, *149*, 184–190. [CrossRef] [PubMed]
39. Kapewangolo, P.; Kandawa-Schulz, M.; Meyer, D. Anti-HIV Activity of *Ocimum labiatum* Extract and Isolated Pheophytin-a. *Molecules* **2017**, *22*, 1763. [CrossRef]
40. Matotoka, M.M.; Ndhlala, A.R.; Masoko, P. In vitro inhibition of HIV-1 reverse transcriptase and anti-inflammatory activities of some herbal concoctions sold in the Limpopo Province. *S. Afr. J. Bot.* **2019**, *126*, 65–69. [CrossRef]
41. Barry, A.L. *Methods for Determining Bactericidal Activity of Antimicrobial Agents: Approved Guideline*; NCCLS Document; NCCLS: Wayne, PA, USA, 1999; ISBN 9781562383848.
42. Kuroda, D.G.; Bauman, J.D.; Challa, J.R.; Patel, D.; Troxler, T.; Das, K.; Arnold, E.; Hochstrasser, R.M. Snapshot of the equilibrium dynamics of a drug bound to HIV-1 reverse transcriptase. *Nat. Chem.* **2013**, *5*, 174–181. [CrossRef] [PubMed]

43. Shelley, J.C.; Cholleti, A.; Frye, L.L.; Greenwood, J.R.; Timlin, M.R.; Uchimaya, M. Epik: A software program for pK a prediction and protonation state generation for drug-like molecules. *J. Comput. Aided Mol. Des.* **2007**, *21*, 681–691. [CrossRef]
44. Kwofie, S.; Dankwa, B.; Odame, E.; Agamah, F.; Doe, L.; Teye, J.; Agyapong, O.; Miller, W.; Mosi, L.; Wilson, M. In Silico Screening of Isocitrate Lyase for Novel Anti-Buruli Ulcer Natural Products Originating from Africa. *Molecules* **2018**, *23*, 1550. [CrossRef]

antibiotics

MDPI

Article

In Vitro Antimicrobial Potential of Portuguese Propolis Extracts from Gerês against Pathogenic Microorganisms

Rafaela Dias Oliveira [1,2,†], **Carina Araújo** [3,†] **and Cristina Almeida-Aguiar** [3,4,*]

[1] Life and Health Sciences Research Institute (ICVS), University of Minho, 4710-057 Braga, Portugal; rafaeladiasoliveira98@gmail.com
[2] ICVS/3B's—PT Government Associate Laboratory, 4710-057 Braga, Portugal
[3] Biology Department, University of Minho, 4710-057 Braga, Portugal; sacarinaaraujo@gmail.com
[4] CBMA—Centre of Molecular and Environmental Biology, University of Minho, 4710-057 Braga, Portugal
[*] Correspondence: cristina.aguiar@bio.uminho.pt; Tel.: +351-253601513
[†] These authors contributed equally to this work.

Abstract: Antimicrobial resistance (AMR) is one of humanity's main health problems today. Despite all the breakthroughs and research over the past few years, the number of microbial illnesses that are resistant to the available antibiotics is increasing at an alarming rate. In this article, we estimated the biomedical potential of Portuguese propolis harvested from the Gerês apiary over five years, evaluating the in vitro antimicrobial effect of five hydroalcoholic extracts prepared from five single propolis samples and of a hydroalcoholic extract obtained from the mixture of all samples. The antimicrobial potential was firstly assessed by determining the minimum inhibitory concentration (MIC) of these extracts against a panel of three Gram-positive (*Bacillus subtilis*, methicillin-sensitive *Staphylococcus aureus*, and methicillin-resistant *Staphylococcus aureus*) and one Gram-negative bacteria (*Escherichia coli*), as well as two yeasts (*Candida albicans* and *Saccharomyces cerevisiae*). As MIC values against each bacterium were consistent across all the evaluated propolis extracts, we decided to further conduct a disk diffusion assay, which included three commercial antibiotics—erythromycin, vancomycin, and amoxicillin/clavulanic acid—for comparison purposes. In addition to displaying a concentration-dependent antibacterial effect, the hydroalcoholic extracts prepared with 70% ethanol exhibited stronger antimicrobial capacity than vancomycin against *B. subtilis* (% of increase ranged between 26 and 59%) and methicillin-sensitive *S. aureus* (% of increase ranged between 63 and 77%). Moreover, methicillin-resistant *S. aureus* (MRSA) showed susceptibility to the activity of the same extracts and resistance to all tested antibiotics. These findings support that propolis from Gerês is a promising natural product with promising antimicrobial activity, representing a very stimulating result considering the actual problem with AMR.

Keywords: Portuguese propolis; antimicrobial activity; pathogenic microorganisms; antibiotic resistance

1. Introduction

Antimicrobial resistance (AMR)—defined as the capacity of a microorganism to survive exposure to a defined dose of an antimicrobial agent—is one of the major current threats to public health [1,2]. Despite all the advancements and research over the past years, the number of microbial illnesses that are resistant to all known and approved antibiotics is increasing at an alarming rate [3]. This problem is exacerbated by the extension of resistance from the restricted hospital environments into open communities and by the ongoing development of novel resistance mechanisms [4]. In 2019, 4.95 million deaths caused by AMR were recorded, among which 1.27 million were caused by bacterial resistance [5]. According to the Organization for Economic Cooperation and Development, infections with resistant microorganisms will cause 2.4 million deaths in Europe, North America, and Australia over the following 30 years [6]. In fact, it is predicted that Southern Europe will have the highest mortality rate and will be the most affected region worldwide [2,6]. In short, the current pattern of increased resistance of microbial pathogens to the available drugs has stimulated the introduction of new alternative agents in the treatment of infectious diseases. Thus, the interest in researching natural products, such as propolis and its isolated compounds, as antimicrobial agents has emerged.

Propolis, a natural resinous product produced by honeybees (mainly *Apis mellifera* L.) is a mixture of resins collected from plant exudates which are digested with ß-glucosidase and mixed with pollen and waxes to form the final product [7–9]. The chemical composition of this natural product is extremely complex and highly variable, depending on distinct factors, such as the bee species, surrounding flora, local climate, and geographic location. In fact, about 800 compounds were already identified in propolis [10]. These compounds belong to distinct classes, namely, phenolic acids (cinnamic and caffeic acids) and their esters, flavonoids (flavones, flavanones, flavonols, and dihydroflavonols chalcones), and terpenes [9,11–13]. Over the years, several biological properties of this natural product have been confirmed, including its antimicrobial capacity, which has been linked to the presence of polyphenolic compounds—namely, flavonoids (chrysin, galangin, pinobanksin, pinocembrin), phenolic acids (caffeic, cinnamic, and ferulic acids) and their esters [7,9,14], and terpenes [9,15]. Notably, some of these propolis compounds, such as chrysin or apigenin, for example, have been shown not only to exhibit antimicrobial effects against microorganisms but also to potentiate the activity of antibiotics [16,17]. According to published data, propolis is a really powerful antibacterial agent that is more effective against Gram-positive bacteria than against Gram-negative bacteria [9,18], and has bactericidal effects which are correlated with several mechanisms, including the inhibition of cell division and protein synthesis, the destruction of cell walls and bacterial cytoplasm, and the inhibition of bacterial mobility and enzyme activity [19,20]. Additionally, propolis displays bacteriostatic activity against diverse bacterial genera [18,21].

Portuguese propolis from Gerês is one of the most studied in Portugal [13,22–29]. Recent results revealed potent antimicrobial and antioxidant activities, regardless of the harvesting year—which seems to be a distinctive and unique feature of this propolis. Indeed, this constancy was also observed in the chemical profile of this national product [22]. Moreover, Peixoto et al. [28] reported that mixtures of propolis samples from Gerês harvested in different years are equally or more effective than the single extracts. This highlights a possible step toward standardization since the combinations of fewer and more active extracts result in similar bioactivities. The use of propolis mixtures is also important to ensure propolis availability, as well as to value this national natural resource, allowing propolis application in food, cosmetic, and medical areas.

Research on Portuguese propolis increased the potential for generating novel and bioactive propolis-based formulations [25,30]. A prior study conducted by Araújo et al. [29] demonstrated the anti-inflammatory capacity of Portuguese propolis from Gerês, both in vivo and in vitro, using single extracts and extracts of mixtures of propolis samples. In the present study, our main goal was to evaluate the antimicrobial effect of the same extracts of Portuguese propolis from Gerês previously studied by Araújo et al. [29]. More precisely, we aimed to assess their effectiveness as antimicrobial agents against a selected panel of susceptibility indicator strains and to compare such effectiveness with the action of three commercial antibiotics—amoxicillin/clavulanic acid (AMC), erythromycin (ERY), and vancomycin (VAN).

2. Results

2.1. In Vitro Antimicrobial Potential of Propolis from Gerês

Globally, none of the six extracts was active against yeast, *Escherichia coli*, and methicillin-resistant *Staphylococcus aureus* at the highest tested concentration in the agar dilution assays (Table 1). However, the same hydroalcoholic extracts were found to be active against *Bacillus subtilis* and methicillin-sensitive *Staphylococcus aureus*. No differences were observed in the antimicrobial capacity of the different extracts against each of the tested bacteria, as indicated by identical MIC values (Table 1).

Table 1. MIC values of Gerês propolis hydroalcoholic extracts against susceptibility indicator strains. Propolis extracts have ethanol 70% (*v*/*v*) (EE_{70}). Results are expressed as the lowest concentration at which no growth occurred. Controls were performed using YPDA/LBA plates without G.EEs and YPDA/LBA plates with ethanol 70%. (G: Gerês; mG: a mixture of propolis samples from Gerês).

	Bacillus subtilis	Methicillin-Sensitive *Staphylococcus aureus* (MSSA)	Methicillin-Resistant *Staphylococcus aureus* (MRSA)	*Escherichia coli*
G11.EE_{70}	50	200	>2000	>2000
G12.EE_{70}	50	200	>2000	>2000
G13.EE_{70}	50	200	>2000	>2000
G14.EE_{70}	50	200	>2000	>2000
G15.EE_{70}	50	200	>2000	>2000
mG.EE_{70}	50	200	>2000	>2000

Since no differences were observed across the antimicrobial activity of the different hydroalcoholic extracts, we decided to pursue a disk diffusion assay to try to distinguish the capacity of each extract against the susceptible panel of microorganisms. Therefore, we evaluated the antimicrobial potential of the hydroalcoholic extracts of propolis from Gerês, either from single samples (G.EE_{70}) or from the resultant mixtures (mG.EE_{70}) (Supplementary Table S1). Additionally, we tested the extracts obtained by diluting the G.EE70 samples (G11.EE35, G12.EE35, G13.EE35, G14.EE35, G15.EE35, and mG.EE35) to assess the effect of dilution on the antimicrobial potential of the propolis extracts. Globally, none of the analyzed extracts was active against yeasts as no growth inhibition zones were observed. On the other hand, antibacterial activity was detected against all the tested bacteria strains, and the G.EE_{70} group displayed invariably inhibition zones with wider diameters than the ones observed within the G.EE_{35} group.

2.1.1. Antimicrobial Effect of Propolis Extracts against *Bacillus subtilis*

The antimicrobial potential of propolis hydroalcoholic extracts from Gerês was evaluated against *Bacillus subtilis*, often used as a Gram-positive model organism [31,32]. Overall, the inhibition zone diameters displayed by the G.EE_{70} samples were generally higher (values ranging from 25.25 ± 2.46 to 31.92 ± 4.48 mm) (Figure 1A) than the ones displayed by the G.EE_{35} samples (values ranging between 16.67 ± 3.62 and 19.92 ± 2.72 mm) (Figure 1B). Additionally, and considering the Zone Diameter Interpretive Standards for each antibiotic–microorganism [33], this species seems to be more susceptible to the antibiotics ERY and VAN and resistant to AMC (Supplementary Table S1 and Figure 1).

The antimicrobial potential did not vary significantly between samples harvested in different years ($ANOVA_{1w}$, $F_{(5,12)} = 2.793$, $p = 0.0673$; Figure 1A; $ANOVA_{1w}$, $F_{(5,12)} = 0.755$, $p = 0.05984$; Figure 1B), suggesting a constancy in antibacterial activity against *Bacillus subtilis*. When comparing with the tested antibiotics, however, the post hoc analysis for the G.EE_{70} group ($ANOVA_{1w}$, $F_{(7,14)} = 5.98$, $p = 0.002$; Figure 1A) showed significant differences between (i) G11.EE and VAN; (ii) G12.EE and VAN; and (iii) G15.EE and VAN. Regarding the G.EE_{35} group, no significant differences were observed across the inhibition zone diameters displayed by propolis extracts and the tested antibiotics ($ANOVA_{1w}$, $F_{(6,12)} = 0.78$, $p = 0.60$; Figure 1B). In fact, the findings point to the limited antibacterial action of G.EE_{35} samples against the tested microorganism. Thus, in summary, G11.EE_{70}, G12.EE_{70}, and G15.EE_{70} have the highest antibacterial activity against *Bacillus subtilis* and are a feasible and potential alternative to the tested antibiotics.

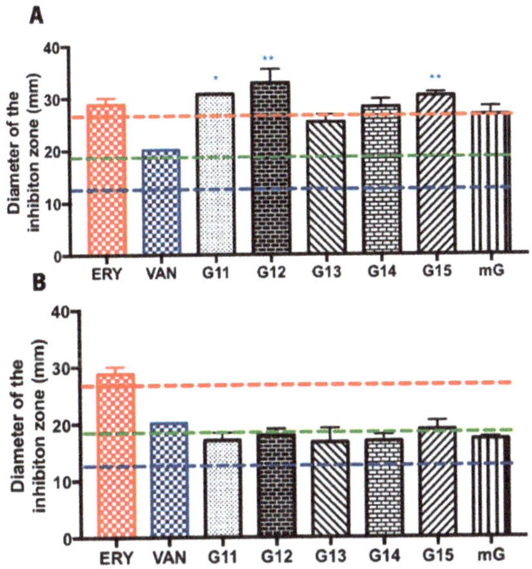

Figure 1. Antibacterial effect of the Gerês propolis hydroalcoholic extracts against *Bacillus subtilis*. Diameter of the inhibition zones displayed by (**A**) G.EE_{70} samples, erythromycin, and vancomycin and by (**B**) G.EE_{35} samples, erythromycin, and vancomycin. Results are expressed as means ± SD (* $0.05 > p \geq 0.01$; ** $0.01 > p \geq 0.001$). No growth inhibition zones were observed for disks containing ethanol 70% or ethanol 35%. Reference values of inhibition zone diameters above which a microorganism is susceptible to ERY, AMC, and VAN [33] are represented with red, green, and blue dashed lines, respectively. (ERY: erythromycin; AMC: amoxicillin/Clavulanic Acid; VAN: vancomycin; G: Gerês, mG: mixture of propolis samples from Gerês).

2.1.2. Antimicrobial Effect of Propolis Extracts against Methicillin-Sensitive *Staphylococcus aureus*

Methicillin-sensitive *Staphylococcus aureus* is a Gram-positive bacterium that is pathogenic for humans, causing soft (e.g., furuncles and cellulitis) and very serious infections (e.g., blood stream infections and pneumonia) [34]. Regarding the evaluation of the Gerês propolis hydroalcoholic extracts' antibacterial capacity against *Staphylococcus aureus*, G.EE_{70} samples exhibited considerably higher inhibition zones (values ranging from 26.93 ± 2.20 to 29.17 ± 2.75 mm) (Figure 2A) than the ones displayed by the G.EE_{35} samples (values ranging from 14.42 ± 0.14 to 18.08 ± 0.62 mm) (Figure 2B). Considering the reference values of the inhibition zone diameters for the tested antibiotics and *Staphylococcus* spp. [33], this MSSA strain was revealed to be susceptible to the action of both VAN and AMC and resistant to ERY (Supplementary Table S1 and Figure 2).

The antimicrobial potential of hydroalcoholic extracts of propolis from Gerês against *S. aureus* is consistent and independent of the harvesting year ($ANOVA_{1w}$, $F_{(5,12)} = 0.877$, $p = 0.5248$; Figure 1A; $ANOVA_{1w}$, $F_{(5,12)} = 1.547$, $p = 0.2478$; Figure 1B). Nevertheless, the G.EE_{70} group post hoc analysis ($ANOVA_{1w}$, $F_{(7,16)} = 5.98$, $p < 0.001$; Figure 2A) revealed significant differences across (i) G11.EE_{70} and VAN; (ii) G12.EE_{70} and VAN; (iii) G13.EE_{70} and VAN; (iv) G14.EE_{70} and VAN; (v) G15.EE_{70} and VAN; and (vi) mG.EE_{70} and VAN. These results suggest the G.EE_{70} samples have stronger antibacterial capacity than vancomycin. In summary, in the G.EE_{70} group, all extracts showed antibacterial activity effectiveness, and the mG.EE_{70} exhibited higher capacity than almost all the individual extracts, indicating that blending propolis with either a high or low antibacterial capacity may enhance this bioactivity.

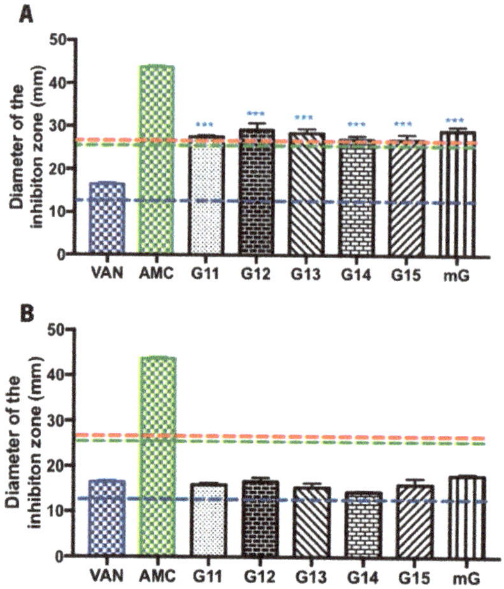

Figure 2. Antibacterial effect of Gerês propolis hydroalcoholic extracts against methicillin-sensitive *Staphylococcus aureus*. Diameter of the inhibition zones displayed by (**A**) G.EE$_{70}$ samples, Vancomycin, and Amoxicillin/Clavulanic acid and by (**B**) G.EE$_{35}$ samples, Vancomycin, and Amoxicillin/Clavulanic acid. Results are expressed as means ± SD (*** $p < 0.001$). No growth inhibition zones were observed for disks containing ethanol 70% or ethanol 35%. Reference values of inhibition zone diameters above which *Staphylococcus* spp. is susceptible to ERY, AMC, and VAN [33] are represented with red, green, and blue dashed lines, respectively. (ERY: erythromycin; AMC: amoxicillin/Clavulanic Acid; VAN: vancomycin; G: Gerês, mG: mixture of propolis samples from Gerês).

Regarding G.EE$_{35}$ samples, significant differences were not observed among the extracts (Figure 2B), and the diameters of the inhibition zones displayed were lower than the ones obtained with the AMC antibiotic. However, they were remarkably similar to the inhibition diameter produced by VAN, and, even though no statistically significant differences were observed (ANOVA$_{1W}$, $F_{(6,13)} = 2.37$, $p = 0.09$; Figure 2B), mG.EE$_{35}$ could be a good option when a low ethanol concentration is required because it is slightly more active than individual extracts (Supplementary Table S1 and Figure 2B).

2.1.3. Antimicrobial Effect of Propolis Extracts against Methicillin-Resistant *Staphylococcus aureus*

Methicillin-resistant *S. aureus* is currently recognized as one of the major multidrug-resistant bacterial pathogens, posing a significant threat to human health [35–37]. Considering the need for new antibacterial compounds to overcome this methicillin resistance, Gerês propolis hydroalcoholic extracts were tested against MRSA (Supplementary Table S1 and Figure 3).

In general, the tested commercial antibiotics (ERY, VAN, and AMC) had no inhibitory effect against MRSA, demonstrating the already known resistance of this bacterium to these antibiotics (Supplementary Table S1). No antibacterial effect was observed after treatment with the hydroalcoholic extracts of Gerês propolis containing ethanol 35% either (Supplementary Table S1). Regarding the G.EE$_{70}$ group, all the tested G.EEs/mG.EEs caused an inhibitory effect in the growth of MRSA, with values ranging between 17.25 ± 1.15 and 21.33 ± 3.17 mm (Supplementary Table S1 and Figure 3). In fact, the antimicrobial potential of these extracts did not vary significantly between samples harvested in different years (ANOVA$_{1W}$, $F_{(5,12)} = 1.51$, $p = 0.26$; Figure 3), revealing a constancy in the antibacterial potential against MRSA. Curiously, the G.EE$_{70}$ group displayed inhibition halos even larger than the reference inhibition zone diameter for vancomycin against *Staphylococcus* spp. (blue line, Figure 3). Although not significantly different, the extract of propolis mixtures—mG.EE$_{70}$—seemed to be the most effective extract against this bacterium.

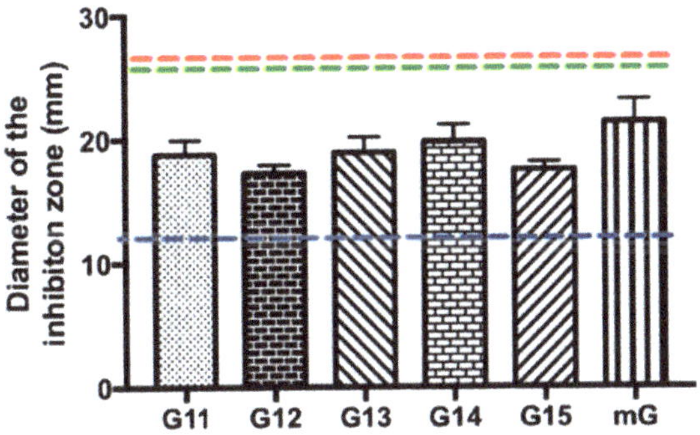

Figure 3. Antibacterial effect of Gerês propolis hydroalcoholic extracts against methicillin-resistant *Staphylococcus aureus*. Diameter of the inhibition zones displayed by G.EE$_{70}$ samples. Results are expressed as means ± SD. No growth inhibition zones were observed for disks containing ethanol 70%. Reference values of inhibition zone diameters above which *Staphylococcus* spp. is susceptible to ERY, AMC, and VAN [33] are represented with red, green, and blue dashed lines, respectively. (ERY: erythromycin; AMC: amoxicillin/Clavulanic Acid; VAN: vancomycin; G: Gerês, mG: mixture of propolis samples from Gerês).

2.1.4. Antimicrobial Effect of Propolis Extracts against *Escherichia coli*

Lastly, the antimicrobial efficacy of Gerês propolis hydroalcoholic extracts was investigated against *Escherichia coli* (*E. coli*), a Gram-negative bacillus commonly found in the intestinal tract of humans and animals [38]. Overall, *E. coli* was only susceptible to the action of the G.EE$_{70}$ group (values ranging between 15.33 ± 1.61 and 22.17 ± 2.16 mm), while the G.EE$_{35}$ group had no inhibitory effect against this enterobacterium (Supplementary Table S1 and Figure 4). Considering the Zone Diameter Interpretive Standards for each antibiotic–microorganism [39] (Figure 4), *E. coli* was revealed to be also susceptible to the action of AMC and resistant to ERY.

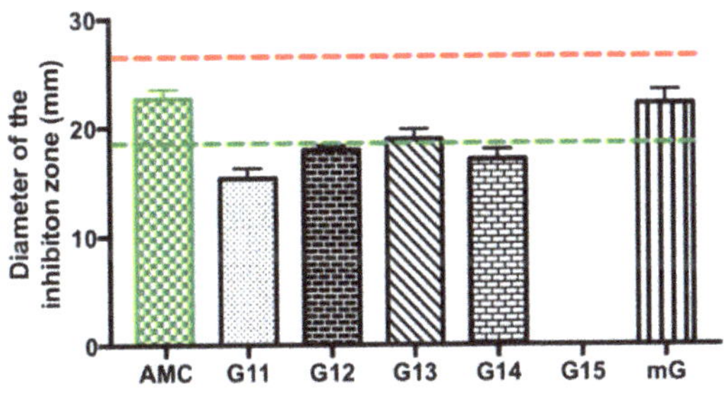

Figure 4. Antibacterial effect of Gerês propolis hydroalcoholic extracts against *Escherichia coli*. Diameter of the inhibition zones displayed by G.EE$_{70}$ samples and AMC. Results are expressed as means ± SD. No growth inhibition zones were observed for disks containing ethanol 70%. Reference values of inhibition zone diameters above which a microorganism is susceptible to ERY and AMC are represented with red and green dashed lines, respectively [33]. Vancomycin has no reference diameter against Gram-negative bacteria because of its inability to breach the outer membrane barrier [40]. (ERY: erythromycin; AMC: amoxicillin/Clavulanic Acid; G: Gerês, mG: mixture of propolis samples from Gerês).

Considering the inhibition diameters displayed by the G.EE_{70} group (Supplementary Table S1 and Figure 4), only the mG.EE_{70} extract displayed an inhibition zone whose diameter was similar to the value obtained for amoxicillin/clavulanic acid ($ANOVA_{1w}$, $F_{(6,14)} = 80.3$, $p < 0.001$; Figure 4) and even exceeded the reference susceptibility value for this antibiotic (green line, Figure 4). Concluding, although not statistically significant, the mixture mG.EE_{70} was proved to be once again the most potent extract, this time against *E. coli*.

3. Discussion

The steady increase in antimicrobial resistance worldwide has become a significant public health concern recently recognized by the World Health Organization [3,41]. In response, there has been a growing effort in recent years to identify and validate new natural and multitargeting approaches as potential replacements for commercial antibiotics given their limited effectiveness against pathogenic microorganisms. Antimicrobial activity has been described for several samples of Portuguese propolis—from Gerês [23,25,28], Caramulo [42], Trás-os-Montes [43], Açores [44], and Bragança and Beja [45]. However, a direct comparison with commercial antibiotics was never reported. Therefore, in this study, we evaluated the antimicrobial potential of extracts of Gerês propolis collected over five consecutive years (G11–G15), as well as of the extract of the mixture (mG.EE), along with the efficacy of three commercial antibiotics—erythromycin, vancomycin, and amoxicillin/clavulanic acid.

The most frequent experimental approaches to estimate the antimicrobial effect of propolis are the disk diffusion and the agar dilution methodologies, either in solid or liquid media [18,46]. Each method can be particularly suitable in certain experimental conditions when chosen accordingly [47]. In this case, the agar dilution methodology was selected to test the extract samples prepared with ethanol 70%, more precisely G11.EE_{70}–G15.EE_{70} and mG.EE_{70}. These propolis extracts lacked antifungal capacity and did not affect MRSA and *E. coli* (Table 1). This aligns with previous studies, which reported no antifungal activity and a lower antibacterial potential against Gram-negative bacteria like *E. coli* [24,25]. Conversely, and as previously published [46,48], ethanol extracts from Gerês propolis are more effective against Gram-positive bacteria. Our results are also consistent with the literature for other extracts of Gerês propolis [22,25,28], although the extracts in the literature were prepared with ethanol 100% instead of ethanol 70%, as all the tested G.EE_{70} samples were effective against *B. subtilis* (MIC value = 50 μg/mL) and MSSA (MIC value = 200 μg/mL) (Table 1). Indeed, *B. subtilis* was the most susceptible strain. Beyond this antibacterial consistency with the antimicrobial spectra of other G.EE_{100} samples, no differences were observed within the antimicrobial activity of the different hydroalcoholic extracts. Therefore, we decided to conduct the disk diffusion assay, aiming to better characterize and distinguish the activity of each G.EE against the selected susceptibility strain panel.

The disk diffusion method was selected to screen a high number of samples, more precisely, 12 extracts (G11.EE_{35}, G12.EE_{35}, G13.EE_{35}, G14.EE_{35}, G15.EE_{35}, G11.EE_{70}, G12.EE_{70}, G13.EE_{70}, G14.EE_{70}, G15.EE_{70}, mG.EE_{35}, and mG.EE_{70}), and to quickly analyze their potential and effectiveness. Additionally, an antibiotic susceptibility test was conducted using antibiotic disks [49]. The results were presented as the diameter of the inhibition halos (Supplementary Table S1). Overall, the disk diffusion assay proved to be the most suitable approach for our study.

None of the Gerês propolis samples extracts (G.EE_{70} and G.EE_{35} groups) showed antifungal capacity against the tested yeasts: *C. albicans* and *S. cerevisiae*. Previous studies demonstrated that the same propolis sample from Gerês extracted with ethanol absolute (100%) had a small inhibition effect against yeasts (MIC = 1500 μg/mL) [28]. These differences can possibly be explained by the chosen solvent, which interferes with the chemical composition of the extracts [50,51]. Indeed, a higher percentage of ethanol culminates in different and/or more phenolic and flavonoid compounds, which are usually associated with an increased antifungal capacity [52,53].

Regarding the antibacterial potential, Gerês propolis extracts displayed a concentration-dependent antibacterial effect, with the most promising results observed in the G.EE$_{70}$ group (Supplementary Table S1). The tested G.EEs exhibited evident activity against Gram-positive bacteria and a weaker effect towards Gram-negative species (Supplementary Table S1), namely, *E. coli* (MIC > 2000 μg/mL), corroborating previously published data [9,18,22,25,28,46]. *E. coli* strains can have different impacts on human health; some of them are harmless and beneficial to humans, while others can lead to a range of illnesses, spanning from mild and self-limiting gastroenteritis to more severe conditions like renal failure and septic shock [44]. Given the rising number of infections caused by this bacterium, there is an increasing need to explore novel approaches to combat this microbial threat. Our results against *E. coli* revealed that mG.EE$_{70}$ (inhibition diameter of 22.17 ± 2.16 mm) presented a similar inhibitory effect to the amoxicillin/clavulanic acid (Figure 4; inhibition diameter of 22.67 ± 1.53 mm), the only commercial antibiotic that affected the Gram-negative bacterium and whose mode of action relies on inhibiting bacterial cell wall synthesis [54]. This finding suggests that effective action against *E. coli* is achievable with higher extract concentration/ethanol proportions. Additionally, our results revealed that mixing propolis samples can enhance the antimicrobial potential of the individual extracts, which might be attributed to a potential synergistic effect among compounds that are differently present in the individual extracts [55], although such synergy was not observed in the calculation of MIC values (Table 1). Recently, Peixoto et al. [27,28] highlighted how interesting the extracts of mixtures of propolis samples are since combining extracts of propolis from Gerês with different levels of bioactivities resulted in comparable and even higher bioactivities. Indeed, these natural mixtures could be a possible way of applying propolis in the future while overcoming the problems of standardization and scarcity. Overall, mixtures of propolis extracts and/or extracts of propolis blends from Gerês may be a promising alternative to antibiotics in combating Gram-negative bacteria, given the results against *E. coli*, and addressing the challenges posed by their distinct cell wall structure and antibiotic resistance.

Analyzing now the activity of G.EEs against the Gram-positive bacteria—*B. subtilis*, MSSA, and MRSA—it becomes evident that extracts prepared with ethanol 70% had comparable or even superior antimicrobial capacity when compared to certain antibiotics (Supplementary Table S1). More specifically, G.EE$_{70}$ samples displayed greater activity against *B. subtilis* (values ranging between 25.25 ± 2.46 and 31.92 ± 4.48 mm; MIC = 50 μg/mL) and *S. aureus* (values ranging between 26.93 ± 2.20 and 29.17 ± 2.75 mm; MIC = 200 μg/mL) than vancomycin (inhibition diameter of 20.08 ± 0.12 and 16.50 ± 0.71 mm, respectively), an extensively utilized and long-standing antibiotic known for its higher potential for Gram-positive bacteria [56]. Vancomycin is a glycopeptide commercial antibiotic that exerts its bactericidal effect by blocking the polymerization of peptidoglycans in the bacterial cell wall [56]. The resistance mechanism of Gram-positive microorganisms against this antibiotic involves alterations in the bacterial cell wall structure through specific gene mutations or the acquisition of resistance genes from other bacteria [19,57]. In light of this, it is plausible to infer that, if propolis extracts exhibit increased antimicrobial effectiveness compared to vancomycin, the mechanism of action of propolis in this specific type of bacteria is a distinct (and more effective) pathway. This outcome emphasizes the potential of Gerês propolis extracts as a potent and more natural substitute for vancomycin in treating Gram-positive bacterial infections. Lastly, it is important to note that the antimicrobial effect of G.EEs against *S. aureus* (Table 1, Figure 2 and Supplementary Table S1), more specifically, MSSA (MIC = 200 μg/mL and inhibition diameter = 26.93–29.17 mm), supports the evidence that propolis from the north and center of Portugal is exceedingly efficient against this microorganism [45], which is currently one of the six leading mortality-causing pathogens [58], making this result extremely important for human health maintenance. Indeed, *S. aureus* is notorious for its capacity to become resistant to antibiotics, reinforcing the importance of the previous result [59,60].

MRSA is a strain of *S. aureus* resistant to methicillin and other β-lactam antibiotics commonly used for treating bacterial infections [61,62]. This resistance is due to the mecA gene, which encodes penicillin-binding protein 2a (PBP2a) within the staphylococcal cassette chromosome mec [63]. MRSA is a major multidrug-resistant pathogen, posing a significant health threat due to its strong resistance, ability to cause severe skin and hospital-acquired infections, and rapid spread in healthcare settings and communities [35–37]. Regarding our results, this microorganism was revealed to be resistant to all the tested commercial antibiotics (no growth inhibition zones) (Supplementary Table S1 and Figure 3). However, although MIC was not determined under the experimental conditions tested herein (MIC > 2000 μg/mL), the $G.EE_{70}$ group exhibited a marked effect against MRSA (inhibition diameter = 17.25–21.33 mm) (Table 1; Supplementary Table S1; and Figure 3), suggesting that propolis from Gerês can be an important source of antimicrobial compounds or lead to the development of antimicrobial drugs against this bacterium. As the mode of action of each antibiotic is specific for a single cellular mechanism—erythromycin is an inhibitor of protein synthesis [64]; vancomycin is an inhibitor of bacterial cell wall synthesis [56]; and amoxicillin/clavulanic acid is an inhibitor of the bacterial wall combined with an inhibitor of β-lactamase [65]—the antibacterial effect of the G.EEs might be related to their chemical diversity, which confers to propolis the possibility to act in different cellular mechanisms. Indeed, several authors [25,55,66] suggested that bioactivities of propolis might be associated with synergistic effects between chemical compounds such as phenylpropanoids, p-coumaric acid, and diterpenic acid [55]. Regarding the $G.EE_{35}$ group, all the tested $G.EE_{35}$ samples did not exhibit an inhibitory effect against *S. aureus*—either MSSA or MRSA—demonstrating that the dilution of extracts and consequently the number of propolis compounds may limit the propolis activity against MRSA.

The chemical composition of propolis is highly dependent on the primary plant source, the season, and the bee species [9]. Consequently, its composition is known for being extremely diverse, complex, and variable [7,9,10]. Propolis is rich in phenolic compounds, particularly in flavonoids and aromatic acids, which are closely associated with its pharmacological properties, including antimicrobial capacity [67,68]. The total phenolic and flavonoid content reported for the six hydroalcoholic extracts (70% ethanol) analyzed in this study—$G11.EE_{70}$, $G12.EE_{70}$, $G13.EE_{70}$, $G14.EE_{70}$, $G15.EE_{70}$, and $mG.EE_{70}$—ranged from 71.00 ± 13.31 to 112.86 ± 19.24 mg GAE/g and from 40.73 ± 5.25 to 57.55 ± 6.69 mg QE/g, respectively [29]. Araújo et al. [29] further illustrated that diluting ethanol extracts from Gerês propolis significantly reduces their phenolic and flavonoid contents. The total phenolic and flavonoid content reported for $G.EE_{35}$ samples and the $mG.EE_{35}$ ranged between 67.61 ± 7.68 and 101.36 ± 10.81 mg GAE/g and between 30.49 ± 3.85 and 38.43 ± 5.72 mg QE/g, respectively [29]. This decrease in the TPC and TFC content affects extracts' antimicrobial capacity (Figures 1 and 2), underscoring the crucial role of these chemicals in propolis bioactivity. They also observed that these Gerês propolis blends exhibited TPC and TFC values surpassing the average values obtained from the individual hydroalcoholic extracts. Considering the known connection between these compounds and the pharmacological properties of propolis [67], it is reasonable to speculate that the heightened antimicrobial efficacy of $G.EE_{70}$ samples (Supplementary Table S1) could be attributed to the greater abundance of flavonoids and phenolics in these hydroalcoholic extracts. Previous studies showed that the main phenolic compounds in Gerês propolis are apigenin, pinobanksin, chrysin, acacetin, galangin, kaempferide, kaempferol, caffeic acid, caffeic acid isoprenyl ester (CAIE), 3,4-dimethyl-caffeic acid (DMCA), p-coumaric acid, p-coumaric acid methyl ester, and ferulic acid [13,22]. Additionally, these studies demonstrated that these compounds are consistently present in propolis samples collected over consecutive years [13,22]. Therefore, considering this constancy of chemical profile observed in ethanol extracts of Gerês propolis [22], along with the relatively consistent antimicrobial spectra of the tested G.EEs, it is feasible to conclude that the G11.EEs–G15.EEs could have a phenolic composition similar to that already described for this type of propolis. In summary, our antimicrobial findings align with the documented chemical composition and known bioactive compounds of the tested hydroalcoholic extracts, emphasizing that higher TFC and TPC values lead to greater antimicrobial potential in this type of propolis.

Overall, and considering all outcomes, it is possible to conclude that propolis extracts are potent antimicrobial agents [69]. Moreover, these extracts represent a compelling and promising alternative, either as the natural product itself or as a source of compounds (lead) for commercial antibiotics, addressing the growing concern of antimicrobial resistance observed in recent years. Araújo et al. [29] also provided evidence of the in vivo and in vitro anti-inflammatory capacity of the same G.EEs used in this work. Thus, this combination of antimicrobial and anti-inflammatory potential underscores the biomedical value of G11–G15 propolis extracts—or the even greater potential of the extract from the mixture of the propolis samples G11, G12, G13, G14, and G15—and emphasizes the importance of further extensive research on Portuguese propolis from Gerês.

4. Materials and Methods

4.1. Propolis Samples

Propolis was harvested in an apiary located next to the Cávado River, between the settlements of Paradela and Sirvozelo, in Montalegre, Gerês, Portugal (41°45′41.62″ N; 7°58′03.34″ W). Raw propolis samples were collected over five years (2011–2015) and identified with the nomenclature criteria adopted in our research group as G11, G12, G13, G14, and G15—samples were named with the capital letter G (referring to their provenance: Gerês) followed by the last two digits of the harvesting year. Propolis samples (G11–G15) were kept at 4 °C for 5, 4, 3, 2, and 1 years, respectively.

4.2. Preparation of Propolis Hydroalcoholic Extracts

Hydroalcoholic extraction of raw propolis was performed according to the methods of previous studies [29]. Succinctly, 10 mL of ethanol 70% (*v*/*v*) (analytical ACS grade (Sigma-Aldrich, St. Louis, MO, USA)) was added individually to each propolis sample—G11, G12, G13, G14, and G15—and incubated under orbital agitation at 100 rpm and 25 °C, in the dark, for 24 h. Suspensions were then centrifuged for 5 min at 5000 rpm and 4 °C. The resultant supernatant was gathered and reserved, and the pellets were extracted again under the same conditions. After the second centrifugation, the resultant supernatant was combined with the previous one, generating stock solutions of 80 mg/mL of each G.EE_{70}. Hydroalcoholic extracts with a final 40 mg/mL concentration were prepared by diluting G.EE_{70} samples in water (1:1), generating the G.EE_{35} samples. The percentage of solvent (ethanol 70% or 35%) was added to each code to help identify the ten G.EEs: G11.EE_{70} and G11.EE_{35}; G12.EE_{70} and G12.EE_{35}; G13.EE_{70} and G13.EE_{35}; G14.EE_{70} and G14.EE_{35}; G15.EE_{70} and G15.$EE3_{35}$.

Hydroalcoholic extraction of the combination of all propolis samples from Gerês (mG) was carried out as stated by Araújo et al. [29]. In a nutshell, 125 mL of ethanol 70% (*v*/*v*) was incubated with G11, G12, G13, G14, and G15 (6.70 g, 6.88 g, 6.57 g, 6.57 g, and 6.67 g, respectively, taking into consideration the extraction yields previously obtained [29]). The mixture was swirled orbitally at 100 rpm for 24 h, in the dark and centrifuged, and the pellet was subsequently re-incubated with ethanol 70% under the conditions mentioned. The resultant supernatants were pooled, giving a stock solution of 80 mg/mL of mG.EE_{70}. Additionally, mG.EE_{35} was prepared by diluting mG.EE_{70} in water (1:1).

4.3. Evaluation of Propolis Antimicrobial Potential

4.3.1. Strains and Culture Conditions

Three Gram-positive bacteria—*Bacillus subtilis* 48886, methicillin-sensitive *Staphylococcus aureus* ATCC 6538 (MSSA), and methicillin-resistant *Staphylococcus aureus* M746665 (MRSA)—and one Gram-negative bacterium—*Escherichia coli* CECT 423—as well as the yeasts *Candida albicans* 53B and *Saccharomyces cerevisiae* BY4741, all from the microbial collection of the Department of Biology of the University of Minho, were selected as susceptibility indicator strains.

Bacteria were cultured in LB broth (Difco—composed of 0.5% (*w*/*v*) yeast extract, 1% (*w*/*v*) tryptone, and 2% (*w*/*v*) NaCl) or in solid LB medium (LBA—LB recipe with 2% (*w*/*v*) agar (Biolife, Milano, Italy)). Bacterial cultures were prepared using LB medium at a ratio of 2:5 (*w*/*v*), and incubation was performed at 37 °C and 200 rpm.

Yeasts were cultured in YPD medium (Difco™ YPD Broth BD—composed of 1% (*w*/*v*) yeast extract, 2% (*w*/*v*) peptone, and 2% (*w*/*v*) dextrose/glucose) or in solid YPD medium (YPDA—YPD recipe with 2% (*w*/*v*) agar). Yeast cultures were prepared using YPD medium at a ratio of 2:5 (*w*/*v*), and incubation was conducted at 30 °C and 200 rpm. Microbial growth was monitored by determining the optical density at 600 nm (OD_{600}) (Thermo Scientific Genesys 20 (Thermo Fischer Scientific, Waltham, MA, USA)).

4.3.2. Antimicrobial Potential of Gerês Propolis: The Agar Dilution Assay

The antimicrobial capacity of propolis extracts from Gerês was evaluated by calculating the minimum inhibitory concentration (MIC) values of 6 hydroalcoholic extracts—G11.EE_{70}, G12.EE_{70}, G13.EE_{70}, G14.EE_{70}, G15.EE_{70}, and mG.EE_{70}—using an adaptation of the agar dilution method [66]. Briefly, yeast and bacterial strains were grown on YPD and LB media (see Section 4.3.1), respectively. Overnight cultures were diluted with fresh medium to an OD_{600} of 0.1 and incubated to the mid-exponential phase of growth (OD_{600} between 0.4–0.6), being 5 µL-drops of each culture inoculated on YPDA or LBA plates containing the propolis extracts at concentrations of 10, 50, 100, 200, 500, 750, 1000, 1500, and 2000 µg/mL. Plates with only LBA and YPDA, and with LBA and YPDA supplemented with ethanol 70% in the same volumes as the extracts, were used as controls. The control using LBA/YPDA alone was included to manage medium contaminations and confirm microbial growth, whereas the control with LBA/YPDA supplemented with ethanol was included to eliminate the influence of the extraction solvent on the observed results. Plates were incubated at 30 °C (Heraeus Incubator (Thermo Fischer Scientific, Waltham, MA, USA)) for 48 h for yeast, and at 37 °C (Incucell MMM MedCenter Incubator, Planegg, Germany) for 24 h in the case of bacteria, and then observed for the presence or absence of growth. MIC values were expressed as the lowest concentrations where no growth was verified.

4.3.3. Antimicrobial Potential of Gerês Propolis: The Disk Diffusion Assay

The antimicrobial activity of propolis from Gerês was evaluated by measuring the diameter of the inhibition zones caused by G.EEs and mG.EEs against the panel of selected microbial indicator strains using an adaptation of the disk diffusion assay [39,70,71]. Three commercial antibiotics—amoxicillin/clavulanic acid (AMC), erythromycin (ERY), and vancomycin (VAN)—were also used for comparison purposes. Bacteria are considered susceptible to the tested antibiotics when the diameter of the inhibition zone is ≥23 mm in the case of ERY; ≥15 mm for VAN; and ≥27 mm (for *Staphylococcus* spp.) or ≥18 mm (for other bacteria) in the case of AMC [33].

Succinctly, mid-exponentially growing yeast and bacterial cultures were obtained as described in Section 4.3.1. A volume of 200 µL of each suspension was then added to 10 mL of 0.8% (*w*/*v*) agar (Biolife, Bothell, WA, USA) and poured on top of an LBA or YPDA plate, depending on the tested microorganism, forming an overlay. Blank paper disks (Thermo Scientific Oxoid™, Hampshire, UK) were placed on top of these microbial layers, and 50 µL of each propolis sample—G.EE_{70} samples/mG.EE_{70} at 80 mg/mL and G.EE_{35} samples/mG.EE_{35} at 40 mg/mL—was added to an individual disk. Antibiotic-containing disks—AMC 30 µg, ERY 15 µg, and VAN 30 µg (BD BBL™ Sensi-Disc™, Porto, Portugal)—were also added on top of the formed microbial layer. Blank paper disks with 50 µL of each solvent (ethanol 70% and 35%) were used as controls. Plates were incubated at 30 °C for 48 h for yeasts, and at 37 °C for 24 h in the case of bacteria, followed by observation of the presence of inhibitions halos and measurement of the respective diameters (in mm) using a ruler.

4.4. Statistical Analysis

Unless otherwise stated experiments, were performed at least three times, each with three replicates per treatment. All data are presented as means ± SD (standard deviation). The statistical analysis was performed using GraphPad Prism 7 software (GraphPad Software, Inc, New York, NY, USA). Differences between results were evaluated using an ANOVA (one-way analysis of variance) followed by a t-test with Bonferroni correction for multiple comparisons. Only the analyses with a $p < 0.05$ were considered statistically significant (* $0.05 > p \geq 0.01$, ** $0.01 > p \geq 0.001$, *** $p < 0.001$).

5. Conclusions

Portuguese propolis from Gerês displays a wide spectrum of bioactivities, making it extremely appealing for several applications, namely, in the medicinal and health sector as an alternative or complement to non-efficient drugs. The G11–G15 hydroalcoholic propolis extracts and resultant mixtures (mG.EEs) (70% ethanol and 35% ethanol) demonstrated a potential antimicrobial capacity against the evaluated pathogens. Gerês propolis extracts prepared with ethanol 70% demonstrated higher antimicrobial potential (larger growth inhibition zones) compared to the tested antibiotics and are thus a promising option for treating infections, especially those caused by the methicillin-resistant *S. aureus* pathogen and Gram-negative bacteria, such as *E. coli*. On the other hand, the $G.EE_{35}$ samples exhibited lower antibacterial activity compared to the $G.EE_{70}$ samples and the tested antibiotics, leading us to deduce that the dilution process could be a constraining factor. This process may result in reduced phenolic and flavonoid contents, thereby potentially decreasing the antibacterial effectiveness of the $G.EE_{35}$ extracts. Furthermore, we provided compelling evidence showcasing the extremely promising application of propolis mixtures as an alternative to individual extracts. Mixtures of Gerês propolis extracts resulted in a synergistic enhancement of antibacterial effectiveness, enabling us to take full advantage of the biomedical potential of Portuguese propolis from Gerês.

We provided further evidence of the potential application of Portuguese propolis as a natural alternative to commercial antibiotics in order to overcome the antimicrobial resistance challenge. We strongly advocate for further exploration of this natural product as a valuable biomedical agent or as a source of compounds with significant biomedical potential.

Supplementary Materials: The following supporting information can be downloaded at: https://www.mdpi.com/article/10.3390/antibiotics13070655/s1, Table S1: Antimicrobial effect of propolis hydroalcoholic extracts from Gerês.

Author Contributions: The first authors, C.A. and R.D.O., conducted and analyzed the presented experiments and were responsible for drafting the initial manuscript. C.A.-A. was responsible for the planning and designing of the experiment, as well as manuscript corrections. Additionally, C.A.-A. played a key role in coordinating the experimental resources and development. All authors have read and agreed to the published version of the manuscript.

Funding: This study had the support of FCT, I.P., which provided national funds through the strategic project 'Financiamento Programático' UIDB/04050/2020 awarded to CBMA (https://doi.org/10.54499/UIDB/04050/2020), and LA/P/0069/2020, awarded to the Associate Laboratory ARNET (https://doi.org/10.54499/LA/P/0069/2020).

Institutional Review Board Statement: Not applicable.

Informed Consent Statement: Not applicable.

Data Availability Statement: Data are contained within the article and Supplementary Materials.

Acknowledgments: The authors would like to express their special thanks to Amadeu Fortunas of Casa do Couto, 5470-362, Montalegre, Portugal, for the propolis sample used in this work.

Conflicts of Interest: The authors declare that there are no conflicts of interest.

References

1. Acar, J.; Röstel, B. Antimicrobial resistance: An overview. *Rev. Sci. Tech.* **2001**, *20*, 797–810. [CrossRef] [PubMed]
2. Hofer, U. The cost of antimicrobial resistance. *Nat. Rev. Microbiol.* **2019**, *17*, 3. [CrossRef]
3. Salatino, A. Perspectives for Uses of Propolis in Therapy against Infectious Diseases. *Molecules* **2022**, *27*, 4594. [CrossRef] [PubMed]
4. Davies, J.; Davies, D. Origins and Evolution of Antibiotic Resistance. *Microbiol. Mol. Biol. Rev.* **2010**, *74*, 417–433. [CrossRef] [PubMed]
5. Murray, C.J.L.; Ikuta, K.S.; Sharara, F.; Swetschinski, L.; Robles Aguilar, G.; Gray, A.; Han, C.; Bisignano, C.; Rao, P.; Wool, E.; et al. Global burden of bacterial antimicrobial resistance in 2019: A systematic analysis. *Lancet* **2022**, *399*, 629–655. [CrossRef] [PubMed]
6. OECD. *Stemming the Superbug Tide: Just a Few Dollars More*; OECD Health Policy Studies; OECD: Paris, France, 2018. [CrossRef]
7. Marcucci, M.C. Propolis: Chemical composition, biological properties and therapeutic activity. *Apidologie* **1995**, *26*, 83–99. [CrossRef]
8. Umthong, S.; Phuwapraisirisan, P.; Puthong, S.; Chanchao, C. In vitro antiproliferative activity of partially purified Trigona laeviceps propolis from Thailand on human cancer cell lines. *BMC Complement. Altern. Med.* **2011**, *11*, 37. [CrossRef] [PubMed]
9. Silva-Carvalho, R.; Baltazar, F.; Almeida-Aguiar, C. Propolis: A Complex Natural Product with a Plethora of Biological Activities That Can Be Explored for Drug Development. *Evid.-Based Complement. Altern. Med.* **2015**, *2015*, 206439. [CrossRef]
10. Kasote, D.; Bankova, V.; Viljoen, A.M. Propolis: Chemical diversity and challenges in quality control. *Phytochem. Rev.* **2022**, *21*, 1887–1911. [CrossRef]
11. Moreira, L.; Dias, L.G.; Pereira, J.A.; Estevinho, L. Antioxidant properties, total phenols and pollen analysis of propolis samples from Portugal. *Food Chem. Toxicol.* **2008**, *46*, 3482–3485. [CrossRef]
12. Šuran, J.; Cepanec, I.; Mašek, T.; Radić, B.; Radić, S.; Tlak Gajger, I.; Vlainić, J. Propolis Extract and Its Bioactive Compounds—From Traditional to Modern Extraction Technologies. *Molecules* **2021**, *26*, 2930. [CrossRef] [PubMed]
13. Caetano, A.R.; Oliveira, R.D.; Celeiro, S.P.; Freitas, A.S.; Cardoso, S.M.; Gonçalves, M.S.T.; Baltazar, F.; Almeida-Aguiar, C. Phenolic Compounds Contribution to Portuguese Propolis Anti-Melanoma Activity. *Molecules* **2023**, *28*, 3107. [CrossRef] [PubMed]
14. Widelski, J.; Okińczyc, P.; Suśniak, K.; Malm, A.; Paluch, E.; Sakipov, A.; Zhumashova, G.; Ibadullayeva, G.; Sakipova, Z.; Korona-Glowniak, I. Phytochemical Profile and Antimicrobial Potential of Propolis Samples from Kazakhstan. *Molecules* **2023**, *28*, 2984. [CrossRef] [PubMed]
15. Popova, M.P.; Chinou, I.B.; Marekov, I.N.; Bankova, V.S. Terpenes with antimicrobial activity from Cretan propolis. *Phytochemistry* **2009**, *70*, 1262–1271. [CrossRef]
16. Krzyżek, P.; Paluch, E.; Gościniak, G. Synergistic Therapies as a Promising Option for the Treatment of Antibiotic-Resistant Helicobacter pylori. *Antibiotics* **2020**, *9*, 658. [CrossRef] [PubMed]
17. Akilandeswari, K.; Ruckmani, K. Synergistic antibacterial effect of apigenin with β-lactam antibiotics and modulation of bacterial resistance by a possible membrane effect against methicillin resistant *Staphylococcus aureus*. *Cell. Mol. Biol.* **2016**, *62*, 74–82. [CrossRef] [PubMed]
18. Fokt, H.; Pereira, A.; Ferreira, A.; Cunha, A.; Almeida Aguiar, C. How do bees prevent hive infections? The antimicrobial properties of propolis. *Curr. Res. Technol. Educ. Top. Appl. Microbiol. Microb. Biotechnol.* **2010**, *1*, 481–493.
19. Reygaert, W.C. An overview of the antimicrobial resistance mechanisms of bacteria. *AIMS Microbiol.* **2018**, *4*, 482–501. [CrossRef] [PubMed]
20. Baran, A.; Kwiatkowska, A.; Potocki, L. Antibiotics and Bacterial Resistance—A Short Story of an Endless Arms Race. *Int. J. Mol. Sci.* **2023**, *24*, 5777. [CrossRef]
21. Anjum, S.I.; Ullah, A.; Khan, K.A.; Attaullah, M.; Khan, H.; Ali, H.; Bashir, M.A.; Tahir, M.; Ansari, M.J.; Ghramh, H.A.; et al. Composition and functional properties of propolis (bee glue): A review. *Saudi J. Biol. Sci.* **2019**, *26*, 1695–1703. [CrossRef]
22. Freitas, A.S.; Cunha, A.; Cardoso, S.M.; Oliveira, R.; Almeida-Aguiar, C. Constancy of the bioactivities of propolis samples collected on the same apiary over four years. *Food Res. Int.* **2019**, *119*, 622–633. [CrossRef] [PubMed]
23. Freitas, A.S.; Cunha, A.; Parpot, P.; Cardoso, S.M.; Oliveira, R.; Almeida-Aguiar, C. Propolis Efficacy: The Quest for Eco-Friendly Solvents. *Molecules* **2022**, *27*, 7531. [CrossRef] [PubMed]
24. Freitas, A.S.; Costa, M.; Pontes, O.; Seidel, V.; Proença, F.; Cardoso, S.M.; Oliveira, R.; Baltazar, F.; Almeida-Aguiar, C. Selective Cytotoxicity of Portuguese Propolis Ethyl Acetate Fraction towards Renal Cancer Cells. *Molecules* **2022**, *27*, 4001. [CrossRef]
25. Freitas, A.S.; Cunha, A.; Oliveira, R.; Almeida-Aguiar, C. Propolis antibacterial and antioxidant synergisms with gentamicin and honey. *J. Appl. Microbiol.* **2022**, *132*, 2733–2745. [CrossRef]
26. Oliveira, R.D.; Celeiro, S.P.; Barbosa-Matos, C.; Freitas, A.S.; Cardoso, S.M.; Viana-Pereira, M.; Almeida-Aguiar, C.; Baltazar, F. Portuguese Propolis Antitumoral Activity in Melanoma Involves ROS Production and Induction of Apoptosis. *Molecules* **2022**, *27*, 3533. [CrossRef] [PubMed]
27. Peixoto, M.; Freitas, A.S.; Cunha, A.; Oliveira, R.; Almeida-Aguiar, C. Mixing Propolis from Different Apiaries and Harvesting Years: Towards Propolis Standardization? *Antibiotics* **2022**, *11*, 1181. [CrossRef] [PubMed]
28. Peixoto, M.; Freitas, A.S.; Cunha, A.; Oliveira, R.; Almeida-Aguiar, C. Antioxidant and antimicrobial activity of blends of propolis samples collected in different years. *LWT* **2021**, *145*, 111311. [CrossRef]
29. Araújo, C.; Oliveira, R.D.; Pinto-Ribeiro, F.; Almeida-Aguiar, C. An Insight on the Biomedical Potential of Portuguese Propolis from Gerês. *Foods* **2022**, *11*, 3431. [CrossRef] [PubMed]

30. Silva-Carvalho, R.; Miranda-Gonçalves, V.; Ferreira, A.M.; Cardoso, S.M.; Sobral, A.J.F.N.; Almeida-Aguiar, C.; Baltazar, F. Antitumoural and antiangiogenic activity of Portuguese propolis in in vitro and in vivo models. *J. Funct. Foods* **2014**, *11*, 160–171. [CrossRef]
31. Errington, J.; van der Aart, L.T. Microbe Profile: Bacillus subtilis: Model organism for cellular development, and industrial workhorse. *Microbiology* **2020**, *166*, 425–427. [CrossRef]
32. Piggot, P.J. *Bacillus subtilis. Encyclopedia of Microbiology*; Elsevier: Amsterdam, The Netherlands, 2009; pp. 45–56.
33. Becton, Dickinson and Company. BBLTM Sensi-DiscTM Antimicrobial Susceptibility Test Discs. 2020; pp. 1–16. Available online: https://dmec.moh.gov.vn/documents/10182/26822970/upload_00027222_1645284052129.pdf?version=1.0&fileId=26844772 (accessed on 1 March 2024).
34. Taylor, T.A.; Unakal, C.G. Staphylococcus aureus Infection. In *StatPearls*; StatPearls Publishing: Treasure Island, FL, USA, 2023.
35. Lee, A.S.; De Lencastre, H.; Garau, J.; Kluytmans, J.; Malhotra-Kumar, S.; Peschel, A.; Harbarth, S. Methicillin-resistant *Staphylococcus aureus*. *Nat. Rev. Dis. Prim.* **2018**, *4*, 18033. [CrossRef] [PubMed]
36. Turner, N.A.; Sharma-Kuinkel, B.K.; Maskarinec, S.A.; Eichenberger, E.M.; Shah, P.P.; Carugati, M.; Holland, T.L.; Fowler, V.G., Jr. Methicillin-resistant *Staphylococcus aureus*: An overview of basic and clinical research. *Nat. Rev. Microbiol.* **2019**, *17*, 203–218. [CrossRef] [PubMed]
37. Stapleton, P.D.; Taylor, P.W. Methicillin Resistance in *Staphylococcus aureus*: Mechanisms and Modulation. *Sci. Prog.* **2002**, *85*, 57–72. [CrossRef] [PubMed]
38. Mueller, M.; Tainter, C.R. Escherichia coli Infection. In *StatPearls*; StatPearls Publishing: Treasure Island, FL, USA, 2023.
39. Zaidan, M.R.S.; Noor Rain, A.; Badrul, A.R.; Adlin, A.; Norazah, A.; Zakiah, I. In vitro screening of five local medicinal plants for antibacterial activity using disc diffusion method. *Trop. Biomed.* **2005**, *22*, 165–170. [PubMed]
40. Nikaido, H. Outer membrane barrier as a mechanism of antimicrobial resistance. *Antimicrob. Agents Chemother.* **1989**, *33*, 1831–1836. [CrossRef] [PubMed]
41. Levy, S.B. Factors impacting on the problem of antibiotic resistance. *J. Antimicrob. Chemother.* **2022**, *49*, 25–30. [CrossRef] [PubMed]
42. Pereira, L.; Cunha, A.; Almeida-Aguiar, C. Portuguese propolis from Caramulo as a biocontrol agent of the apple blue mold. *Food Control* **2022**, *139*, 109071. [CrossRef]
43. Dias, L.G.; Pereira, A.P.; Estevinho, L.M. Comparative study of different Portuguese samples of propolis: Pollinic, sensorial, physicochemical, microbiological characterization and antibacterial activity. *Food Chem. Toxicol.* **2012**, *50*, 4246–4253. [CrossRef] [PubMed]
44. Lourenço, T.; Oliveira, T.; Ferreira, A.M.; Oliveira, R.; Bento, F.; Geraldo, D.; Almeida-Aguiar, C.; Cunha, A. Antimicrobial and antioxidant properties of propolis ethanol extracts from Terceira Island (Azores, Portugal). *Planta Med.* **2014**, *80*, P1L17. [CrossRef]
45. Silva, J.C.; Rodrigues, S.; Feás, X.; Estevinho, L.M. Antimicrobial activity, phenolic profile and role in the inflammation of propolis. *Food Chem. Toxicol.* **2012**, *50*, 1790–1795. [CrossRef]
46. Kujumgiev, A.; Tsvetkova, I.; Serkedjieva, Y.; Bankova, V.; Christov, R.; Popov, S. Antibacterial, antifungal and antiviral activity of propolis of different geographic origin. *J. Ethnopharmacol.* **1999**, *64*, 235–240. [CrossRef]
47. Jorgensen, J.H.; Ferraro, M.J. Antimicrobial Susceptibility Testing: A Review of General Principles and Contemporary Practices. *Clin. Infect. Dis.* **2009**, *49*, 1749–1755. [CrossRef] [PubMed]
48. Przybyłek, I.; Karpiński, T.M. Antibacterial Properties of Propolis. *Molecules* **2019**, *24*, 2047. [CrossRef]
49. Truong, W.R.; Hidayat, L.; Bolaris, M.A.; Nguyen, L.; Yamaki, J. The antibiogram: Key considerations for its development and utilization. *JAC-Antimicrob. Resist.* **2021**, *3*, dlab060. [CrossRef] [PubMed]
50. Sforcin, J.M.; Bankova, V. Propolis: Is there a potential for the development of new drugs? *J. Ethnopharmacol.* **2011**, *133*, 253–260. [CrossRef]
51. Cunha, I.B.S.; Sawaya, A.C.H.F.; Caetano, F.M.; Shimizu, M.T.; Marcucci, M.C.; Drezza, F.T.; Povia, G.S.; Carvalho, P.O. Factors that influence the yield and composition of Brazilian propolis extracts. *J. Braz. Chem. Soc.* **2004**, *15*, 964–970. [CrossRef]
52. Al Aboody, M.S.; Mickymaray, S. Anti-Fungal Efficacy and Mechanisms of Flavonoids. *Antibiotics* **2020**, *9*, 45. [CrossRef]
53. Teodoro, G.R.; Ellepola, K.; Seneviratne, C.J.; Koga-Ito, C.Y. Potential Use of Phenolic Acids as Anti-Candida Agents: A Review. *Front. Microbiol.* **2015**, *6*, 1420. [CrossRef] [PubMed]
54. Singh, N.S.; Singhal, N.; Kumar, M.; Virdi, J.S. Exploring the genetic mechanisms underlying amoxicillin-clavulanate resistance in waterborne *Escherichia coli*. *Infect. Genet. Evol.* **2021**, *90*, 104767. [CrossRef]
55. Abdullah, N.A.; Ja'afar, F.; Yasin, H.M.; Taha, H.; Petalcorin, M.I.R.; Mamit, M.H.; Kusrini, E.; Usman, A. Physicochemical analyses, antioxidant, antibacterial, and toxicity of propolis particles produced by stingless bee Heterotrigona itama found in Brunei Darussalam. *Heliyon* **2019**, *5*, e02476. [CrossRef]
56. Patel, S.; Preuss, C.V.; Bernice, F. Vancomycin. In *StatPearls*; StatPearls Publishing: Treasure Island, FL, USA, 2023.
57. Selim, S. Mechanisms of gram-positive vancomycin resistance (Review). *Biomed. Rep.* **2021**, *16*, 7. [CrossRef] [PubMed]
58. Gajic, I.; Kabic, J.; Kekic, D.; Jovicevic, M.; Milenkovic, M.; Mitic Culafic, D.; Trudic, A.; Ranin, L.; Opavski, N. Antimicrobial Susceptibility Testing: A Comprehensive Review of Currently Used Methods. *Antibiotics* **2022**, *11*, 427. [CrossRef]
59. Grundmann, H.; Aires-de-Sousa, M.; Boyce, J.; Tiemersma, E. Emergence and resurgence of meticillin-resistant *Staphylococcus aureus* as a public-health threat. *Lancet* **2006**, *368*, 874–885. [CrossRef]
60. Chambers, H.F.; DeLeo, F.R. Waves of resistance: *Staphylococcus aureus* in the antibiotic era. *Nat. Rev. Microbiol.* **2009**, *7*, 629–641. [CrossRef]

61. Bæk, K.T.; Gründling, A.; Mogensen, R.G.; Thøgersen, L.; Petersen, A.; Paulander, W.; Frees, D. β-Lactam resistance in methicillin-resistant *Staphylococcus aureus* USA300 is increased by inactivation of the ClpXP protease. *Antimicrob. Agents Chemother.* **2014**, *58*, 4593–4603. [CrossRef] [PubMed]
62. Ali Alghamdi, B.; Al-Johani, I.; Al-Shamrani, J.M.; Musamed Alshamrani, H.; Al-Otaibi, B.G.; Almazmomi, K.; Yusnoraini Yusof, N. Antimicrobial resistance in methicillin-resistant *Staphylococcus aureus*. *Saudi J. Biol. Sci.* **2023**, *30*, 103604. [CrossRef] [PubMed]
63. Martínez-Meléndez, A.; Morfín-Otero, R.; Villarreal-Treviño, L.; González-González, G.; Llaca-Díaz, J.; Rodríguez-Noriega, E.; Camacho-Ortíz, A.; Garza-González, E. Staphylococcal Cassette Chromosome mec (SCCmec) in coagulase negative staphylococci. *Med. Univ.* **2015**, *17*, 229–233. [CrossRef]
64. Weisblum, B. Insights into erythromycin action from studies of its activity as inducer of resistance. *Antimicrob. Agents Chemother.* **1995**, *39*, 797–805. [CrossRef]
65. Todd, P.A.; Benfield, P. Amoxicillin/Clavulanic Acid. *Drugs* **1990**, *39*, 264–307. [CrossRef]
66. Stepanović, S.; Antić, N.; Dakić, I.; Švabić-Vlahović, M. In vitro antimicrobial activity of propolis and synergism between propolis and antimicrobial drugs. *Microbiol. Res.* **2003**, *158*, 353–357. [CrossRef]
67. Wieczorek, P.P.; Hudz, N.; Yezerska, O.; Horčinová-Sedláčková, V.; Shanaida, M.; Korytniuk, O.; Jasicka-Misiak, I. Chemical Variability and Pharmacological Potential of Propolis as a Source for the Development of New Pharmaceutical Products. *Molecules* **2022**, *27*, 1600. [CrossRef] [PubMed]
68. Miguel, M.G.; Nunes, S.; Dandlen, S.A.; Cavaco, A.M.; Antunes, M.D. Phenols and antioxidant activity of hydro-alcoholic extracts of propolis from Algarve, South of Portugal. *Food Chem. Toxicol.* **2010**, *48*, 3418–3423. [CrossRef] [PubMed]
69. Vică, M.L.; Glevitzky, M.; Heghedus-Mîndru, R.C.; Glevitzky, I.; Matei, H.V.; Bâlici, Ș.; Popa, M.; Teodoru, C.A. Potential Effects of Romanian Propolis Extracts against Pathogen Strains. *Int. J. Environ. Res. Public Health* **2022**, *19*, 2640. [CrossRef] [PubMed]
70. Bauer, A.W.; Kirby, W.M.M.; Sherris, J.C.; Turck, M. Antibiotic Susceptibility Testing by a Standardized Single Disk Method. *Am. J. Clin. Pathol.* **1966**, *45*, 493–496. [CrossRef]
71. Wilkins, T.D.; Holdeman, L.V.; Abramson, I.J.; Moore, W.E.C. Standardized Single-Disc Method for Antibiotic Susceptibility Testing of Anaerobic Bacteria. *Antimicrob. Agents Chemother.* **1972**, *1*, 451–459. [CrossRef]

antibiotics

Article

Green Synthesis of Silver Nanoparticle Using Black Mulberry and Characterization, Phytochemical, and Bioactivity

Yoo-Na Jeon [1], Su-Ji Ryu [1], Ha-Yeon Lee [1], Jang-Oh Kim [2] and Jong-Suep Baek [1,3,4,*]

1 Department of Bio-Health Convergence, Kangwon National University, Chuncheon 24341, Republic of Korea
2 Department of Radiological Science, Kangwon National University, Samcheok 25949, Republic of Korea
3 Department of Bio-Functional Materia, Kangwon National University, Samcheok 25949, Republic of Korea
4 BeNatureBioLab, Chuncheon 24206, Republic of Korea
* Correspondence: jsbaek@kangwon.ac.kr; Tel.: +82-33-257-5908

Abstract: Synthesis of silver nanoparticles (AgNPs) using plant extracts has been proposed as a more advantageous and environmentally friendly alternative compared to existing physical/chemical methods. In this study, AgNPs were synthesized from silver nitrate using black mulberry (BM) extract. The biosynthesized AgNPs were characterized through an UV-visible spectrometer, X-ray diffraction, and transmission electron microscopy. Additionally, BM-AgNPs were subjected to antioxidant, antibacterial, anti-inflammatory, and anticancer activities. AgNPs biosynthesized from BM extract were dark brown in color and showed a strong peak at 437 nm, confirming that AgNPs were successfully synthesized. The size of AgNPs was 170.17 ± 12.65 nm, the polydispersity index was 0.281 ± 0.07, and the zeta potential value was −56.6 ± 0.56 mV, indicating that the particles were stable. The higher total phenol, flavonoid, and anthocyanin content of BM-AgNPs compared to BM extract indicates that the particles contain multiple active substances due to the formation of AgNPs. The DPPH and ABTS assays showed decreased IC_{50} values compared to BM extract, demonstrating improved antioxidant activity. AgNPs inhibited the growth of *S. aureus* and *E. coli* at 600 μg/mL, with minimum bactericidal concentrations determined to be 1000 and 1200 μg/mL, respectively. The anti-inflammatory activity was 64.28% at a BM-AgNPs concentration of 250 μg/mL. As the concentration increased, the difference from the standard decreased, indicating the inhibitory effect of AgNPs on bovine serum albumin denaturation. The viability of MCF-7 cells treated with BM-AgNPs was found to be significantly lower than that of cells treated with BM extract. The IC_{50} value of BM-AgNPs was determined to be 96.9 μg/mL. This study showed that BM-AgNPs have the potential to be used in the pharmaceutical industry as antioxidant, antibacterial, anti-inflammatory, and anticancer agents.

Keywords: silver nanoparticles; black mulberry; antioxidant; antibacterial; anti-inflammation; anticancer

1. Introduction

The field of nanotechnology research has made significant progress over the past few decades and has emerged as a promising technology in the scientific field, particularly in the development of materials with potential medical applications [1]. Metal nanoparticles, including gold, silver, copper, platinum, iridium, osmium, and palladium are being increasingly used in biomedical applications due to their biological activity [2,3].

Silver (Ag) is the metal most often used in nanoparticle preparation due to its antifungal, antibacterial, antiviral, anticancer, and antioxidant properties [4,5]. Silver nanoparticles (AgNPs) have been used as an antibacterial agent due to their effectiveness in sterilizing the growth inhibition of various microorganisms, such as *Vibrio cholerae*, *Candida albicans*, *Pseudomonas aeruginosa*, *Escherichia coli* (*E. coli*), and *Staphylococcus aureus* (*S. aureus*) [6]. AgNPs have been reported to induce selective death of cancer cells, and can play a role in drug delivery and act as a promising therapeutic agent against cancer [7]. AgNPs are synthesized using organisms such as algae, plants, bacteria, fungi, and their metabolites, which act as reducing agents, stabilizers, and capping agents [8]. A synthetic method using such biological materials is cost-effective, efficient, and environmentally friendly [9]. AgNPs prepared using biological materials have the properties of a high surface area, small size, and high dispersion [10]. Among them, plants are considered the preferred source because they are highly productive and easily available [11]. Plants such as leaves, fruits, and seeds used in green synthesis contain a variety of natural phytochemicals such as terpenoids, flavonoids, alkaloids, phenols, saponins, tannins, and amino acids [12–14]. By these phytochemicals, Ag ions are stabilized and reduced to nanoparticles, and improve the biological activity of nanomaterials [15]. The method of synthesis with plant extracts is effective in preventing oxidation and aggregation of synthesized AgNPs while reducing cost and chemical usage [16].

Mulberry, belonging to the genus Morus, has been distributed worldwide in diverse climatic conditions from temperate to tropical [17]. The three types of mulberry fruit are red mulberry (*Morus rubra* L.), black mulberry (*Morus nigra* L.) (BM), and white mulberry (*Morus alba* L.) [18]. Among them, BM belongs to the Moraceae family and is a purple to black fruit with a distinctive, slightly acidic taste [19]. BM is a fruit traditionally known for its use in natural medicine because of its nutritional properties and taste as well as its high content of various phytochemicals, including flavonoids and phenolic compounds, especially anthocyanins [20]. The major anthocyanins identified in BM are cyanidine-3-glucoside and cyanidine-3-lutinoside (C3G) [21], and the C3G is considered the most commonly occurring and important molecule among anthocyanins due to its diverse biological activities [22]. BM has been reported for various biological activities including antioxidant, anticancer, anti-inflammatory, antibacterial, neuroprotective, hyperlipidemia, antiatherosclerosis, and anti-obesity agents due to the bioactive content of polyphenols and anthocyanins [23–25].

This study aimed to synthesize AgNPs from BM extracts and evaluate their various biological activities. The anti-inflammatory and anticancer activities of BM-AgNPs are reported here for the first time. A UV-visible spectrometer (UV-vis spectra), dynamic light scattering (DLS), X-ray diffraction (XRD), transmission electron microscopy (TEM) and energy dispersive X-ray spectroscopy (EDS) were performed to characterize the size, shape, dispersion, crystallinity, stability, and surface morphology of the biosynthesized nanoparticles. Through the total phenolic content (TPC) and total flavonoid contents (TFC) of BM-AgNPs, the content of phytochemicals such as polyphenols and flavonoids contained in BM-AgNPs was measured, and the C3G content was measured through TAC. In addition, 2,2-diphenyl-1-picrylhydrazide (DPPH) and '2,2'-azino-bis (3-ethylbenzothiazoline-6-sulfonic acid) (ABTS) experiments were performed to confirm the antioxidant capacity of BM-AgNPs. The antibacterial activity of the prepared BM-AgNPs was analyzed on Gram-positive (*E. coli*) and Gram-negative (*S. aureus*). The anti-inflammatory effect of BM-AgNPs was evaluated through protein denaturation experiments using bovine serum almumin (BSA), while the anti-cancer properties of BM-AgNPs were investigated using a breast cancer (MCF-7) cell line.

2. Results and Discussion

2.1. Characterization of BM-AgNPs

2.1.1. UV-Vis Absorbance Spectra and Color Change

The color change observed in AgNPs confirms the reduction of Ag ions in silver nitrate ($AgNO_3$) [26]. The reaction mixture containing $AgNO_3$ and BM extract turned dark brown, confirming the formation of BM-AgNPs (Figure 1a). UV-vis spectral analysis is used to obtain information about AgNPs, such as their stability with regard to size and shape, in aqueous suspensions [27]. In the absorption spectrum of AgNPs, the surface plasmon resonance band appears in a broad range of 300–700 nm, with a strong absorption band observed at 437 nm (Figure 1b). This absorption spectrum broadened with an increasing absorption peak up to 24 h. This is due to the distinct size distribution of BM-AgNPs and indicates the presence of polydisperse AgNPs in the solution [28,29]. The polydisperse presence of AgNPs in the solution may be due to the slow reduction rate [30].

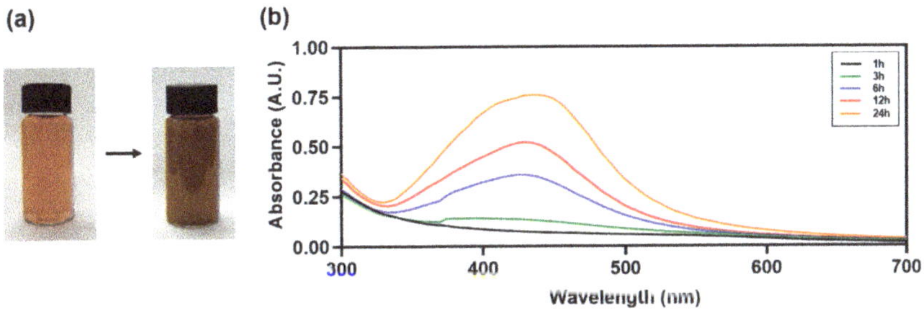

Figure 1. (**a**) Image of Ag colloids synthesized with BM extract after 24 h, (**b**) UV-Vis spectra of BM-AgNPs synthesized from different incubation times (1, 3, 6, 12, and 24 h).

2.1.2. TEM

TEM images provide information about the size, morphology, and shape of the biosynthesized BM-AgNPs. Through TEM analysis of BM-AgNPs, it was confirmed that the particles were spherical below 100 nm (Figure 2a,b). Spherical particles have a large surface area to prevent aggregation and are relatively evenly distributed [31]. In Figure 2c, EDS mapping confirmed the presence of Ag ions in the AgNPs. In the EDS spectrum, Ag was identified by a strong and distinct peak at approximately 3 keV (Figure 2d). This finding is consistent with a previous study, which also confirmed that AgNPs exhibit a strong peak at 3 keV, aligning with the EDS spectrum results of BM-AgNPs [32].

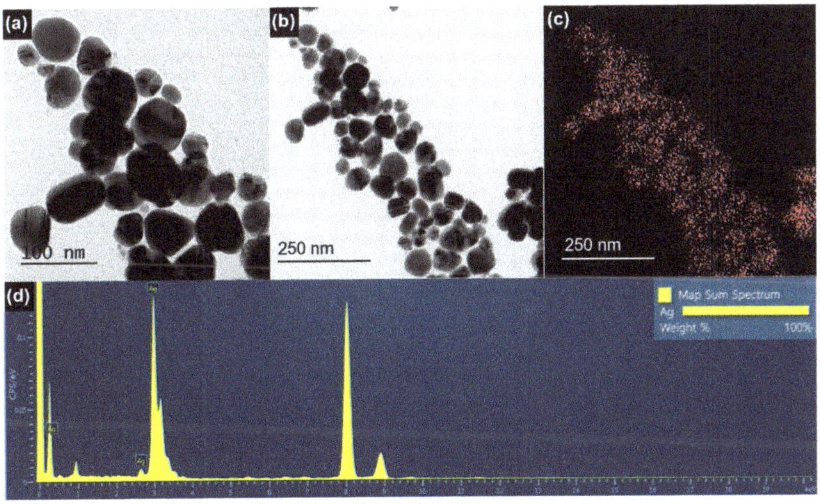

Figure 2. (**a**,**b**) TEM images of BM-AgNPs with different magnification, (**c**) EDS mapping of Ag, and (**d**) EDS spectrum of BM-AgNPs.

2.1.3. Particle Size, PDI and Zeta Potential

The particle size, PDI, and zeta potential of BM-AgNPs were measured (Table 1). DLS was performed to measure the size distribution of AgNPs, and the particle size of BM-AgNPs was measured to be 170.17 ± 12.65 nm. DLS is an indirect ensemble technique based on the luminosity signal of particles, which can bias particle size measurements toward larger sizes [33,34]. Therefore, diameters obtained through DLS are generally larger than those measured through TEM [35]. PDI values less than 0.1 are considered monodisperse [36]. The PDI value of BM-AgNPs is 0.281 ± 0.07, which is considered polydisperse, and the broad absorption peak in Figure 1 mentioned above supports the existence of polydisperse AgNPs in the solution. BM-AgNPs were measured with a PDI of 0.281 ± 0.07, which is relatively monodisperse. The determination of the zeta potential serves as a key factor in assessing the stability of AgNPs. Zeta potential is an indicator of the magnitude of electrostatic interactions between dispersed particles and can indirectly evaluate the stability of nanoparticles in colloidal suspensions by reflecting their ability to electrostatically repel each other [37,38]. Particles with zeta potential values exceeding +30 mV or dropping below −30 mV are considered stable [39]. The zeta potential value of BM-AgNPs was −56.6 ± 0.56 mV, confirming that the particles were in a stable condition. Therefore, BM-AgNPs can prevent aggregation of the particles and maintain stability in the solution due to the electrostatic repulsion between them.

Table 1. Particle size (nm), PDI and Zeta potential (mV) of BM-AgNPs.

	Particle Size (nm)	PDI (Index)	Zeta Potential (mV)
BM-AgNPs	170.17 ± 12.65	0.281 ± 0.07	−56.6 ± 0.56

2.1.4. XRD

The crystalline nature of BM-AgNPs was confirmed through analysis of the XRD pattern (Figure 3). The XRD analysis revealed the presence of four prominent diffraction peaks at specific 2θ values: 38.12° (111), 44.12° (200), 64.40° (220), and 77.26° (311). The high intensity diffraction peak at (111) shows the growth direction of the nanocrystals [40]. It was confirmed that these peaks are due to the crystallographic plane of the BM-AgNPs by face-centered cubic (fcc) information provided in JCPDS (Joint Committee Powder Diffraction Standards) [41].

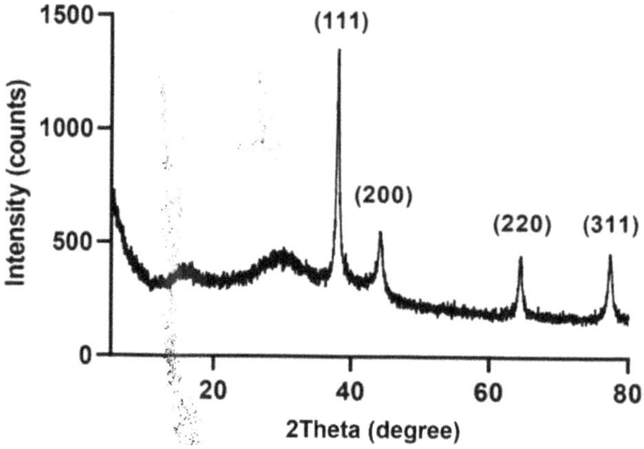

Figure 3. XRD pattern of BM-AgNPs.

2.2. *Phytochemical of BM-AgNPs*

Table 2 shows the TPC, TFC, TAC, and antioxidant results of BM extract and BM-AgNPs. The TPC of BM extract was 24.57 ± 0.17 mg/g, BM-AgNPs was 148.42 ± 2.33 mg/g, showing the higher phenol content of BM-AgNPs. The TFC was 8.07 ± 0.04 mg/g for BM extract and 12.80 ± 0.65 mg/g for BM-AgNPs, confirming that BM-AgNPs had a higher flavonoid content. Anthocyanins are a group of water-soluble flavonoids that relieve oxidant stress responses by scavenging free radicals [42,43]. As a result of the TAC, it was confirmed that the C3G content of BM-AgNPs was higher at 1.04 ± 0.24 mg/L for the BM extract and 1.63 ± 0.60 mg/L for BM-AgNPs, which showed that anthocyanin was used to obtain metal nanoparticles in the process. Previous studies have reported the reduction of metal ions and stabilization of AgNPs by polyphenols and anthocyanins present in blackcurrant extract [44].

Table 2. TPC, TFC, TAC, and IC_{50} values of antioxidant activities of BM and BM-AgNPs. Data are expressed as means ± standard deviation (n = 3). Different letters (a–c) in each samples (BM extract, BM-AgNPs, and Ascorbic acid) ($p < 0.05$) indicate significant differences.

Sample	TPC	TFC	TAC	ABTS	DPPH
	GAE mg/g	QE mg/g	C3G mg/L	IC_{50} μg/mL	IC_{50} μg/mL
BM extract	24.57 ± 0.17 [a]	8.07 ± 0.04 [a]	1.04 ± 0.24 [a]	312.18 ± 15.34 [a]	335.35 ± 39.33 [a]
BM-AgNPs	148.42 ± 2.33 [b]	12.80 ± 0.65 [b]	1.63 ± 0.60 [a]	250.53 ± 24.28 [a]	134.90 ± 26.29 [b]
Ascorbic acid	-	-	-	12.76 ± 0.54 [c]	11.36 ± 0.27 [c]

In ABTS, IC_{50} values were 575.48 ± 21.69 μg/mL for the BM extract and 257.09 ± 35.44 μg/mL for BM-AgNPs. In DPPH, IC_{50} values were 334.63 ± 23.98 μg/mL for the BM extract and 100.94 ± 15.01 μg/mL for BM-AgNPs. It was found that BM-AgNPs have a stronger effect on neutralizing free radicals. The DPPH and ABTS radical scavenging of BM extracts and BM-AgNPs are shown in Figure 4. At all concentrations, the ABTS radical scavenging value increased in a concentration-dependent manner and was higher for BM-AgNPs than for BM extracts. The DPPH radical scavenging value also increased in a concentration-dependent manner at all concentrations and was higher for BM-AgNPs than for BM extracts. Due to the different radical scavenging mechanisms of DPPH and ABTS, significant differences between extracts and AgNPs in DPPH radical scavenging were only identified at 200 μg/mL [45]. The higher TFC, TPC, and TAC of BM-AgNPs in comparison to the BM extract may be associated with an enhanced antioxidant activity. This indicates that the active molecules present in the extract utilized as a reducing agent may elevate the antioxidant activity of BM-AgNPs [46].

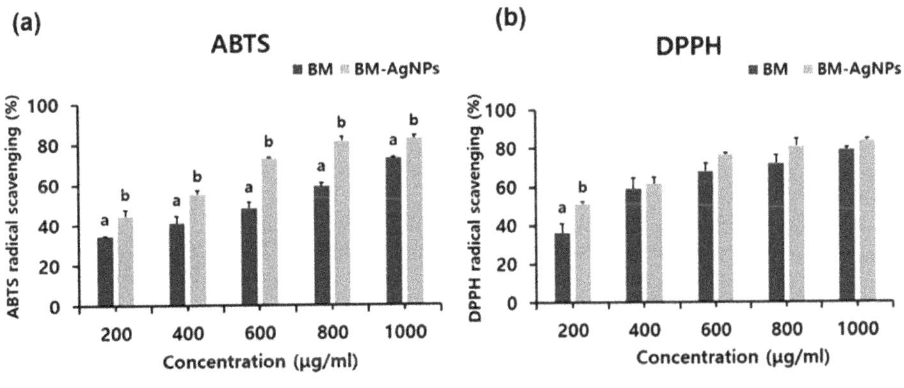

Figure 4. Antioxidant activities of BM extract and BM-AgNPs. (**a**) DPPH radical scavenging activity, (**b**) ABTS radical scavenging activity. Data are expressed as means ± standard deviation (n = 3). Different letters (a, b) in the same concentrations ($p < 0.05$) indicate significant differences.

2.3. Antibacterial Activities

The results of minimum inhibitory concentration (MIC) and minimum bactericidal concentration (MBC) on the BM extract and BM-AgNPs at various concentrations are shown in Table 3. In the MIC assay, the lowest antimicrobial concentration preventing growth is shown, while the MBC assay is determined by the minimal concentration causing microbial death [47]. A specific MIC indicating inhibition implies the possibility of organism proliferation, unlike MBC, which causes death [48]. Antibacterial effects against *S. aureus* and *E. coli* were observed using the BM extract and BM-AgNPs. The BM extract did not show antibacterial activity up to 1000 μg/mL, and the MIC of BM-AgNPs were 600 μg/mL in *S. aureus* and 600 μg/mL in *E. coli*. The MBC for BM-AgNPs were 1000 μg/mL in *S. aureus* and 1200 μg/mL in *E. coli*. These results are attributed to differences in the thickness, structure, and composition of bacterial cell walls and membranes [49]. The cell wall of Gram-negative bacteria is thinner and more sensitive to AgNPs than positive bacteria [50], while the thick peptidoglycan layer of Gram-positive bacteria constitutes a solid barrier that makes it difficult for AgNPs to adhere and penetrate [51]. AgNPs have a small particle size and a large surface area that can contact bacteria and easily cause cell membrane damage when silver ions penetrate bacterial cells through cell wall destruction [52,53]. Although antibacterial activity was not confirmed in the BM extract, BM-AgNPs showed a notable antibacterial effect. The results obtained were consistent with previous studies that found that nanoparticles synthesized using plant extracts, such as mulberry leaf [54], *Curcuma longa* [55], and *Carduus crispus* [56], exhibited high antibacterial properties.

Table 3. MIC and MBC value of BM extract and BM-AgNPs.

	BM Extract		**BM-AgNPs**	
	MIC (μg/mL)	**MBC (μg/mL)**	**MIC (μg/mL)**	**MBC (μg/mL)**
S. aureus	-	-	600	1000
E. coli	-	-	600	1200

- (growth bacteria).

2.4. Anti-Inflammatory

In this study, BSA assay was used to investigate the anti-inflammatory activity of the BM extract and BM-AgNPs at different concentrations of 10, 50, 100, and 250 μg/mL. Inflammation is caused by the denaturation of proteins, a process in which the secondary and tertiary structures of proteins are damaged by stress compounds such as strong acids, bases, and heat [57]. BSA accounts for 60% of the total protein in animal serum, and when heated, it denatures into protein and begins to express antigens associated with type 3 hypersensitivity reactions associated with diseases such as rheumatoid arthritis and serum sickness [58]. The anti-inflammatory effect of BM-AgNPs was shown to inhibit BSA protein denaturation to a higher extent than that of the BM extract, with the level of inhibition increasing with the increasing concentration (Figure 5). The BM extract showed an albumin denaturation inhibition effect of 4.21, 6.52, 7.81, and 14.52% at each concentration, and BM-AgNPs showed an albumin denaturation inhibition effect of 14.29, 28.37, 47.90, and 64.28%. The standard, diclofenac sodium showed an inhibitory effect of 70.84, 75.26, 75.32, and 82.09%. It was confirmed that the difference in inhibitory activity between the BM extract and BM-AgNPs increased as the concentration increased. The inhibitory activity of BM-AgNPs became similar to that of Diclofenac sodium as the concentration increased. These results suggest that AgNPs may exhibit anti-inflammatory properties similar to diclofenac sodium at concentrations above 250 μg/mL. These results affirmed the regulation of autoantigen production responsible for protein denaturation in BM-AgNPs and demonstrated the inhibitory effect of AgNPs on BSA denaturation.

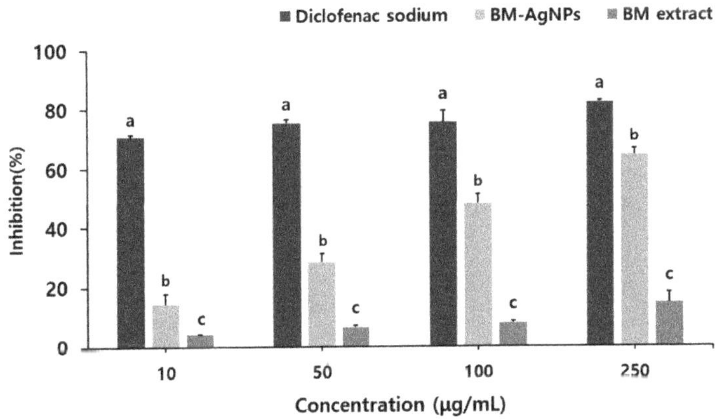

Figure 5. Anti-inflammatory activity of BM extract and BM-AgNPs. Data are expressed as means ± standard deviation (n = 3). Different letters (a–c) in the same concentrations ($p < 0.05$) indicate significant differences.

2.5. Anti-Cancer

The *in-vitro* cytotoxic activity of the BM extract and BM-AgNPs against MCF-7 cells was evaluated using the MTT assay (Figure 6). The BM extract and BM-AgNPs exhibited cytotoxic effects in a dose-dependent manner, with cell viability significantly decreasing with increasing concentration of AgNPs tested compared to the extract. In the BM extract, cytotoxicity was 79.70% at 250 μg/mL, while BM-AgNPs showed a high cytotoxic effect of 22.99%. The concentration that can reduce the cell viability of MCF-7 cells by 50% is the IC_{50}, which is 96.9 μg/mL for BM-AgNPs. AgNPs were shown to stimulate and induce reactive oxygen species (ROS) by inhibiting the synthesis of intracellular antioxidant systems [59]. This damage extends to cellular components such as cell membranes and DNA, resulting in cell death [60].

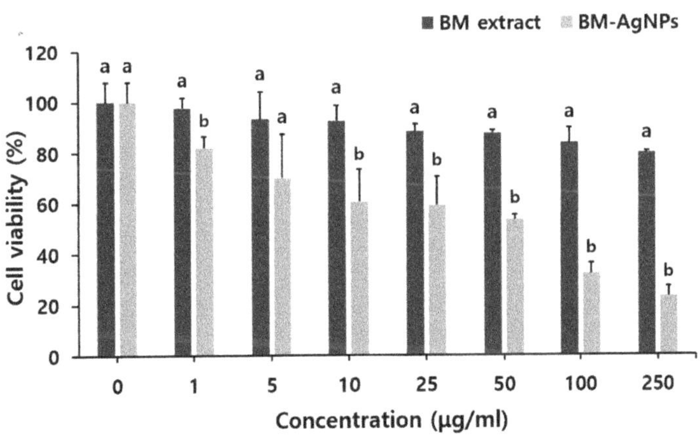

Figure 6. Anti-cancer activity of BM extract and BM-AgNPs. Data are expressed as means ± standard deviation (n = 3). Different letters (a, b) in the same concentrations ($p < 0.05$) indicate significant differences.

3. Materials and Methods

3.1. Materials

Potassium acetate was obtained from Junsei Chemical (Tokyo, Japan). Folin-Ciocalteu's phenol reagent (F9252), $AgNO_3$, and sodium carbonate were purchased from Daejung

Chemical (Siheung, Republic of Korea). Phosphate-buffered saline (PBS) was used without a further purification system. Gallic acid (G7384), quercetin (Q4951), and bovine serum albumin (A2153) were obtained from Sigma-Aldrich (St. Louis, MO, USA). Diclofenac sodium salt (D2508) was purchased from TCI (Tokyo, Japan). Mueller hinton broth (MHB) and agar were bought from Kisanbio (Seoul, Republic of Korea). Dimethyl sulfoxide (DMSO) for cell culture was purchased from Gen DEPOT (Altair, TX, USA). Dulbecco's Modification of Eagle's Medium (DMEM), penicillin-streptomycin, and fetal bovine serum (FBS) were purchased from Thermo Fisher Scientific (Waltham, MA, USA).

3.2. *Preparation of BM Extract*

BM powder was obtained by purchasing fresh mulberry fruit from a farmer in Wanju, Republic of Korea, and freeze-drying it at −50 °C. 1 g of BM powder was placed into 50 mL of distilled water (DW), followed by ultrasonic extraction at 40 °C for 30 min. The extract was filtrated through Whatman 6 filter paper at room temperature. The aqueous extract was stored at 4 °C, intended for later applications.

3.3. *Synthesis of BM-AgNPs*

The solution of $AgNO_3$ at a concentration of 3 mM was mixed with the BM extract in a ratio of 9:1 ($AgNO_3$: BM extract). The mixture was stored at 60 °C for 24 h. Afterward, the mixture was centrifuged at 13,000 rpm for 20 min. DW was then added, and the pellet was washed three times to eliminate any remaining residues of Ag ions and extract.

3.4. *Characterization of BM-AgNPs*

The analysis of AgNPs synthesis involved the use of a UV-Vis spectrometer (BioMate 160, Thermo Fisher Scientific, Waltham, MA, USA) at 300–700 nm, employing a cuvette cell containing 1 mL of the sample. The particle size, polydispersity index (PDI), and zeta potential were measured in a Mastersizer 2000 (Malvern Instruments, Malvern, UK). Morphological details were observed by loading AgNPs samples onto copper grids and employing TEM with EDS capabilities (JEM-2100F, JEOL, Tokyo, Japan). The crystalline structure of AgNPs was analyzed using XRD patterns over 2θ angles ranging from 5 to 80°, employing an X-ray diffractometer (MPD, PAN Analytical, Almelo, The Netherlands).

3.5. *Phytochemical of BM-AgNPs*

3.5.1. TFC

A modified aluminum chloride colorimetric method was used to measure the TFC of BM-AgNPs [61]. 400 μL of each sample was added along with 20 μL of 10% aluminum chloride and 20 μL of 1 M potassium acetate. 1 mL of DW was added to the mixture and reacted for 30 min. The mixture's absorbance was evaluated at a wavelength of 415 nm. For calibration, a range of quercetin standard solutions (31.25, 62.5, 125, 250, and 500 μg/mL) was utilized.

3.5.2. TPC

Quantification of the TPC in BM-AgNPs was performed using the Folin-Ciocalteu colorimetric method, which operates according to the oxidation−reduction principle, with slight modifications [62]. A total of 0.2 mL of the sample and 0.6 mL of the DW was mixed in a test tube. A total of 1 mL Folin-Ciocalteu reagent 0.8 mL and 7.5% sodium carbonate solution of were mixed and added to the test tube. Room temperature incubation was maintained for 45 min and the mixture's absorbance was evaluated at a wavelength of 760 nm. The standard used was a gallic acid solution.

3.5.3. TAC

TAC quantification of BM-AgNPs was performed using the pH difference method with slight modifications [63]. A potassium chloride buffer (pH 1.0) was prepared by combining a 0.2 M hydrochloric acid (HCl) solution with a 0.2 N potassium chloride (KCl) solution. A sodium acetate buffer (pH 4.5) was created using a 0.4 M sodium acetate solution, with pH adjustment achieved through the addition of acetic acid. The sample (1 mg/mL) was diluted ten-fold with a buffer solution. The absorbance was evaluated at 510 to 700 nm. The calculation of the sample's TAC (mg of C3G/L) was performed utilizing the relevant Equation (1):

$$\text{TAC (mg/L)} = \text{A} \times \text{M} \times \text{DF} \times 1000/(\varepsilon \times \text{d}) \tag{1}$$

A (absorbance) = $(A_{510nm} - A_{700nm})_{pH = 1.0} - (A_{500nm} - A_{700\,nm})_{pH = 4.5}$
M (molecular weight) = 449.2 g/mol for C3G
DF (dilution factor)
1000 = conversion from g to mg
ε (molar absorptivity coefficient in L/mol/cm for C3G) = 26,900
d (path length) = 1 cm

3.6. *Antioxidant Activity*

3.6.1. ABTS

The scavenging activity of ABTS radicals was measured based on published methods [64]. Potassium persulfate and $ABTS^+$ were adjusted to 2.45 mM and 7 mM, respectively, and then mixed in a 1:1 ratio. Darkness was maintained for 16 h during the reaction of the mixture. The solution was suitably diluted using 0.1 M PBS (pH 7.4) until an absorbance of 0.9 at 730 nm was achieved. The BM extract and BM-AgNPs were diluted to various concentrations (200, 400, 600, 800, and 1000 µg/mL). A mixture consisting of 30 µL of the extract and 150 µL of the ABTS solution was prepared and reacted for 25 min in a dark environment. The absorbance was measured at 730 nm. Ascorbic acid solutions were employed for calibration purposes.

3.6.2. DPPH

The scavenging activity of DPPH radicals was measured with a slightly modified method [65]. To measure the scavenging activity of DPPH radicals, the BM extract and BM-AgNPs were diluted to various concentrations (200, 400, 600, 800, and 1000 µg/mL) and then mixed with 0.4 mM DPPH solution in a 1:1 ratio. The mixture was incubated in a dark condition (room temperature, 20 min) and the mixture's absorbance was evaluated at a wavelength of 517 nm. Ascorbic acid was used as a standard and the concentrations were 20, 40, 60, 80, and 100 µg/mL.

3.7. *Antibacterial Activity*

The strain of *E. coli* and *S. aureus* was used for the test. The strain of *E. coli* and *S. aureus* was grown in MHB and incubated overnight at 37 °C. Each sample was diluted to an appropriate concentration. DW was used as a negative control. Bacteria were diluted with MHB until 0.5 at 600 nm. 100 µL of the sample and bacteria were treated in a 96-well plate, respectively. The absorbance was evaluated at a wavelength of 600 nm at 0 and 24 h. The experiment was conducted three times. The bactericidal effect of the biosynthesized AgNPs was determined through a solid medium (Mueller Hinton Agar (MHA)) culture. Dispense 50 µL of the sample onto MHA and incubate at 37 °C for 24 h to calculate the colony forming unit (CFU) value.

3.8. *Anti-Inflammatory*

The anti-inflammatory test of BM-AgNPs was measured using bovine serum albumin (BSA) with some modifications [66]. A total of 0.45 mL of 1% BSA was mixed with 0.5 mL of BM extract and BM-AgNPs of various concentrations. The reaction was incubated at 37 °C for 20 min, then the temperature was gradually increased to 60 °C and heated for

20 min. Diclofenac sodium at different concentrations of 0.5 mL was used as a positive control. Protein inhibition by AgNPs was calculated using the following Equation (2).

$$\text{Inhibition (\%)} = [(\text{Abs Control} - \text{Abs Sample})/\text{Abs Control}] \times 100 \quad (2)$$

3.9. *Anti-Cancer*

3.9.1. Cell Culture

MCF-7 cells were cultured in DMEM with 10% FBS and 1% penicillin-streptomycin. MCF-7 cells were incubated at 37 °C in 5% CO_2, and the medium was changed every two days.

3.9.2. Cell Viability (%)

The cytotoxic activity of BM-AGNPs against MCF-7 was performed through an MTT assay. MCF-7 cells were cultured and 1×10^5 cells/well were seeded into the wells and incubated overnight for 24 h at 37 °C in a 5% CO_2 environment for cell attachment. The attached cells were treated with the BM extract and BM-AgNPs at various concentrations (1, 5, 10, 25, 50, 100, and 250 µg/mL) and cultured overnight for 24 h. The MTT solution dissolved in PBS was adjusted to 2 mg/mL, and then 50 µL was added to each well containing the cell and incubated for an additional 4 h. After inhaling the MTT solution, the formed purple formazan crystals were dissolved by adding 100 µL of DMSO. The absorbance was recorded at 540 nm using a microplate reader. Cell viability (%) was determined by substituting the absorbance value using the following Formula (3):

$$\text{Cell viability (\%)} = (\text{Abs of samples}/\text{Abs of controls}) \times 100 \quad (3)$$

3.10. *Statistical Analysis*

Analysis data were expressed as mean ± standard deviation using SAS version 9.4 (SAS Institute Inc., Cary, NC, USA) and were performed in triplicate. One-way analysis of variance (ANOVA) with Duncan's multiple range test (DMRT) was used to compare significant differences between samples at the 5% level ($p < 0.05$).

4. Conclusions

In this study, biosynthesis of AgNPs was performed from $AgNO_3$ using BM extracts serving as capping and reducing agents, and an environmentally friendly and cost-effective approach was adopted. The study results suggest the efficient application of green-synthesized nanoparticles in pharmaceuticals using BM extracts. The prepared AgNPs were characterized in structural and morphological aspects with antioxidant activity, anti-inflammatory, and dose-dependent cytotoxicity against MCF-7 breast cancer cells. UV-vis spectra confirmed the formation of AgNPs by showing an absorption peak for Ag nanoparticles at 437 nm. The crystalline structure of the biosynthesized AgNPs was confirmed through the XRD study. The particle size of AgNPs was determined to be less than 100 nm through TEM. The zeta potential value of −56.6 ± 0.56 mV obtained from DLS indicates the stability of AgNPs, and the particle size was measured to be 170.17 ± 12.65 nm. The diameter obtained via DLS can typically be higher than the diameter measured through TEM. The PDI was measured at 0.281 ± 0.07, indicating relative monodispersity. BM-AgNPs were all higher than the BM extract in TFC, TPC, and TAC. The relatively high content of active molecules can suggest a correlation with improved antioxidant and antibacterial activity. In potential antioxidant activity and antibacterial studies, AgNPs showed activity compared to the extract and could be used to treat bacteria-related diseases. In addition, BM-AgNPs exhibited suppression of protein denaturation in BSA, showing an anti-inflammatory effect comparable to the control group treated with declofenac sodium as the concentration increased. MTT analysis confirmed significant anticancer properties of BM-AgNPs against human breast cancer, exhibiting cytotoxic even at low concentrations. These findings suggest that BM-AgNPs possess potential therapeutic effects on antioxidant, antibacterial, anti-inflammatory, and anticancer properties, making them suitable candidates for nano-drug formulations in the pharmaceutical industry.

Author Contributions: Conceptualization, Y.-N.J., S.-J.R. and J.-S.B.; methodology, Y.-N.J., J.-O.K. and J.-S.B.; formal analysis, H.-Y.L. and S.-J.R.; investigation, Y.-N.J. and J.-O.K.; data curation, Y.-N.J.; writing—original draft preparation, Y.-N.J. and J.-S.B.; writing—review and editing, J.-S.B.; visualization, Y.-N.J. and S.-J.R.; supervision, J.-S.B.; project administration, J.-S.B.; funding acquisition, J.-S.B. All authors have read and agreed to the published version of the manuscript.

Funding: This research was funded by a grant (PJ015570012023) from the National Institute of Agricultural Sciences, Rural Development Administration, Korea.

Institutional Review Board Statement: Not applicable.

Informed Consent Statement: Not applicable.

Data Availability Statement: All data generated for this study are contained within the article.

Conflicts of Interest: Author Jong-Suep Baek was employed by the company BeNatureBioLab. The remaining authors declare that the research was conducted in the absence of any commercial or financial relationships that could be construed as a potential conflict of interest.

References

1. Saxena, S.K.; Nyodu, R.; Kumar, S.; Maurya, V.K. Current advances in nanotechnology and medicine. In *NanoBioMedicine*; Springer: Singapore, 2020; pp. 3–16.
2. Mani, M.; Okla, M.K.; Selvaraj, S.; Ram Kumar, A.; Kumaresan, S.; Muthukumaran, A.; Kaviyarasu, K.; El-Tayeb, M.A.; Elbadawi, Y.B.; Almaary, K.S.; et al. A novel biogenic Allium cepa leaf mediated silver nanoparticles for antimicrobial, antioxidant, and anticancer effects on MCF-7 cell line. *Environ. Res.* **2021**, *198*, 111199. [CrossRef] [PubMed]
3. Marinescu, L.; Ficai, D.; Oprea, O.; Marin, A.; Ficai, A.; Andronescu, E.; Holban, A.-M. Optimized synthesis approaches of metal nanoparticles with antimicrobial applications. *J. Nanomater.* **2020**, *2020*, 6651207. [CrossRef]
4. Alahmad, A.; Feldhoff, A.; Bigall, N.C.; Rusch, P.; Scheper, T.; Walter, J.-G. Hypericum perforatum L.-mediated green synthesis of silver nanoparticles exhibiting antioxidant and anticancer activities. *Nanomaterials* **2021**, *11*, 487. [CrossRef] [PubMed]
5. Kumkoon, T.; Srisaisap, M.; Boonserm, P. Biosynthesized Silver Nanoparticles Using *Morus alba* (White Mulberry) Leaf Extract as Potential Antibacterial and Anticancer Agents. *Molecules* **2023**, *28*, 1213. [CrossRef] [PubMed]
6. Wongpreecha, J.; Polpanich, D.; Suteewong, T.; Kaewsaneha, C.; Tangboriboonrat, P. One-pot, large-scale green synthesis of silver nanoparticles-chitosan with enhanced antibacterial activity and low cytotoxicity. *Carbohydr. Polym.* **2018**, *199*, 641–648. [CrossRef] [PubMed]
7. Haque, S.; Norbert, C.C.; Acharyya, R.; Mukherjee, S.; Kathirvel, M.; Patra, C.R. Biosynthesized silver nanoparticles for cancer therapy and in vivo bioimaging. *Cancers* **2021**, *13*, 6114. [CrossRef] [PubMed]
8. Chugh, D.; Viswamalya, V.S.; Das, B. Green synthesis of silver nanoparticles with algae and the importance of capping agents in the process. *J. Genet. Eng. Biotechnol.* **2021**, *19*, 126. [CrossRef]
9. Kumar, S.; Basumatary, I.B.; Sudhani, H.P.K.; Bajpai, V.K.; Chen, L.; Shukla, S.; Mukherjee, A. Plant extract mediated silver nanoparticles and their applications as antimicrobials and in sustainable food packaging: A state-of-the-art review. *Trends Food Sci. Technol.* **2021**, *112*, 651–666. [CrossRef]
10. Liu, Y.; Hussain, M.; Memon, H.; Yasin, S. Solar irradiation and Nageia nagi extract assisted rapid synthesis of silver nanoparticles and their antibacterial activity. *Dig. J. Nanomater. Biostruct.* **2015**, *10*, 1019–1024.
11. Ghatage, M.M.; Mane, P.A.; Gambhir, R.P.; Parkhe, V.S.; Kamble, P.A.; Lokhande, C.D.; Tiwari, A.P. Green synthesis of silver nanoparticles via *Aloe barbadensis* miller leaves: Anticancer, antioxidative, antimicrobial and photocatalytic properties. *Appl. Surf. Sci. Adv.* **2023**, *16*, 100426. [CrossRef]
12. Jain, S.; Mehata, M.S. Medicinal plant leaf extract and pure flavonoid mediated green synthesis of silver nanoparticles and their enhanced antibacterial property. *Sci. Rep.* **2017**, *7*, 15867. [CrossRef] [PubMed]
13. Raji, P.; Samrot, A.V.; Keerthana, D.; Karishma, S. Antibacterial activity of alkaloids, flavonoids, saponins and tannins mediated green synthesised silver nanoparticles against *Pseudomonas aeruginosa* and *Bacillus subtilis*. *J. Clust. Sci.* **2019**, *30*, 881–895. [CrossRef]
14. Farshori, N.N.; Al-Oqail, M.M.; Al-Sheddi, E.S.; Al-Massarani, S.M.; Saquib, Q.; Siddiqui, M.A.; Wahab, R.; Al-Khedhairy, A.A. Green synthesis of silver nanoparticles using *Phoenix dactylifera* seed extract and its anticancer effect against human lung adenocarcinoma cells. *J. Drug Deliv. Sci. Technol.* **2022**, *70*, 103260. [CrossRef]
15. Pirtarighat, S.; Ghannadnia, M.; Baghshahi, S. Green synthesis of silver nanoparticles using the plant extract of *Salvia spinosa* grown in vitro and their antibacterial activity assessment. *J. Nanostruct. Chem.* **2018**, *9*, 1–9. [CrossRef]
16. Ahmed, S.; Ahmad, M.; Swami, B.L.; Ikram, S. A review on plants extract mediated synthesis of silver nanoparticles for antimicrobial applications: A green expertise. *J. Adv. Res.* **2016**, *7*, 17–28. [CrossRef] [PubMed]

17. Sheikh, S.; Siddique, F.; Ameer, K.; Ahmad, R.S.; Hameed, A.; Ebad, A.; Mohamed Ahmed, I.A.; Shibli, S. Effects of white mulberry powder fortification on antioxidant activity, physicochemical, microbial and sensorial properties of yogurt produced from buffalo milk. *Food Sci. Nutr.* **2023**, *11*, 204–215. [CrossRef] [PubMed]
18. Suriyaprom, S.; Kaewkod, T.; Promputtha, I.; Desvaux, M.; Tragoolpua, Y. Evaluation of Antioxidant and Antibacterial Activities of White Mulberry (*Morus alba* L.) Fruit Extracts. *Plants* **2021**, *10*, 2736. [CrossRef] [PubMed]
19. Turan, E.; Simsek, A. Effects of lyophilized black mulberry water extract on lipid oxidation, metmyoglobin formation, color stability, microbial quality and sensory properties of beef patties stored under aerobic and vacuum packaging conditions. *Meat Sci.* **2021**, *178*, 108522. [CrossRef] [PubMed]
20. Sánchez-Salcedo, E.M.; Mena, P.; García-Viguera, C.; Martínez, J.J.; Hernández, F. Phytochemical evaluation of white (*Morus alba* L.) and black (*Morus nigra* L.) mulberry fruits, a starting point for the assessment of their beneficial properties. *J. Funct. Foods* **2015**, *12*, 399–408. [CrossRef]
21. Zhou, M.; Chen, Q.; Bi, J.; Wang, Y.; Wu, X. Degradation kinetics of cyanidin 3-O-glucoside and cyanidin 3-O-rutinoside during hot air and vacuum drying in mulberry (*Morus alba* L.) fruit: A comparative study based on solid food system. *Food Chem.* **2017**, *229*, 574–579. [CrossRef] [PubMed]
22. Wang, S.; Zhang, X.; Ai, J.; Yue, Z.; Wang, Y.; Bao, B.; Tian, L.; Bai, W. Interaction between black mulberry pectin-rich fractions and cyanidin-3-O-glucoside under in vitro digestion. *Food Hydrocoll.* **2023**, *134*, 108110. [CrossRef]
23. Deniz, G.; Laloglu, E.; Koc, K.; Nadaroglu, H.; Geyikoglu, F. The effect of black mulberry (*Morus nigra*) extract on carbon tetrachloride-induced liver damage. *Arch. Biol. Sci.* **2018**, *70*, 371–378. [CrossRef]
24. Chen, Q.; Li, Z.; Bi, J.; Zhou, L.; Yi, J.; Wu, X. Effect of hybrid drying methods on physicochemical, nutritional and antioxidant properties of dried black mulberry. *LWT* **2017**, *80*, 178–184. [CrossRef]
25. Manzoor, S.; Qayoom, K. *Morus nigra* L. (*Black mulberry*): A plant with potent medicinal value. *South Asian J. Agric. Sci.* **2023**, *3*, 60–63. [CrossRef]
26. Kanimozhi, S.; Durga, R.; Sabithasree, M.; Kumar, A.V.; Sofiavizhimalar, A.; Kadam, A.A.; Rajagopal, R.; Sathya, R.; Azelee, N.I.W. Biogenic synthesis of silver nanoparticle using *Cissus quadrangularis* extract and its invitro study. *J. King Saud Univ. Sci.* **2022**, *34*, 101930. [CrossRef]
27. Baghizadeh, A.; Ranjbar, S.; Gupta, V.K.; Asif, M.; Pourseyedi, S.; Karimi, M.J.; Mohammadinejad, R. Green synthesis of silver nanoparticles using seed extract of *Calendula officinalis* in liquid phase. *J. Mol. Liq.* **2015**, *207*, 159–163. [CrossRef]
28. Alsammarraie, F.K.; Wang, W.; Zhou, P.; Mustapha, A.; Lin, M. Green synthesis of silver nanoparticles using turmeric extracts and investigation of their antibacterial activities. *Colloids Surf. B Biointerfaces* **2018**, *171*, 398–405. [CrossRef] [PubMed]
29. Mittal, J.; Jain, R.; Sharma, M.M. Phytofabrication of silver nanoparticles using aqueous leaf extract of *Xanthium strumerium* L. and their bactericidal efficacy. *Adv. Nat. Sci. Nanosci. Nanotechnol.* **2017**, *8*, 025011. [CrossRef]
30. Mittal, J.; Singh, A.; Batra, A.; Sharma, M.M. Synthesis and characterization of silver nanoparticles and their antimicrobial efficacy. *Part. Sci. Technol.* **2017**, *35*, 338–345. [CrossRef]
31. Azmi, S.N.H.; Al-Jassasi, B.M.H.; Al-Sawafi, H.M.S.; Al-Shukaili, S.H.G.; Rahman, N.; Nasir, M. Optimization for synthesis of silver nanoparticles through response surface methodology using leaf extract of Boswellia sacra and its application in antimicrobial activity. *Environ. Monit. Assess.* **2021**, *193*, 497. [CrossRef]
32. Chand, K.; Jiao, C.; Lakhan, M.N.; Shah, A.H.; Kumar, V.; Fouad, D.E.; Chandio, M.B.; Maitlo, A.A.; Ahmed, M.; Cao, D. Green synthesis, characterization and photocatalytic activity of silver nanoparticles synthesized with Nigella Sativa seed extract. *Chem. Phys. Lett.* **2021**, *763*, 138218. [CrossRef]
33. Wilson, B.K.; Prud'homme, R.K. Nanoparticle size distribution quantification from transmission electron microscopy (TEM) of ruthenium tetroxide stained polymeric nanoparticles. *J. Colloid. Interface Sci.* **2021**, *604*, 208–220. [CrossRef] [PubMed]
34. Al-Khafaji, M.A.; Gaal, A.; Wacha, A.; Bota, A.; Varga, Z. Particle Size Distribution of Bimodal Silica Nanoparticles: A Comparison of Different Measurement Techniques. *Materials* **2020**, *13*, 3101. [CrossRef] [PubMed]
35. Scherer, M.D.; Sposito, J.C.; Falco, W.F.; Grisolia, A.B.; Andrade, L.H.; Lima, S.M.; Machado, G.; Nascimento, V.A.; Gonçalves, D.A.; Wender, H. Cytotoxic and genotoxic effects of silver nanoparticles on meristematic cells of *Allium cepa* roots: A close analysis of particle size dependence. *Sci. Total Environ.* **2019**, *660*, 459–467. [CrossRef] [PubMed]
36. Tantra, R.; Schulze, P.; Quincey, P. Effect of nanoparticle concentration on zeta-potential measurement results and reproducibility. *Particuology* **2010**, *8*, 279–285. [CrossRef]
37. Pochapski, D.J.; Carvalho dos Santos, C.; Leite, G.W.; Pulcinelli, S.H.; Santilli, C.V. Zeta potential and colloidal stability predictions for inorganic nanoparticle dispersions: Effects of experimental conditions and electrokinetic models on the interpretation of results. *Langmuir* **2021**, *37*, 13379–13389. [CrossRef] [PubMed]
38. Riaz, M.; Mutreja, V.; Sareen, S.; Ahmad, B.; Faheem, M.; Zahid, N.; Jabbour, G.; Park, J. Exceptional antibacterial and cytotoxic potency of monodisperse greener AgNPs prepared under optimized pH and temperature. *Sci. Rep.* **2021**, *11*, 2866. [CrossRef] [PubMed]
39. Chinni, S.V.; Gopinath, S.C.B.; Anbu, P.; Fuloria, N.K.; Fuloria, S.; Mariappan, P.; Krusnamurthy, K.; Veeranjaneya Reddy, L.; Ramachawolran, G.; Sreeramanan, S.; et al. Characterization and Antibacterial Response of Silver Nanoparticles Biosynthesized Using an Ethanolic Extract of *Coccinia indica* Leaves. *Crystals* **2021**, *11*, 97. [CrossRef]

40. Yi, Z.; Xu, X.; Fang, Q.; Wang, Y.; Li, X.; Tan, X.; Luo, J.; Jiang, X.; Wu, W.; Yi, Y. Fabrication of silver nanosheets on quartz glass substrates through electroless plating approach. *Appl. Phys. A* **2014**, *114*, 485–493. [CrossRef]
41. Ryu, S.; Nam, S.H.; Baek, J.S. Green Synthesis of Silver Nanoparticles (AgNPs) of Angelica Gigas Fabricated by Hot-Melt Extrusion Technology for Enhanced Antifungal Effects. *Materials* **2022**, *15*, 7231. [CrossRef] [PubMed]
42. Noori, A.; Donnelly, T.; Colbert, J.; Cai, W.; Newman, L.A.; White, J.C. Exposure of tomato (*Lycopersicon esculentum*) to silver nanoparticles and silver nitrate: Physiological and molecular response. *Int. J. Phytoremediation* **2020**, *22*, 40–51. [CrossRef] [PubMed]
43. Shavalibor, A.; Esmaeilzadeh Bahabadi, S. Effect of biologically synthesized silver nanoparticles on *Melissa officinalis* L. Evaluation of growth parameters, secondary metabolites and antioxidant enzymes. *Iran. J. Plant Physiol.* **2021**, *11*, 3799–3809.
44. Krajewska, J.B.; Dlugosz, O.; Salaga, M.; Banach, M.; Fichna, J. Silver nanoparticles based on blackcurrant extract show potent anti-inflammatory effect in vitro and in DSS-induced colitis in mice. *Int. J. Pharm.* **2020**, *585*, 119549. [CrossRef] [PubMed]
45. Campos, A.M.; Lissi, E.A. Kinetics of the reaction between 2, 2′-azinobis (3-ethylbenzothiazoline-6-sulfonic acid (ABTS) derived radical cations and phenols. *Int. J. Chem. Kinet.* **1997**, *29*, 219–224. [CrossRef]
46. Siakavella, I.K.; Lamari, F.; Papoulis, D.; Orkoula, M.; Gkolfi, P.; Lykouras, M.; Avgoustakis, K.; Hatziantoniou, S. Effect of Plant Extracts on the Characteristics of Silver Nanoparticles for Topical Application. *Pharmaceutics* **2020**, *12*, 1244. [CrossRef] [PubMed]
47. Reyes-Blas, M.; Maldonado-Luna, N.M.; Rivera-Quiñones, C.M.; Vega-Avila, A.L.; Roman-Velázquez, F.R.; Perales-Perez, O.J. Single Step Microwave Assisted Synthesis and Antimicrobial Activity of Silver, Copper and Silver-Copper Nanoparticles. *J. Mater. Sci. Chem. Eng.* **2020**, *8*, 13–29. [CrossRef]
48. Sukeri, S.; Karem, A.A.; Kamarudin, E.; Bahari, M. Antimicrobial Activity of Methanolic and Aqueous Extract of *Rhodomyrtus tomentosa* Leaves against *Staphylococcus aureus* and *Escherichia coli*. *J. Pure Appl. Microbiol.* **2021**, *15*, 186–193. [CrossRef]
49. Naghmouchi, S.; Al-Zaban, M.I.; Al-Zaben, M.; Alharbi, N.; Bahatheq, A.; Gnilitskyi, I. Generation and Characterization of Silver Nanoparticles in *Mentha pulegium* Extract and Evaluation of Biological Activities of the Prepared Extract. *J. Nanomater.* **2022**, *2022*, 5410274. [CrossRef]
50. Nogueira, S.S.; de Araujo-Nobre, A.R.; Mafud, A.C.; Guimaraes, M.A.; Alves, M.M.M.; Placido, A.; Carvalho, F.A.A.; Arcanjo, D.D.R.; Mascarenhas, Y.; Costa, F.G.; et al. Silver nanoparticle stabilized by hydrolyzed collagen and natural polymers: Synthesis, characterization and antibacterial-antifungal evaluation. *Int. J. Biol. Macromol.* **2019**, *135*, 808–814. [CrossRef] [PubMed]
51. Ontong, J.C.; Singh, S.; Nwabor, O.F.; Chusri, S.; Voravuthikunchai, S.P. Potential of antimicrobial topical gel with synthesized biogenic silver nanoparticle using *Rhodomyrtus tomentosa* leaf extract and silk sericin. *Biotechnol. Lett.* **2020**, *42*, 2653–2664. [CrossRef] [PubMed]
52. Mahadevan, S.; Vijayakumar, S.; Arulmozhi, P. Green synthesis of silver nano particles from *Atalantia monophylla* (L) Correa leaf extract, their antimicrobial activity and sensing capability of H(2)O(2). *Microb. Pathog.* **2017**, *113*, 445–450. [CrossRef] [PubMed]
53. Mortazavi-Derazkola, S.; Hosseinzadeh, M.; Yousefinia, A.; Naghizadeh, A. Green Synthesis and Investigation of Antibacterial Activity of Silver Nanoparticles Using Eryngium bungei Boiss Plant Extract. *J. Polym. Environ.* **2021**, *29*, 2978–2985. [CrossRef]
54. Dong, C.; Cheng, F.; Zhang, X.; Wang, X.; Yang, X.; Bin, Y. Rapid and green synthesis of monodisperse silver nanoparticles using mulberry leaf extract. *Rare Met. Mater. Eng.* **2018**, *47*, 1089–1095.
55. Khan, M.A.; Moghul, N.B.; Butt, M.A.; Kiyani, M.M.; Zafar, I.; Bukhari, A.I. Assessment of antibacterial and antifungal potential of *Curcuma longa* and synthesized nanoparticles: A comparative study. *J. Basic Microbiol.* **2021**, *61*, 603–611. [CrossRef] [PubMed]
56. Urnukhsaikhan, E.; Bold, B.-E.; Gunbileg, A.; Sukhbaatar, N.; Mishig-Ochir, T. Antibacterial activity and characteristics of silver nanoparticles biosynthesized from *Carduus crispus*. *Sci. Rep.* **2021**, *11*, 21047. [CrossRef] [PubMed]
57. Salve, P.; Vinchurkar, A.; Raut, R.; Chondekar, R.; Lakkakula, J.; Roy, A.; Hossain, M.J.; Alghamdi, S.; Almehmadi, M.; Abdulaziz, O.; et al. An Evaluation of Antimicrobial, Anticancer, Anti-Inflammatory and Antioxidant Activities of Silver Nanoparticles Synthesized from Leaf Extract of Madhuca longifolia Utilizing Quantitative and Qualitative Methods. *Molecules* **2022**, *27*, 6404. [CrossRef] [PubMed]
58. Ahiwale, A.V.; Anandh, S.; Gosavi, P. Effect of Therapeutic Exercises on Pre-Menopausal Symptoms in Sedentary Life Style Women. *Indian J. Public Health Res. Dev.* **2019**, *10*, 1. [CrossRef]
59. Sathishkumar, G.; Gobinath, C.; Wilson, A.; Sivaramakrishnan, S. *Dendrophthoe falcata* (L.f) Ettingsh (Neem mistletoe): A potent bioresource to fabricate silver nanoparticles for anticancer effect against human breast cancer cells (MCF-7). *Spectrochim. Acta A Mol. Biomol. Spectrosc.* **2014**, *128*, 285–290. [CrossRef] [PubMed]
60. Karuppaiah, A.; Siram, K.; Selvaraj, D.; Ramasamy, M.; Babu, D.; Sankar, V. Synergistic and enhanced anticancer effect of a facile surface modified non-cytotoxic silver nanoparticle conjugated with gemcitabine in metastatic breast cancer cells. *Mater. Today Commun.* **2020**, *23*, 100884. [CrossRef]
61. Chang, C.-C.; Yang, M.-H.; Wen, H.-M.; Chern, J.-C. Estimation of total flavonoid content in propolis by two complementary colorimetric methods. *J. Food Drug Anal.* **2002**, *10*, 3.
62. Zhang, Q.; Zhang, J.; Shen, J.; Silva, A.; Dennis, D.A.; Barrow, C.J. A simple 96-well microplate method for estimation of total polyphenol content in seaweeds. *J. Appl. Phycol.* **2006**, *18*, 445–450. [CrossRef]
63. Nguyen, N.-T.; Vo, T.-L.-H. Fabrication of Silver Nanoparticles Using *Cordyline fruticosa* L. Leave Extract Endowing Silk Fibroin Modified Viscose Fabric with Durable Antibacterial Property. *Polymers* **2022**, *14*, 2409. [CrossRef] [PubMed]
64. Re, R.; Pellegrini, N.; Proteggente, A.; Pannala, A.; Yang, M.; Rice-Evans, C. Antioxidant activity applying an improved ABTS radical cation decolorization assay. *Free. Radic. Biol. Med.* **1999**, *26*, 1231–1237. [CrossRef] [PubMed]

65. Brand-Williams, W.; Cuvelier, M.-E.; Berset, C. Use of a free radical method to evaluate antioxidant activity. *LWT-Food Sci. Technol.* **1995**, *28*, 25–30. [CrossRef]
66. Govindappa, M.; Farheen, H.; Chandrappa, C.; Rai, R.V.; Raghavendra, V.B. Mycosynthesis of silver nanoparticles using extract of endophytic fungi, Penicillium species of Glycosmis mauritiana, and its antioxidant, antimicrobial, anti-inflammatory and tyrokinase inhibitory activity. *Adv. Nat. Sci. Nanosci. Nanotechnol.* **2016**, *7*, 035014. [CrossRef]

antibiotics

MDPI

Article

Unlocking the Skin-Protective Effects of *Ferulago* spp. Essential Oils: A Focus on Antidermatophytic and Wound-Healing Potential

Ceyda Sibel Kılıç [1,†], Inês Catarino [2,†], Jorge Alves-Silva [2,3,4,5], Betül Demirci [6], Damla Kırcı [7], Lígia Salgueiro [4,8] and Mónica Zuzarte [2,3,4,5,8,*]

1 Ankara University, Faculty of Pharmacy, Department of Pharmaceutical Botany, 06560 Ankara, Türkiye; erdurak@pharmacy.ankara.edu.tr
2 Univ Coimbra, Coimbra Institute for Clinical and Biomedical Research (iCBR), Faculty of Medicine, Azinhaga de S. Comba, 3000-548 Coimbra, Portugal; uc2024232528@student.uc.pt (I.C.); jmasilva@fmed.uc.pt (J.A.-S.)
3 Univ Coimbra, Center for Innovative Biomedicine and Biotechnology (CIBB), 3000-548 Coimbra, Portugal
4 Univ Coimbra, Faculty of Pharmacy, Azinhaga de S. Comba, 3000-548 Coimbra, Portugal; ligia@ff.uc.pt
5 Clinical Academic Centre of Coimbra (CACC), 3000-548 Coimbra, Portugal
6 Anadolu University, Faculty of Pharmacy, Department of Pharmacognosy, 26470 Tepebaşı, Eskişehir, Türkiye; bdemirca@anadolu.edu.tr
7 İzmir Katip Çelebi University, Faculty of Pharmacy, Department of Pharmacognosy, 35620 Çiğli, İzmir, Türkiye; damla.kirci@ikc.edu.tr
8 Univ Coimbra, Chemical Engineering and Renewable Resources for Sustainability (CERES), Department of Chemical Engineering, 3030-790 Coimbra, Portugal
* Correspondence: mzuzarte@uc.pt
† These authors contributed equally to this work.

Abstract: Background/Objectives: Fungal infections have significant health risks due to virulence factors like biofilms that lead to chronic and persistent infections. To decrease associated rates of morbidity and mortality, it is crucial to develop effective antifungal treatments. The present study aims to evaluate the bioactive potential of *Ferulago* spp. essential oils by assessing their effect on dermatophytes and wound healing, as these fungi are often associated with wound infection. Methods: The minimal inhibitory concentration (MIC) and the minimal lethal concentration (MLC) of five essential oils were assessed on planktonic dermatophytes and the most promising were used to evaluate their effect on the formation and disruption of biofilms, by quantifying biofilm mass using crystal violet and extracellular matrix deposition using safranin staining. Alterations in fungal morphology were confirmed by optical microscopy and a cell migration assay was used to assess wound-healing capacity. Results: *Ferulago silaifolia* essential oil characterized by high amounts of α-pinene (45.4%) and *cis*-crysanthenyl acetate (39.1%) was the most active, particularly against *Microsporum canis* and *Trichophyton rubrum* (MIC = 50 μg/mL). Regarding biofilm assays, *Trichophyton rubrum* was the most susceptible strain, with both biofilm mass and extracellular matrix being highly compromised with an evident decrease in hyphal growth and mycelial density. In addition, this essential oil significantly increased fibroblast migration at 25 μg/mL, indicating a wound-healing effect that could prevent systemic infections. Conclusions: The present study provides new insights into the treatment of dermatophytosis by highlighting the antivirulent effects of *F. silaifolia* essential oil and its potential wound-healing properties.

Keywords: dermatophytosis; biofilm; virulence; cell migration

https://doi.org/10.3390/antibiotics14040343

1. Introduction

The Global Action Fund for Fungal Infections (GAFFI) estimates that over 6 million individuals develop life-threatening fungal infections annually, of which 3.75 million do not survive [1]. Indeed, fungal infections are becoming a worldwide public health issue, especially in immunocompromised individuals, including patients with chronic lung disease, prior tuberculosis, HIV, cancer and diabetes mellitus [2]. These infections can manifest in various forms, ranging from localized cutaneous to invasive and life-threatening fungal infections [3].

The present study focuses on dermatophytes that are known to present an infection rate of around 20–25% worldwide [4] and are divided into nine genera: *Arthroderma*, *Lopophyton*, *Nannizia*, *Ctenomyces*, *Guarromyces*, *Paraphyton*, *Microsporum*, *Epidermophyton*, and *Trichophyton*, the last three being the ones that can cause infections in humans [5,6]. These filamentous fungi affect the keratin present in the skin, hair, and nails, leading to diseases known as dermatophytosis and onychomycosis (nails), which are the the most disseminated type of mycosis. Dermatophytes represent a category of primary pathogenic fungi that rely on keratin cleavage for nutritional sustenance [3].

In the early stages of infection, dermatophytes prompt keratinocytes to generate various cytokine patterns, depending on their classification, that mediate the inflammatory response and the accumulation of leukocytes in the infected tissue [3]. Dermatophytes can be categorized as anthropophilic, geophilic or zoophilic species, depending on their normal habitat (humans, soil or animals, respectively). Clinically, dermatophytosis is classified according to the site of infection as follows: tinea capitis (affects the scalp and hair), tinea faciei (affects the face), tinea barbae (affects the beard), tinea corporis (affects the body), tinea manuum (affects the hands), tinea cruris (affects the groin), tinea pedis (affects the feet, known as athlete's foot) and tinea unguium (affects the nails). Other clinical variants include tinea imbricata, pseudoimbricata, and Majocchi's granuloma, the latter being mostly caused by *Trichophyton rubrum* and occurring when a long-standing superficial fungal infection causes progressive dissemination into subcutaneous tissue [7].

An important virulence factor in dermatophytes is their ability to form biofilms, a more resistant structure than planktonic forms, formed by a larger group of fungal cells and extracellular matrix. The fungal cells first adhere to a substrate, either a mucosal or abiotic surface, and proliferate until they form an extracellular matrix composed of polysaccharides, lipids, proteins, and nucleic acids. The main function of this organization is to increase fungi survival, leading to a strong resistance to conventional therapy and host defenses [8].

The most commonly used treatments for dermatophytosis include topical and oral agents. Topical antifungal agents like butenafine, clotrimazole and miconazole are recommended for dermatophytosis that presents as localized skin infections. However, oral drugs such as terbinafine, itraconazole, griseofulvin and fluconazole are recommended for more extensive infections, especially in combination therapy, to prevent the emergence of resistance and ensure wide coverage [9]. Nevertheless, despite the use of numerous antifungal drugs, addressing these infections remains a challenge, as virulence factors such as biofilms are known to confer high resistance [8]. Furthermore, it is commonplace for patients to neglect and discontinue treatment, primarily due to its long-term duration. These barriers to conventional treatments have driven the emergence of resistant strains, highlighting the need for effective alternatives such as plant metabolites like essential oils (EOs). These volatile extracts have been the target of many researchers as alternatives and/or complementary strategies in healthcare.

The genus *Ferulago* W.D.J.M. Koch (Apiaceae) is considered the eighth largest botanical family in Türkiye. Of the 48 taxa in this genus [10], 35 grow naturally in this country and

19 are endemic, thus positioning Türkiye as the gene center for *Ferulago* [11]. Several *Ferulago* species have been traditionally used for a variety of medicinal purposes, including the treatment of skin wound infections, ulcers, snake bites, headaches, and spleen problems. They are also used as sedatives, immunostimulants, carminatives, and remedies for peptic issues, intestinal worms, hemorrhoids, and even as aphrodisiacs. Moreover, some species are used as spices and to increase the shelf life of vegetable oils and dairy products [12,13]. Although the antimicrobial properties of several *Ferulago* species have been identified, regarding fungal infections, the majority of studies focus on the effect of *Ferulago* EOs on *Candida* spp. growth [14,15]. Indeed, as far as we know, only one study assessed the antifungal effect of *F. capillaris* EO against dermatophytes [16] but failed to address its impact on relevant virulence factors such as biofilms. Therefore, the present study aims to evaluate the antifungal potential of five *Ferulago* species on planktonic dermatophytes, selecting the most promising one to assess its ability to eradicate and/or prevent biofilm formation and promote wound healing.

2. Results

2.1. Chemical Composition

Table 1 summarizes the main chemical compounds found in the EOs obtained from different species of *Ferulago* and confirms a similar chemical profile to that previously reported [17].

Table 1. Main compounds of *Ferulago* spp. essential oil.

Species	Composition	
	Main Compounds	**Amount (%)**
F. cassia Boiss.	α-pinene	61.0
	cis-crysanthenyl acetate	21.3
	α-phellandrene	4.5
F. isaurica Peşmen	α-pinene	17.3
	α-phellandrene	36.5
	limonene	16.3
F. setifolia K. Koch	α-pinene	39.1
	2,3,6-trimethyl benzaldehyde	35.1
	sabinene	8.2
F. silaifolia (Boiss.) Boiss.	α-pinene	45.4
	cis-crysanthenyl acetate	39.1
	myrcene	1.7
F. syriaca Boiss.	myrcene	16.8
	α-phellandrene	8.8
	cubenol	9.1

2.2. Minimal Inhibitory and Minimal Lethal Concentrations of *Ferulago* spp. Essential Oils

The MIC and MLC of the EOs obtained from the species referred to in Table 1 were determined, with *F. setifolia* and *F. silaifolia* being the most active species. Indeed, these species showed very promising antifungal effects, the latter being very effective (Table 2), especially against *M. canis* and *T. rubrum*, with the lowest recorded values of MIC = 50 μg/mL and MLC = 100 μg/mL. Moreover, a fungicidal effect was observed for *T. mentagrophytes* in

F. setifolia and for *M. gypseum* and *T. mentagrophytes* in *F. silaifolia*, as the MIC and MLC values were equivalent. For the remaining species, MIC values were quite high (≥400 μg/mL), suggesting a very mild antifungal effect. For this reason, these species were disregarded and not considered in the following studies. To further confirm its antifungal potential, we selected *F. silaifolia*, which presented the most promising effects, to perform antibiofilm assays against the most common dermatophyte species, representing the three different genera: *Epidermophyton floccosum*, *Microsporum canis* and *Trichophyton rubrum*, as detailed in the next section.

Table 2. Antifungal effect of *Ferulago silaifolia* and *F. setifolia* essential oils.

Dermatophyte Strains	*Ferulago silaifolia*		*Ferulago setifolia*	
	MIC	MLC	MIC	MLC
Epidermophyton floccosum FF9	100	100–200	100–200	200–400
Microsporum canis FF1	50	100	200	200–400
Microsporum gypseum CECT 2908	200	200	200	400–600
Trichophyton mentagrophytes FF7	100–200	100–200	200	200
Trichophyton mentagrophytes var. *interdigitale* CECT 2958	200	400	200	600
Trichophyton rubrum CECT 2794	50	100	50–100	100
Trichophyton verrucosum CECT 2992	200–400	600	200	400

MIC—Minimal Inhibitory Concentration; MLC—Minimal Lethal Concentration; MIC and MLC determined by the macrodilution method and expressed in μg/mL.

2.3. *Effect of Ferulago silaifolia Essential Oil on the Formation of Dermatophyte Biofilms*

In the present study, the effect of *F. silaifolia* EO on the formation and disruption of biofilms formed by the three selected fungal strains was assessed.

Considering the effect of the EO on biofilm formation, it is possible to observe a decrease in biofilm biomass and extracellular matrix deposition in the tested strains (Figure 1A–C), especially at the MIC value, as expected. It is notable that *T. rubrum* was the most susceptible strain to the EOs' activity, being susceptible even at concentrations as low as MIC/4 (Figure 1C). Indeed, the EO was able to significantly decrease extracellular matrix deposition at 12.5 μg/mL (MIC/4), and at 25 μg/mL (MIC/2), it was also effective in reducing biofilm biomass (Figure 1C). On the other hand, *M. canis* was the least susceptible strain, with all parameters exhibiting a slight decrease but without statistical significance between the positive control (DMSO) and the different EO concentrations.

2.4. *Effect of Ferulago silaifolia Essential Oil on the Disruption of Dermatophyte Pre-Formed Biofilms*

The effect of *F. silaifolia* EO on disrupting pre-formed biofilms was also assessed as these structures are highly resistant to both therapy and host defenses. As shown in Figure 2, a significant susceptibility of *T. rubrum* to the EO (Figure 2C) is noteworthy, with a statistically significant reduction seen in extracellular matrix deposition up to MIC/8 (6.25 μg/mL). In addition, *E. floccosum* was quite susceptible at MIC (50 μg/mL) and MIC/2 (25 μg/mL), particularly in the biomass assay (Figure 2A). In contrast, and similarly to the observations in the biofilm formation inhibition assay, *M. canis* was the least susceptible strain to the EO's activity (Figure 2B).

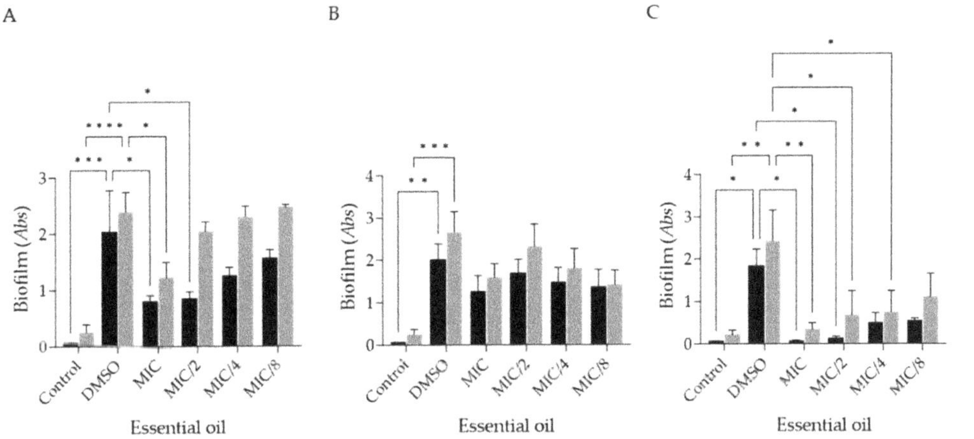

Figure 1. *Ferulago silaifolia* essential oil's effect on biofilm formation mass (represented by dark bars) and extracellular matrix (represented by grey bars) of *Epidermophyton fluccosum* (**A**), *Microsporum canis* (**B**) and *Trichophyton rubrum* (**C**). Results expressed as absorbance (*Abs*) values and comparisons relative to DMSO (mean ± SEM values; * $p < 0.05$, ** $p < 0.01$, *** $p < 0.001$, **** $p < 0.0001$). DMSO—positive control containing EO-free inoculated medium; control—negative control containing non-inoculated medium.

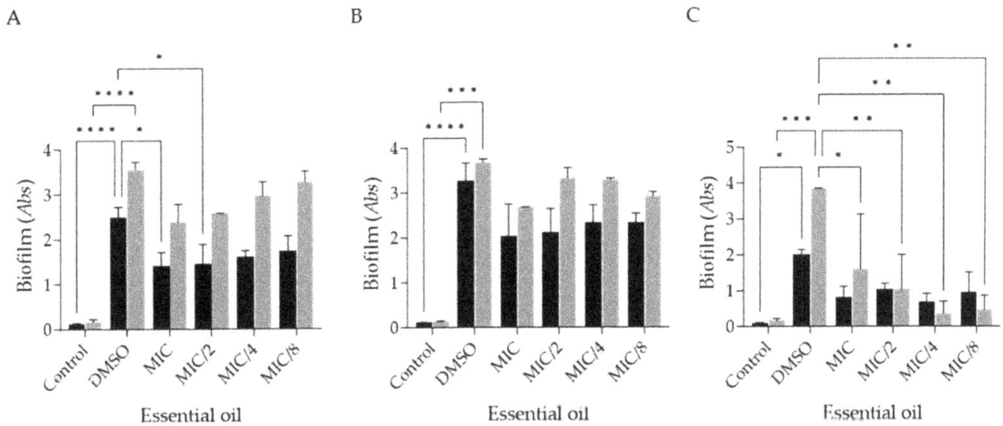

Figure 2. *Ferulago silaifolia* essential oil's effect on pre-formed biofilm mass (represented by dark bars) and extracellular matrix (represented by grey bars) of *Epidermophyton fluccosum* (**A**), *Microsporum canis* (**B**) and *Trichophyton rubrum* (**C**). Results expressed as absorbance (*Abs*) values and comparisons relative to DMSO (mean ± SEM values; * $p < 0.05$, ** $p < 0.01$, *** $p < 0.001$, **** $p < 0.0001$). DMSO—positive control containing EO-free inoculated medium; control—negative control containing non-inoculated medium.

2.5. Morphological Alterations Induced by Ferulago silaifolia Essential Oil on Dermatophytes

Optical microscopy observations were performed following biofilm assays to assess the morphological alterations induced by *F. silaifolia* EO (MIC) on three fungal strains (*Epidermophyton floccosum*, *Microsporum canis* and *Trichophyton rubrum*). As illustrated in Figures 3 and 4, the EO produced significant morphological changes in both biofilm formation and pre-formed biofilms. Regarding biomass (Figure 3), there is a noticeable increase in septate hyphae (arrows) and less dense mycelia, particularly in *T. rubrum* in the presence of the EO; whereas, in the positive control (DMSO), an evident clustering of hyphae and spores in an organized structure was observed. Indeed, the exposure of the fungi to the EO resulted in hypha growth inhibition or destruction with an increase in planktonic fungi. This effect was more evident on biofilm formation compared to pre-formed biofilms, corroborating the results above. Extracellular matrix deposition (Figure 4) was also significantly compromised with very effective results observed once again in *T. rubrum*, in which no extracellular matrix deposition was observed in the presence of the EO.

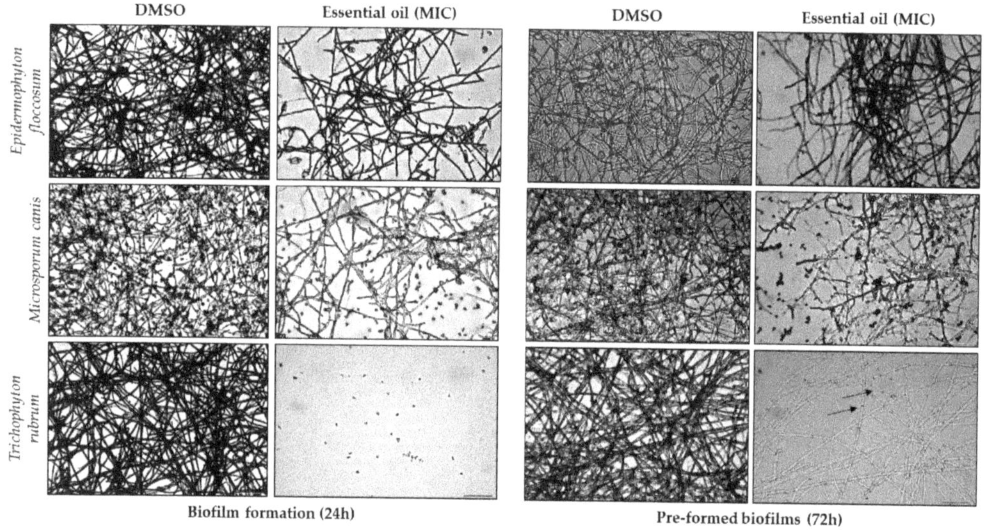

Figure 3. Optical microscopy observations of biofilm formation and pre-formed biofilms in *Epidermophyton floccosum*, *Microsporum canis* and *Trichophyton rubrum*, both untreated (DMSO) and treated with the MICs of *F. silaifolia* essential oil, following biomass quantification using a crystal violet assay. Arrows indicate septate hyphae. Scale bar = 100 µm.

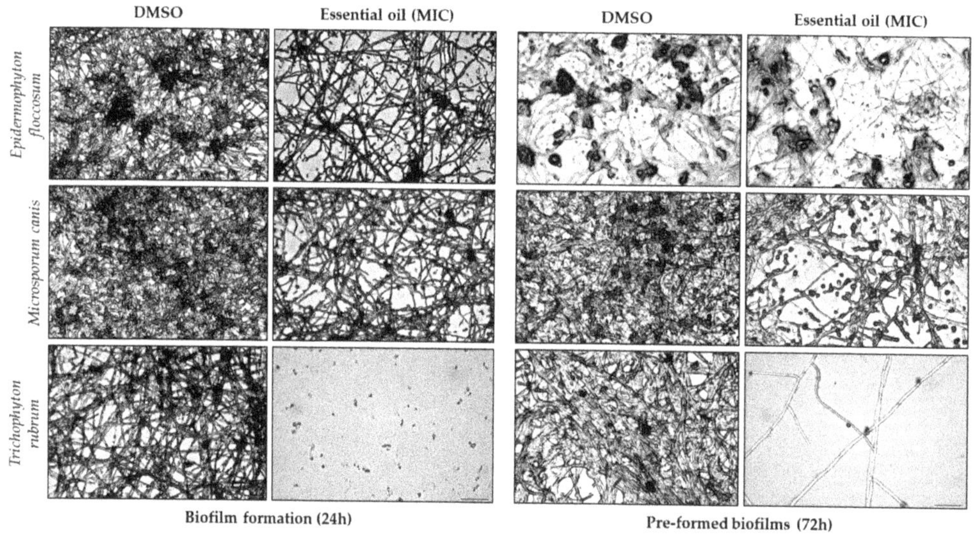

Figure 4. Optical microscopy observations of biofilm formation and pre-formed biofilms in *Epidermophyton floccosum*, *Microsporum canis* and *Trichophyton rubrum*, both untreated (DMSO) and treated with the MICs of *F. silaifolia* essential oil, following biomass quantification using a safranin assay. Scale bar = 100 µm.

2.6. Effect of *Ferulago setifolia* and *F. silaifolia* Essential Oil on Fibrablast Migration

As both *F. setifolia* and *F. silaifolia* EOs were effective in inhibiting dermatophyte growth, their effect on cell migration was also assessed. As shown in Figures 5 and 6, both oils increased cell migration at the highest concentration devoid of toxicity, although statistical significance was only attained for *F. silaifolia* EO at 25 μg/mL (Figure 6B).

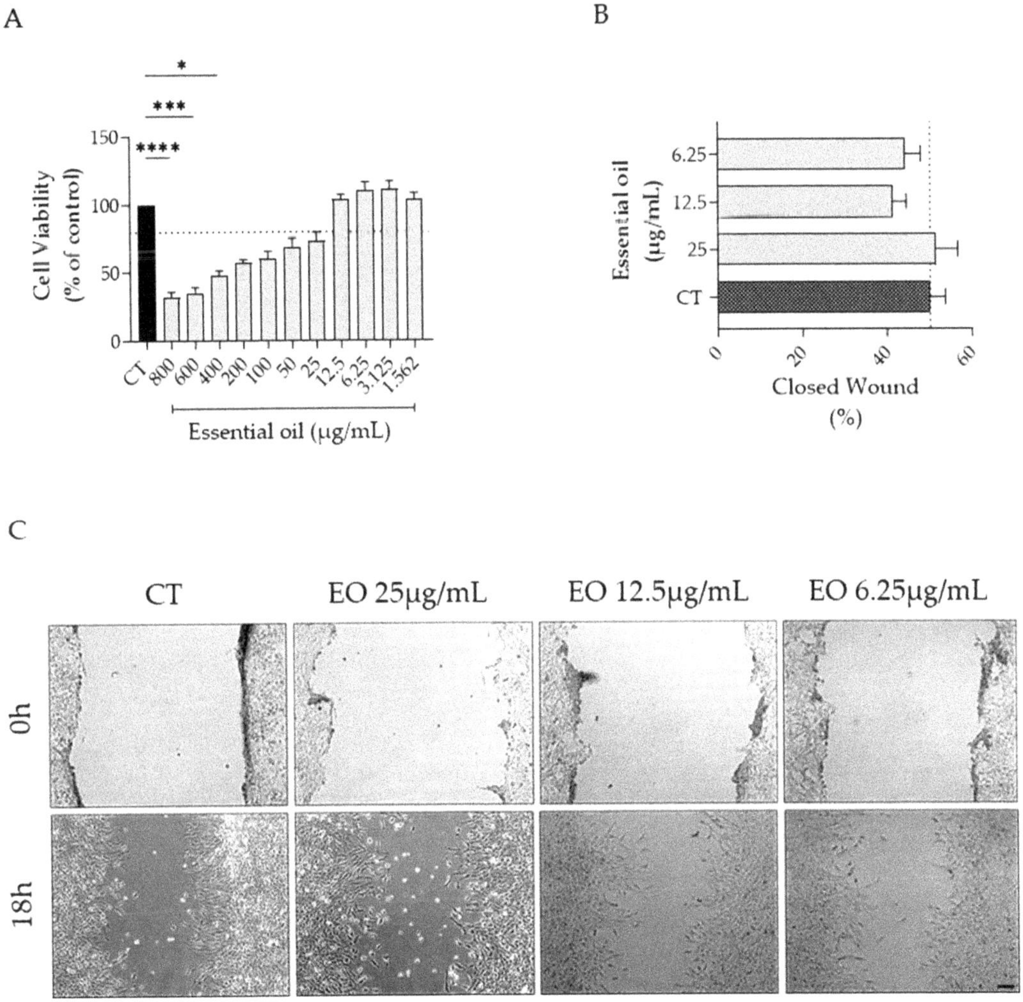

Figure 5. NIH/3T3 fibroblasts migration assessed using a scratch wound healing assay. (**A**) Cell viability following 24h of treatment with *F. setifolia* essential oil. (**B**) Percentage of closed wound at 18h post-scratch; the dashed line represents the percentage of wound closure in control (CT) cells. (**C**) Representative bright-field images of NIH/3T3 fibroblasts 0h and 18h post-scratch. Results as mean ± SEM of three independent assays performed in duplicate. * $p < 0.05$, *** $p < 0.001$ and **** $p < 0.0001$, compared to control (CT). Scale bar = 100 μm.

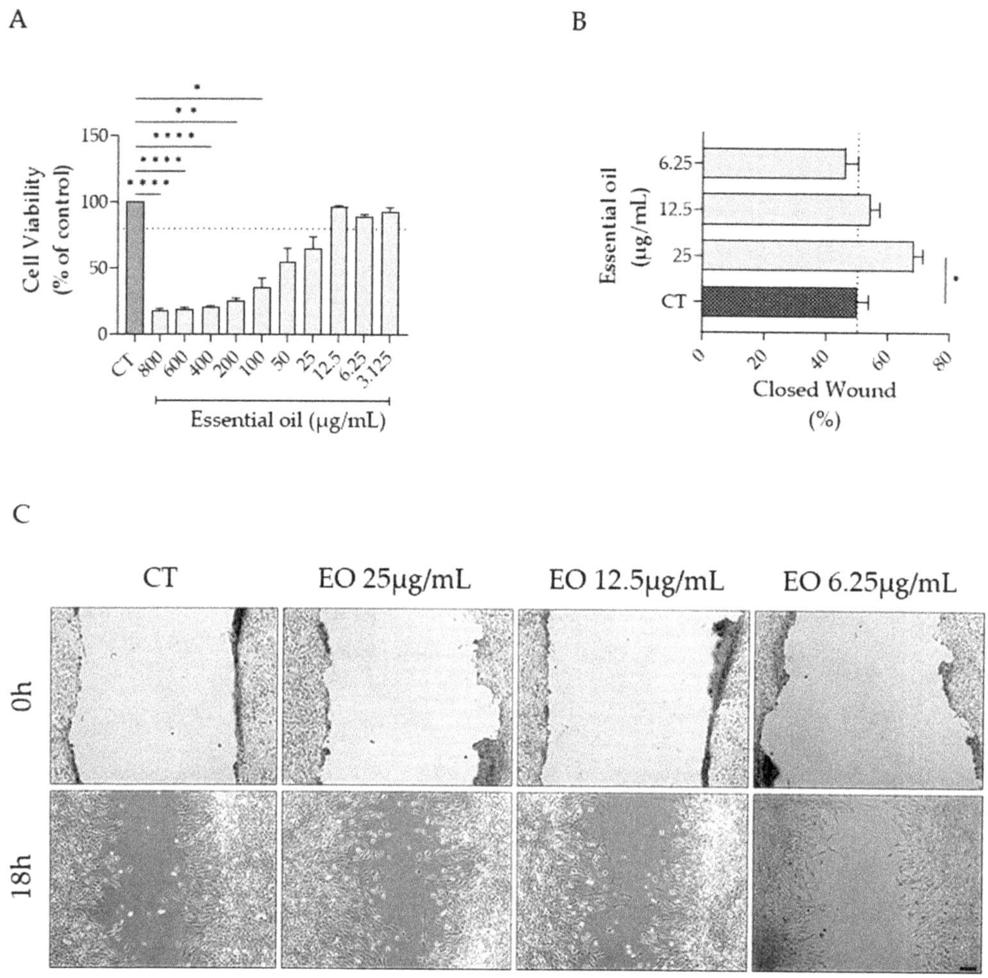

Figure 6. NIH/3T3 fibroblasts migration assessed by the scratch wound healing assay. (**A**) Cell viability following 24 h of treatment with *F. silaifolia* essential oil. (**B**) Percentage of closed wound at 18 h post-scratch; the dashed line represents the percentage of wound closure in control (CT) cells. (**C**) Representative bright-field images of NIH/3T3 fibroblasts 0h and 18h post-scratch. Results as mean ± SEM of three independent assays performed in duplicate. * $p < 0.05$, ** $p < 0.01$ and **** $p < 0.0001$, compared to control (CT). Scale bar = 100 μm.

3. Discussion

Currently, there is significant neglect regarding dermatophytosis, despite the fact that these infections can progress to severe chronic conditions and become invasive, affecting the dermis and, in some cases, internal organs. This is often due to treatment challenges, as conventional antifungal drugs require long-term administration to be effective. Otherwise, the infection tends to relapse, and the fungi develop resistance to antifungal agents. Given this issue, aromatic plants and their extracts, particularly EOs, have emerged as promising candidates for antifungal treatments, due to their high content in bioactive compounds.

Herein, the antifungal effect of EOs obtained from five *Ferulago* species was assessed using both collection and clinical strains of dermatophytes. Regarding the planktonic forms of dermatophytes, both *F. setifolia* and *F. silaifolia* EOs were quite effective, especially against *M. canis*, *T. mentagrophytes* and *T. rubrum*, although the latter was more active in presenting lower MICs. This aligns with previous research indicating that EOs, especially those rich in monoterpene hydrocarbons, exhibit strong antifungal properties. For example, a previous study conducted by Pinto et al. [16], the only one focusing on *Ferulago* and dermatophytes, demonstrated that the EO of *F. capillaris*, which is rich in α-pinene and limonene, exhibited significant antifungal activity against various pathogens, particularly dermatophytes. However, the activity was significantly lower than that observed in the present study.

Another interesting aspect to point out is that, although several studies have reported the antimicrobial effects of *Ferulago* EOs, the majority focus on bacteria [18] and direct comparisons across studies are quite challenging due to the use of different methodologies. On the other hand, regarding fungi, *Candida* species are generally targeted e.g., [14,15], with *C. albicans* being the most studied strain. In this regard, the EO obtained from the roots of *F. campestris*, with high amounts of α-pinene (58.3–75.0%), showed relevant antifungal effects with a MIC value of 78 μg/mL [14].

It is notable that α-pinene is quite relevant in the genus *Ferulago*, as several species are rich in this compound. Other relevant compounds include cis-chrysanthenyl acetate, and 2,3,6-trimethyl benzaldehyde. A literature review compiled by Pandey and colleagues supports that these compounds, among others, have the capacity to disrupt fungal cell membranes, increasing permeability and leading to cell lysis, making these EOs promising antifungal agents [19]. The high antifungal activity of *F. silaifolia* is likely attributed to the presence of major constituents such as α-pinene, which has been shown to have a potent effect on dermatophytes [20] as well as other fungal strains [21,22]. Additionally, synergistic interactions among various compounds, including minor ones, may also play a role in enhancing the overall activity of the EO and, therefore, cannot be disregarded.

Regarding biofilms, for most strains, an inhibitory effect of *F. silaifolia* EO on biofilm formation was observed, with greater reductions in biofilm biomass and extracellular matrix at MIC concentrations and, in some cases, even at lower concentrations. However, this trend was not so evident in pre-formed biofilms, as these tend to be more resistant. Nevertheless, for *T. rubrum*, a significant reduction in extracellular matrix deposition was observed at very low concentrations, with a statistical significance at MIC/8 (6.25 μg/mL). These results were supported by microscopic observations, which revealed notable morphological changes in biofilms treated with this EO, namely, an evident reduction in hyphae and mycelia density. Also, it was evident that the EO disrupted the biofilm structure, causing the fungal cells to disperse into the surrounding medium. These structural disruptions are critical as they compromise the integrity and functionality of the biofilm, making it more susceptible. In general, significant inhibitions of fungal growth and biofilm, particularly in *T. rubrum*, in the presence of *F. silaifolia* EO are noteworthy and given the importance of biofilms in enhancing fungal survival and resistance to therapy and host defenses, this finding is particularly significant. By eradicating this critical virulence factor, it is possible to control the pathogenesis of dermatophytes. Importantly, this study is the first to report the effect of *Ferulago* EOs on dermatophytes biofilms, providing a foundational basis for future studies. Indeed, studies on the effect of EOs on dermatophyte virulence factors are quite scarce. However, a recent study by Alves-Silva et al. provided new insights into the effects of EOs on dermatophytes, demonstrating that the EO of *Lavandula multifida* has a strong inhibitory effect on the formation of biofilms of dermatophytes and *Candida albicans* [23]. Additionally, the EO was effective in eradicating pre-formed biofilms in all tested strains, with dermatophytes showing greater susceptibility compared to *C. albicans*. Notably, the species *E. floccosum*, *T. mentagrophytes*, *M. gypseum*, and *T. rubrum* exhibited the highest sensitivity. These results support the strong antibiofilm potential of some EOs, endorsing their use in the development of medicinal plant-based antifungal products.

In the present study, *F. silaifolia* was also effective in promoting wound closure at 25 µg/mL. As far as we know, this is the first study to assess the effect of *Ferulago* EOs on fibroblast migration. However, a previous study highlighted the enzyme inhibitory potential, relevant in wound healing, of ethanolic extracts of *Ferulago* spp. with *F. macrosciadia* and *F. syriaca* exhibiting the most potent inhibitory effects on collagenase and elastase enzymes [24].

4. Materials and Methods

4.1. Plant Material and Essential Oils

Plants were collected from the locations summarized in Table 3 and identified by Prof. Dr. Hayri Duman from Gazi University's Faculty of Science, Department of Biology. Voucher specimens were deposited at the Herbarium of Ankara University, Faculty of Pharmacy (AEF) with the accession numbers referred to in Table 3.

Table 3. Date and site of collection of *Ferulago* species.

Species	Local Name	Collection Date	Site of Collection	Accession Number
F. cassia (1)	şeytankişnişi	20 June 2016	Konya: Beyşehir, Tınaztepe-Bozkır old road, 1st km, serpentine areas, in gladelike areas under *Pinus nigra*, 1555 m	AEF 28775
F. isaurica (2)	kargıkişnişi	21 June 2016	Antalya-Alanya, between Durbannaz-Banlıca, under *Pinus brutia* forest, calcaerous rocks, 837 m	AEF 28778
F. setifolia (3)	kılkişniş	14 July 2017	Erzincan: Üzümlü district, Karakaya Town, between Karakaya-Tekçam Highland, high mountain steppe, 2047 m	AEF 28770
F. silaifolia (4)	ulukişniş	7 June 2016	Bursa, Uludağ road, 100 km to National Park, under *Castanea sativa* Mill. and *Pinus nigra* forest, 837 m	AEF 28771
F. syriaca (5)	kırkişnişi	22 June 2016	Hatay: Harbiye-Şenköy roa, ca. 5th–6th km, around maquis, 452 m	AEF28779

The EOs were isolated from the fruits of different *Ferulago* species and analyzed using gas chromatography and gas chromatography associated with mass spectrometry, as previously described [17]. Dilutions of each EO were prepared in dimethyl sulfoxide (DMSO) and added to the respective experimental conditions to achieve final concentrations ranging from 50 to 800 µg/mL, with the DMSO concentration in the experimental setups not exceeding 1%.

4.2. Antifungal Effect of Essential Oils on Planktonic Fungi

The antifungal activity of *Ferulago* EOs was assessed against dermatophytes from the Colección Española de Cultivos Tipo (CECT), namely, *Microsporum gypseum* CECT 2908, *Trichophyton mentagrophytes* var. *interdigitale* CECT 2958, *T. rubrum* CECT 2794 and *T. verrucosum* CECT 2992, as well as three clinical strains, which are part of the fungal collection of the laboratory: *Epidermophyton floccosum* FF9, *M. canis* FF1 and *T. mentagrophytes* FF7, isolated from patients' nails and skin.

First, the fungal strains were cultured on Sabouraud dextrose agar (SDA) to ensure optimal growth characteristics and purity. To determine the minimal inhibitory concentration (MIC) and the minimal lethal concentration (MLC) of the oils, the macrodilution method was used, according to Clinical and Laboratory Standards Institute guidelines M38-A2 for filamentous fungi [25]. Dilutions of each EO were prepared in DMSO (50 to 800 μg/mL). Then, using recent cultures of each strain, a fungal suspension was prepared and adjusted to 1 McFarland. The test tubes, containing the respective inoculum and EO concentration, were incubated aerobically at 30 °C for 7 days and MICs were determined by visualizing the turbidity of each test tube. The MIC was determined as the lowest concentration of EO capable of inhibiting the growth of the fungus. After MIC readings, aliquots (20 μL) of broth were taken from each negative tube to evaluate MLCs. Plates containing SDA medium and the inoculated aliquots were incubated at 30 °C for 7 days. The MLC was considered to be the lowest concentration of EO that, after incubation, showed no fungal growth on the plate. For each strain tested, the growing conditions and the sterility of the medium were checked (negative control). The innocuity of DMSO (positive control) was also checked at the highest concentration tested. All experiments were performed in triplicate and repeated if the results differed.

4.3. Evaluation of Essential Oils' Effect on Biofilm Formation

After MICs and MLCs of the fungi were determined, the most promising *Ferulago* species was chosen to evaluate the effect of its EO on the formation and disruption of biofilms formed by the most common species to cause infection, namely, *M. canis*, *E. floccosum* and *T. rubrum*. To evaluate the effect of the EO on the formation of dermatophyte biofilms, 100 μL of conidia suspension adjusted to 1 McFarland was added to sterile 96-well flat-bottomed polystyrene microtiter plates and incubated at 37 °C for 3 h for conidia to adhere. The saline was then removed, and the wells were washed with PBS (0.8% NaCl, 0.02% KH_2PO_4. 0.31% $Na_2HPO_4 \cdot 12\ H_2O$ and 0.02% KCl; pH 7.4) to remove non-adherent cells. Finally, 200 μL of RPMI medium containing different concentrations of the selected EO were added and incubated at 37 °C for 72 h. A negative control containing non-inoculated medium and a positive control with 1% DMSO in inoculated medium (without EO) were also included. To evaluate the effect of the EOs on the formation of dermatophyte biofilms, both biofilm mass and extracellular matrix deposition were quantified, in the presence of EOs' MIC and lower concentrations (up to MIC/8), using crystal violet and safranin assays, respectively.

- Biofilm mass:

After discarding the medium, the biofilms were washed with PBS to remove the non-adherent cells. Then, to fix the biofilms, methanol was added for 10 min, followed by 100 μL of a 0.5% crystal violet solution for 15 min. After removing the solution, the biofilms were washed twice with sterile water and 150 μL of 33% acetic acid was added to dissolve the crystals. Finally, the volume was transferred to another well and the absorbance (*Abs*) was measured at 620 nm [26].

- Extracellular matrix:

After discarding the medium, the biofilms were washed with PBS to remove the non-adherent cells. Then 100 μL of a 0.5% safranin red solution was added for 5 min. After removing the solution, the biofilms were washed twice with sterile PBS to remove the unbound dye. Then, 150 μL of 33% acetic acid was added to dissolve the crystals and, finally, the volume was transferred to a new well and the absorbance was measured at 520 nm [27].

4.4. Evaluation of Essential Oils' Effect on Biofilm Disruption

To evaluate the effect of the EO on pre-formed dermatophyte biofilm disruption, suspensions of dermatophyte conidia (200 μL) adjusted to 1 McFarland were added to sterile 96-well flat-bottomed polystyrene microtiter plates and incubated at 37 °C for 3 h to allow the conidia to adhere. After removing the saline, the cells were washed with PBS to remove the non-adherent cells and 200 μL of sterile RPMI medium was added to the plates and incubated at 37 °C for 72 h. The medium was removed, followed by a washing step with PBS. Finally, 200 μL of RPMI with different concentrations of EO (6.25–50 μg/mL) was added to each well and incubated at 37 °C for 24 h. A negative control containing non-inoculated medium and a positive control with 1% DMSO in inoculated medium (without EO) were also included. To evaluate the effect of EO on disrupting the dermatophyte's biofilm, both biofilm mass and extracellular matrix deposition were quantified, as referred to in Section 4.3.

4.5. Effect of the Essential Oil on Fungal Morphology

Fungal morphological alterations in the presence of the most effective EO were evaluated using brightfield microscopy (Motic AE 2000) with a 20× objective (LWD 20×/0.3 Ph).

4.6. Cell Viability Assay

The two most effective EOs in inhibiting dermatophytes growth were used to assess their effect on the viability of NIH/3T3 fibroblasts (mouse embryonic fibroblast cell line—ATCC CRL-1658), prior to investigating their impact on cell migration. For that, 50,000 cells/mL were plated on 48-well microplates in DMEM (Dulbecco's Modified Eagle Medium) with 10% FBS and 1% Penicillin/Streptomycin and allowed to stabilize for 12 h in a culture chamber (37 °C, 5% CO_2). Then, different concentrations of the EOs (1.562–800 μg/mL) were added to the respective wells for 24 h, after which the medium was removed and fresh medium containing resazurin (1:10) was added for 4 h. The absorbance at 570 nm with a reference filter of 620 nm was registered in an automated plate reader (SLT, Wels, Austria). Cell viability was determined using the following equation:

$$\textit{Cell viability}\ (\%) = \textit{Abs T}/\textit{Abs Ct} \times 100$$

where *Abs T* is the absorbance (difference between 570 and 620 nm) in the different experimental conditions and *Abs Ct* is the absorbance in the control cells (no EO added).

4.7. Scratch Wound Assay

Cell migration was assessed using a scratch wound assay, as previously reported. Briefly, fibroblasts were plated at 300,000 cells/mL in 12-well plates and allowed to grow for 12 h. Then, a scratch was induced with a 200 μL pipette tip, and the detached cells were washed off with PBS. New culture medium with or without the EOs (6.25–25 μg/mL) was added and images were acquired at 0h and 18h post-scratch using a phase-contrast Zeiss Axio HXP IRE 2 microscope (Carl Zeiss AG, Jena, Germany) at 10× magnification (EC Plan-Neofluar 10×/0.3 Ph1) [28,29]. The wound area was measured using ImageJ/Fiji v1.54p software and the results were obtained using the following equation:

$$\textit{Wound closure}\ (\%) = A\ 0\text{h} - A\ x\text{h}/A\ 0\text{h} \times 100$$

where *A* 0h is the area of the wound 0h after the scratch and *A* xh is the area 18h post-scratch.

4.8. Statistical Analysis

All statistical analysis were performed using GraphPad Prism version 9.5.0 (GraphPad Software, San Diego, CA, USA) and the results are presented as mean values ± SEM (standard error of the mean) from at least three independent experiments performed in duplicate. Statistical significance was determined using a two-way ANOVA followed by Šídák's or Dunnet's multiple comparison test. Results with a $p \leq 0.05$ were considered statistically significant.

5. Conclusions

This study investigated the antifungal potential of EOs extracted from *Ferulago* species from Türkiye in inhibiting and disrupting biofilms formed by dermatophytes. The EO of *F. silaifolia* demonstrated significant antifungal activity, particularly against *M. canis* and *T. rubrum*, with MIC values of 50 µg/mL. Notably, *T. rubrum* was the most susceptible fungal species, exhibiting marked morphological changes, including reduced hyphal growth and mycelial density. These findings are promising and suggest that *Ferulago* EOs could serve as alternative antifungal therapies by complementing or replacing conventional treatments while reducing the dosage of synthetic drugs and their associated side effects. This study is the first to highlight the potential of *Ferulago* EOs as effective antifungal agents against dermatophytes, particularly in targeting biofilms, offering a novel approach to treating resistant fungal infections. Moreover, their wound-healing potential adds additional relevance as it may help avoid systemic infection.

Author Contributions: Conceptualization, C.S.K., L.S. and M.Z.; methodology, I.C., J.A.-S., B.D. and D.K., validation, C.S.K., L.S. and M.Z.; formal analysis, M.Z.; investigation, I.C., J.A.-S., B.D. and D.K.; resources, C.S.K., L.S. and M.Z.; writing—original draft preparation, I.C.; writing—review and supervision, L.S. and M.Z.; funding acquisition, M.Z. All authors have read and agreed to the published version of the manuscript.

Funding: This research was funded by Fundação para a Ciência e Tecnologia, under the project PATCH 022.05810.PTDC and projects UIDB/00102/2020 and UIDP/00102/2020.

Institutional Review Board Statement: Not applicable.

Informed Consent Statement: Not applicable.

Data Availability Statement: The data presented in this study are available on request from the corresponding author.

Conflicts of Interest: The authors declare no conflicts of interest.

Abbreviations

The following abbreviations are used in this manuscript:

MIC	Minimal Inhibitory Concentration
MLC	Minimal Lethal Concentration

References

1. Gaffi Global Action for Fungal Infections Fungal Disease Frequency. Available online: https://gaffi.org/why/fungi-fungal-infections/ (accessed on 1 December 2024).
2. Bongomin, F.; Gago, S.; Oladele, R.; Denning, D. Global and Multi-National Prevalence of Fungal Diseases—Estimate Precision. *J. Fungi* **2017**, *3*, 57. [CrossRef]

3. Martinez-Rossi, N.M.; Bitencourt, T.A.; Peres, N.T.A.; Lang, E.A.S.; Gomes, E.V.; Quaresemin, N.R.; Martins, M.P.; Lopes, L.; Rossi, A. Dermatophyte Resistance to Antifungal Drugs: Mechanisms and Prospectus. *Front. Microbiol.* **2018**, *9*, 1108. [CrossRef]
4. Deng, R.; Wang, X.; Li, R. Dermatophyte infection: From fungal pathogenicity to host immune responses. *Front. Immunol.* **2023**, *14*, 1285887. [CrossRef] [PubMed]
5. Mercer, D.K.; Stewart, C.S. Keratin hydrolysis by dermatophytes. *Med. Mycol.* **2019**, *57*, 13–22. [CrossRef] [PubMed]
6. Jartarkar, S.R.; Patil, A.; Goldust, Y.; Cockerell, C.J.; Schwartz, R.A.; Grabbe, S.; Goldust, M. Pathogenesis, Immunology and Management of Dermatophytosis. *J. Fungi* **2021**, *8*, 39. [CrossRef] [PubMed]
7. Sahoo, A.; Mahajan, R. Management of tinea corporis, tinea cruris, and tinea pedis: A comprehensive review. *Indian Dermatol. Online J.* **2016**, *7*, 77. [CrossRef]
8. Markantonatou, A.-M.; Samaras, K.; Vyzantiadis, T.-A. Dermatophytic Biofilms: Characteristics, Significance and Treatment Approaches. *J. Fungi* **2023**, *9*, 228. [CrossRef]
9. Aditya, K.G.; Jennifer, E.R.; Melody, C.; Elizabeth, A.C. Dermatophytosis: The Management of Fungal Infections. *SKINmed Dermatol. Clin.* **2005**, *4*, 305–310. [CrossRef]
10. Dabaghian, F.; Aalinezhad, S.; Kesheh, A.R.; Azargashb, N.; Ansari, R.; Ardekani, M.R.S.; Emami, S.A.; Khanavi, M.; Delnavazi, M.R. A review of the ethnomedicinal, phytochemical, and pharmacological properties of the *Ferulago* genus based on Structure–Activity Relationship (SAR) of coumarins. *DARU J. Pharm. Sci.* **2024**, *32*, 825–899. [CrossRef]
11. Karakaya, S.; Göğer, G.; Kılıç, C.S.; Demirci, B. Türkiye'de Yetişen Ferulago blancheana Post. (Apiaceae) Türünün Toprak Üstü, Çiçek ve Köklerinden Elde Edilen Uçucu Yağların İçeriklerinin ve Antimikrobiyal Aktivitesinin Biyootografi Yöntemiyle Tanımlanması. *Turk. J. Pharm. Sci.* **2016**, *13*, 51–59. [CrossRef]
12. Badalamenti, N.; Ilardi, V.; Rosselli, S.; Bruno, M. The ethnobotany, phytochemistry and biological properties of genus *Ferulago*–A review. *J. Ethnopharmacol.* **2021**, *274*, 114050. [CrossRef] [PubMed]
13. Rahimpour, Y.; Delazar, A.; Asnaashari, S.; Asgharian, P. The Genus *Ferulago*: A Review on Ethnopharmacology, Phytochemistry, and Pharmacology. *Iran. J. Pharm. Res. IJPR* **2021**, *20*, 352–377. [CrossRef]
14. Cecchini, C.; Coman, M.M.; Cresci, A.; Tirillini, B.; Cristalli, G.; Papa, F.; Sagratini, G.; Vittori, S.; Maggi, F. Essential oil from fruits and roots of *Ferulago campestris* (Besser) Grecescu (Apiaceae): Composition and antioxidant and anti-*Candida* activity. *Flavour Fragr. J.* **2010**, *25*, 493–502. [CrossRef]
15. Demirci, B.; Dilmaç, E.; Kırcı, D.; Demirci, F.; Kılıç, C.S.; Duman, H.; Gürbüz, İ. Chemical and antimicrobial characterization of essential oils obtained from aerial part, root and fruit of *Ferulago longistylis* Boiss., an endemic species. *Nat. Volatiles Essent. Oils* **2020**, *7*, 18–25. [CrossRef]
16. Pinto, E.; Hrimpeng, K.; Lopes, G.; Vaz, S.; Gonçalves, M.J.; Cavaleiro, C.; Salgueiro, L. Antifungal activity of *Ferulago capillaris* essential oil against *Candida, Cryptococcus, Aspergillus* and dermatophyte species. *Eur. J. Clin. Microbiol. Infect. Dis.* **2013**, *32*, 1311–1320. [CrossRef]
17. Kırcı, D.; Demirci, B.; Kılıç, C.S.; Gürbüz, İ.; Duman, H. Essential Oil Composition of Fruits of Eight *Ferulago* Species Growing in Türkiye and Multivariate Statistical Analyses. *Chem. Biodivers.* **2023**, *20*, e202301098. [CrossRef]
18. Ghasemi Pirbalouti, A.; Izadi, A.; Malek Poor, F.; Hamedi, B. Chemical composition, antioxidant and antibacterial activities of essential oils from *Ferulago angulata*. *Pharm. Biol.* **2016**, *54*, 2515–2520. [CrossRef]
19. Pandey, A.K.; Singh, P. The Genus *Artemisia*: A 2012–2017 Literature Review on Chemical Composition, Antimicrobial, Insecticidal and Antioxidant Activities of Essential Oils. *Medicines* **2017**, *4*, 68. [CrossRef]
20. Cavaleiro, C.; Salgueiro, L.; Gonçalves, M.-J.; Hrimpeng, K.; Pinto, J.; Pinto, E. Antifungal activity of the essential oil of *Angelica major* against *Candida, Cryptococcus, Aspergillus* and dermatophyte species. *J. Nat. Med.* **2015**, *69*, 241–248. [CrossRef]
21. Nóbrega, J.R.; Silva, D.d.F.; Andrade Júnior, F.P.d.; Sousa, P.M.S.; de Figueiredo, P.T.R.; Cordeiro, L.V.; Lima, E.d.O. Antifungal action of α-pinene against *Candida* spp. isolated from patients with otomycosis and effects of its association with boric acid. *Nat. Prod. Res.* **2021**, *35*, 6190–6193. [CrossRef]
22. de Barros, D.B.; Lima, L.d.O.e.; Silva, L.A.; Fonseca, M.C.; Diniz-Neto, H.; Rocha, W.P.d.S.; Beltrão, G.V.d.M.; Castellano, L.R.C.; Guerra, F.Q.S.; da Silva, M.V. Efeito antifúngico de α-pineno isolado e em associação com antifúngicos frente às cepas de *Candida albicans*. *Res. Soc. Dev.* **2022**, *11*, e58711427748. [CrossRef]
23. Alves-Silva, J.; Zuzarte, M.; Cavaleiro, C.; Salgueiro, L. Antibiofilm Effect of *Lavandula multifida* Essential Oil: A New Approach for Chronic Infections. *Pharmaceutics* **2023**, *15*, 2142. [CrossRef] [PubMed]
24. Akkol, E.K.; Günbatan, T.; Gürbüz, İ.; Duman, H.; Kılıç, C.S.; İlhan, M. In Vitro Enzyme Inhibitory Activity of Ten *Ferulago* W. Koch Species Growing in Turkey. *Braz. Arch. Biol. Technol.* **2022**, *65*, e22210207. [CrossRef]
25. HR, J. Reference Method for Broth Dilution Antifungal Susceptibility Testing of Filamentous Fungi; Approved Standard. Second Edition. M38-A2. *Clin. Lab. Stand. Inst.* **2008**, *28*, 1–35.
26. Castelo-Branco, D.d.S.C.M.; de Aguiar, L.; Araújo, G.d.S.; Lopes, R.G.P.; Sales, J.d.A.; Pereira-Neto, W.A.; Pinheiro, A.d.Q.; Paixão, G.C.; Cordeiro, R.d.A.; Sidrim, J.J.C.; et al. In vitro and ex vivo biofilms of dermatophytes: A new panorama for the study of antifungal drugs. *Biofouling* **2020**, *36*, 783–791. [CrossRef] [PubMed]

27. Costa-Orlandi, C.B.; Sardi, J.C.O.; Santos, C.T.; Fusco-Almeida, A.M.; Mendes-Giannini, M.J.S. In vitro characterization of *Trichophyton rubrum* and *T. mentagrophytes* biofilms. *Biofouling* **2014**, *30*, 719–727. [CrossRef]
28. Martinotti, S.; Ranzato, E. Scratch Wound Healing Assay. In *Epidermal Cells: Methods and Protocols;* Humana: New York, NY, USA, 2019; pp. 225–229.
29. Suarez-Arnedo, A.; Torres Figueroa, F.; Clavijo, C.; Arbeláez, P.; Cruz, J.C.; Muñoz-Camargo, C. An image J plugin for the high throughput image analysis of in vitro scratch wound healing assays. *PLoS ONE* **2020**, *15*, e0232565. [CrossRef]

antibiotics

MDPI

Article

Agave amica (Medik.) Thiede & Govaerts (Asparagaceae)—Insights into Its Valuable Phenolic Profile and *In Vitro* Antimicrobial, Antibiofilm, Antioxidative, and Antiproliferative Properties

Mihaela Niculae [1], Daniela Hanganu [2,*], Ilioara Oniga [2], Sergiu-Alexandru Burcă [2], Brîndușa Tiperciuc [3], Irina Ielciu [4], Emoke Pall [1], Timea Bab [2,5], Ramona Flavia Burtescu [5], Mihaela Andreea Sava [6] and Daniela Benedec [2]

1 Department of Clinical Sciences, University of Agricultural Sciences and Veterinary Medicine Cluj-Napoca, 400372 Cluj-Napoca, Romania; mihaela.niculae@usamvcluj.ro (M.N.); emoke.pall@usamvcluj.ro (E.P.)
2 Department of Pharmacognosy, Faculty of Pharmacy, "Iuliu Hattieganu" University of Medicine and Pharmacy, 400010 Cluj-Napoca, Romania; ioniga@umfcluj.ro (I.O.); burca.sergiu.alex@elearn.umfcluj.ro (S.-A.B.); bab.timea.henrietta@elearn.umfcluj.ro (T.B.); dbenedec@umfcluj.ro (D.B.)
3 Department of Pharmaceutical Chemistry, "Iuliu Hațieganu" University of Medicine and Pharmacy, 41 V. Babeş Street, 400012 Cluj-Napoca, Romania; btiperciuc@umfcluj.ro
4 Department of Pharmaceutical Botany, Faculty of Pharmacy, "Iuliu Haţieganu" University of Medicine and Pharmacy, 400337 Cluj-Napoca, Romania; irina.ielciu@umfcluj.ro
5 PlantExtrakt Ltd., 407059 Cluj-Napoca, Romania; ramona.burtescu@plantextrakt.ro
6 Transylvania School of Botanic Art & Illustration, Copșa Mare, 557046 Sibiu, Romania; mihaelaandreea.sava@yahoo.com
* Correspondence: dhanganu@umfcluj.ro

Abstract: Background/Objectives: *Agave amica* (Medik.) Thiede & Govaerts (Asparagaceae family) is an ornamental bulbous species, widely used for its fragrance but less studied as a medicinal species. This study is aimed at assessing the phenolic profile and selected biological properties of ethanolic extracts obtained from the aerial parts and bulbs of *A. amica* cultivated in Romania. **Methods**: The phenolic composition was characterized by spectrophotometric methods and LC/MS analysis. The antioxidant activity was evaluated by DPPH (2,2-diphenyl-1-picrylhydrazyl radical scavenging capacity) and FRAP (Ferric reducing antioxidant power) tests, while the *in vitro* antimicrobial capacity was investigated by the agar-well diffusion, the broth microdilution, and the antibiofilm assays. Cytotoxicity was tested on a colorectal adenocarcinoma cell line (DLD-1) by a CCK-8 assay. **Results**: Both ethanolic extracts showed important polyphenol content and caffeic acid as their main compound. Significantly higher amounts of total polyphenols (44.25 ± 1.08 mg/g), tannins (12.55 ± 0.34 mg/g), flavonoids (9.20 ± 0.19 mg/g), caffeic acid derivatives (19.95 ± 0.05 mg/g), and also antioxidant activity (IC_{50} = 0.82 ± 0.02 mg/mL, and 79.75 ± 1.80 μM TE/g, respectively) were found for the aerial parts extract compared to the bulbs one ($p < 0.001$). Notable anti-*Candida albicans* activity and moderate efficacy against *Listeria monocytogenes* and *Staphylococcus aureus* were displayed by both extracts against planktonic cells and biofilm. A dose-dependent cytotoxicity towards the colorectal adenocarcinoma cell line was recorded as well. **Conclusions**: This study brings novelty to the scientific literature by characterizing the phenolic profile and *in vitro* antimicrobial, antibiofilm, antioxidant, and antiproliferative activities of the ethanolic extracts obtained from *A. amica*, thus highlighting this herbal species's medicinal potential.

Keywords: *A. amica*; phenols; antibacterial; antifungal; antibiofilm; antioxidant; cytotoxic

https://doi.org/10.3390/antibiotics14070638

1. Introduction

According to a 2022 World Health Organization (WHO) report, resistance to antimicrobial agents remains a major threat to global health due to an increasing number of microbial infections that are refractory to antimicrobial therapy, together with the rapid spread of multidrug-resistant strains at different community levels and settings. Several mechanisms entitle organisms to manifest resistance against conventional antibiotics as well as towards the host's innate immune responses [1]. Among them, biofilms are an established hallmark of microbial adaptation, persistence, and virulence [1,2]. Furthermore, biofilm-forming ability was demonstrated for numerous pathogenic and opportunistic organisms [2]. To tackle this worrying and complex phenomenon, a multifaceted approach that includes targeted strategies aiming to reduce antimicrobial usage and implement surveillance and monitoring systems, as well as promoting the search and discovery of alternative therapies, is recommended [3]. In fact, developing new antimicrobial drugs and alternative therapies has crucial importance, with plant-derived extracts suggested by numerous studies as a potential source of antimicrobial bioactive components that can address several limitations and disadvantages of conventional antibiotics [4]. Certain bioactive metabolites produced by plants have shown not only antimicrobial efficacy but also antibiofilm activity against different bacterial strains [5].

Agave amica (Medik.) Thiede & Govaerts, tuberose, is a member of the Asparagaceae family, an ornamental bulbous species especially known for its pleasant fragrance and use in the perfumery and cosmetic industry [6–10]. Formerly known as *Polianthes tuberosa* L., based on recent phylogenetic studies and molecular data, it has been removed from the Agavaceae and Amaryllidaceae families systematically and included in the family Asparagaceae and subfamily Agavoideae [8–11]. This species is the only taxon of the former genus *Polianthes*, cultivated across the globe and used for medicinal, ornamental, and ceremonial practices since ancient times [11,12]. It is a perennial monocotiledonate species, native to Mexico and cultivated in China, Egypt, South Africa, India, and Japan for its fragrant flowers and the essential oil it yields [6,13–15]. The bulbs of the species produce flowers after four years of growth, only once, in late summer, when spikes develop from them [16]. The flowers are white, night-blooming, with single or double fragrant tepals, and are funnel-shaped while the leaves are sessile, elongated, arched, and ribbon-shaped [16–18]. Nowadays, it is cultivated widely in tropical, subtropical, and subtemperate areas for the fragrance of its flowers, used as flower decorations, but especially for their fragrance, with tuberose fragrant oils being one of the most highly demanded perfumes globally [15,19]. Tuberose flowers are completely edible and can be used as ingredients in different ways [16,17]. Traditionally, it is described not only as an ornamental species but also as a medicinal plant with flowers and bulbs used for its diuretic, emetic, anti-gonorrhea, anti-insomnia, antispasmodic, and anti-rashes effects [13,20]. The scientific literature suggests that this species is a promising source of bioactive compounds, such as tuberolactone and three flavonoids (kaempferol, kaempferol-3-O-xyloside, and kaempferol-3-4′-O-dixyloside), polysaccharides, polianthosides B-G, spirostanols, furostanols, steroidal glycosides, and saponins [14,21–23]. Volatile metabolites belong to the classes of terpenoids, benzenoids, phenylpropanoids, and fatty acid derivatives [24]. All these bioactive metabolites were linked to anti-inflammatory, antioxidant, antimicrobial, cytotoxic, membrane stabilizing, and thrombolytic activities [14,19,25,26].

Previously performed studies on this species aimed to evaluate the antibacterial, antitumor, and anti-inflammatory activities of its flower extracts including in silver, gold, and zinc nanoparticles [20,27–29]. The flower extracts also exhibited antimicrobial activity [21,30]. There are few other studies focusing on other parts of the species, such as the bulbs, attributing the antioxidant and antimicrobial activity to alkaloids [13] or phenols [18], or the leaves, assessing their antioxidant or antibacterial capacity attributed to polyphenols [14,31]. Only one study aims to compare the antibacterial activity of flowers, stems, and bulbs [32]. At the same time, some studies focus on the volatile metabolites of the species, studying their flower extracts [6,17,33].

In Romania, the *A. amica* species is cultivated and perfectly adapted to the pedoclimatic conditions in the whole country but research on it is very limited [34–36]. Taking all these factors into consideration, it appears there is currently little information on the primary and secondary metabolites of *A. amica*, and evidence of its biological activities is also reduced [16,17]. To the best of our knowledge, no previous studies on the *A. amica* polyphenolic profile and the antioxidant, antimicrobial, antibiofilm, and cytotoxic/antiproliferative properties of ethanolic extracts have been performed in Romania. Moreover, no previously performed study has aimed to study the comparison between extracts obtained from the aerial parts (Pta) and bulbs (Ptb). In this context, the present study's main purpose is to bring novelty to the scientific literature by characterizing the ethanolic extracts obtained from *A. amica* cultured in Romania, with a special focus on the phenolic profile and selected activities, namely antibacterial, antifungal, antibiofilm, antioxidant, and cytotoxic, aiming therefore to provide scientific arguments for the medicinal uses of the species.

2. Results

2.1. Quantitative Evaluation of Phenolic Compounds and Antioxidant Activity

The contents of total polyphenols (TPC) expressed in mg GAE/g, flavonoids (FC) expressed in mg RE/g, and caffeic acid derivatives (CADC) expressed in mg CAE/g, as well as the results obtained for the evaluation of the antioxidant activity determined by DPPH and FRAP methods of the two *A. amica* extracts are presented in Table 1. It was found that the extract obtained from the aerial parts of *A. amica* (Pta) contained a significantly higher amount ($p < 0.001$) of active polyphenolic principles: total polyphenols (44.25 ± 1.08 mg/g), tannins (12.55 ± 0.34 mg/g), flavonoids (9.20 ± 0.19 mg/g), and caffeic acid derivatives (19.95 ± 0.05 mg/g) compared to the bulb extract (TPC: 29.60 ± 0.89 mg/g, TC: 9.85 ± 0.15 mg/g, FC: 5.65 ± 0.13 mg/g, and CADC: 8.08 ± 0.12 mg/g, respectively). Comparative analysis of these quantitative results showed highly statistically significant differences ($p < 0.001$) between the two *A. amica* samples ($p < 0.001$).

Table 1. Polyphenolic content and antioxidant activity of *A. amica* extracts.

A. amica Extracts	TPC (mg GAE/g)	TC (mg GAE/g)	FC (mg RE/g)	CADC (mg CAE/g)	DPPH (IC_{50} mg/mL)	FRAP (μM TE/g)
Pta	44.25 ± 1.08 [e]	12.55 ± 0.34	9.20 ± 0.19	19.95 ± 0.05	0.82 ± 0.02 [c]	79.75 ± 1.80
Ptb	29.60 ± 0.89 [a,e]	9.85 ± 0.15 [a]	5.65 ± 0.13 [a]	8.08 ± 0.12 [a]	1.65 ± 0.01 [b,c]	29.62 ± 0.37 [d]
Trolox	-	-	-	-	0.011 ± 0.002	-

Notes: Each value is the mean ± SD of three independent measurements; TPC: total polyphenolic content; TC: tannin content; FC: flavonoid content; CADC: caffeic acid derivatives content; GAE, RE, CAE, and TE: gallic acid, rutin, caffeic acid, and Trolox equivalents. Lowercase letters in the same row indicate significant differences: [a] $p < 0.001$ (Ptb vs. Pta); [b] $p < 0.001$ (Ptb vs. Pta); [c] $p < 0.001$ (Trolox vs. Ptb and Pta); [d] $p < 0.001$ (Ptb vs. Pta); [e] $p < 0.001$ (TPC vs. Ptb and Pta).

The results of the antioxidant activity assays performed on the *A. amica* extracts obtained from aerial parts and bulbs showed that the Pta extract presented higher antioxidant activity (IC_{50} = 0.82 ± 0.02 mg/mL, and 79.75 ± 1.80 μM TE/g, respectively) than the Ptb extract (IC_{50} = 1.65 ± 0.01 mg/mL, and 29.62 ± 0.37 μM TE/g, respectively) by both methods used. Statistical analysis of the results from the two antioxidant activity assays showed very significant differences ($p < 0.001$) between the *A. amica* extracts ($p < 0.001$), and in the case of the DPPH test, the results were less effective than the Trolox standard ($p < 0.001$).

2.2. LC–MS Analysis

The results obtained using LC–MS identification and quantification of the individual components of the two *A. amica* samples can be found in Table 2 and Supplementary Materials Figures S1–S16.

Table 2. Phenolic compounds identified by LC–MS analysis in the composition of the tested *A. amica* extracts (μg/g).

Compound	Retention Time, min		*m/z* and Main Transition		*A. amica* Content (μg/g)	
	Reference	Separated Compound	Reference	Separated Compound	Ptb	Pta
Caffeic acid	13.8	13.6	179.0 > 135.0	179.0 > 135.0	4230 ± 10.40	8890 ± 22.20
Chlorogenic acid	12.0	12.0	353.0 > 191.0	353.0 > 191.0	1501 ± 3.70	4000 ± 9.90
Salicylic acid	23.5	23.4	137.0 > 93.0	137.0 > 93.0	-	3400 ± 8.40
Carnosol	30.7	30.4	329.1 > 285.1	329.1 > 285.1	37 ± 0.20	34 ± 0.20
Chrysin	29.7	29.8	253.0 > 143.0	253.0 > 253.0	1380 ± 3.40	1920 ± 4.70
Hyperoside	20.3	20.4	463.1 > 300.0	463.1 > 300.0	1510 ± 3.70	1640 ± 4.10
Luteolin-7-*O*-glucoside	19.9	19.8	447.0 > 284.9	447.0 > 284.9	72 ± 0.40	2290 ± 5.70
Luteolin	26.9	26.8	287.0 > 153.0	287.0 > 153.0	-	2300 ± 5.70
Naringenin	26.3	26.1	271.0 > 119.0	271.0 > 119.0	-	45 ± 0.20
Quercetin	25.4	25.4	300.9 > 151.0	300.9 > 151.0	-	20 ± 0.09

Note: values represent the mean ± standard deviations of three independent measurements.

The compositions of these samples highlight that caffeic acid was the main compound, with its highest amount being found in the Pta samples (8890 ± 22.2 μg/g). This compound is accompanied by other phenolic acids important for biological effects, namely chlorogenic acid and salicylic acid (present only in the Pta sample). Flavonoids are represented by both glycosylated and free structures (aglycons), such as hyperoside, luteolin-7-*O*-glucoside, chrysin, luteolin, naringenin, and quercetin. For the Pta sample, luteolin was the flavonoidic compound found with the highest amount (2300 ± 5.7 μg/g), followed by luteolin-7-*O*-glucoside (2290 ± 5.7 μg/g), chrysin (1920 ± 4.7 μg/g), hyperoside (1640 ± 4.1 μg/g), naringenin (45 ± 0.2 μg/g), and quercetin (20 ± 0.09 μg/g). For the Ptb sample, only glycosylated structures, namely hyperoside and luteolin-7-*O*-glucosid, in important amounts (1510 ± 3.7 μg/g and, respectively, 72 ± 0.4 μg/g) were found. The LC–MS analysis also confirms the lower polyphenol content in the Ptb sample compared to the Pta sample.

2.3. Antiproliferative Assays

The antiproliferative effects of *A. amica* extracts (Pta and Ptb) were evaluated by measuring the absorbance of colorectal adenocarcinoma cells (DLD-1) after treatment. The CAbs (absolute control) group exhibited the highest absorbance, indicating minimal interference with cell viability. In contrast, the cultures treated with ethanol showed only a marginal reduction in absorbance, suggesting that ethanol did not significantly affect cell viability relative to the baseline control. Treatment with the highest concentration of the Pta extract resulted in a dose-dependent decrease in absorbance, with Pta5 demonstrating the most pronounced effect, as evidenced by the lowest absorbance values. The DOX (doxorubicin) group, serving as the positive control, consistently exhibited the lowest absorbance values, reflecting its potent cytotoxicity. The observed reductions in absorbance for the Pta groups, particularly for Pta5, indicate a notable antiproliferative effect; however, the efficacy of Pta extracts remained weaker compared to doxorubicin (Figure 1).

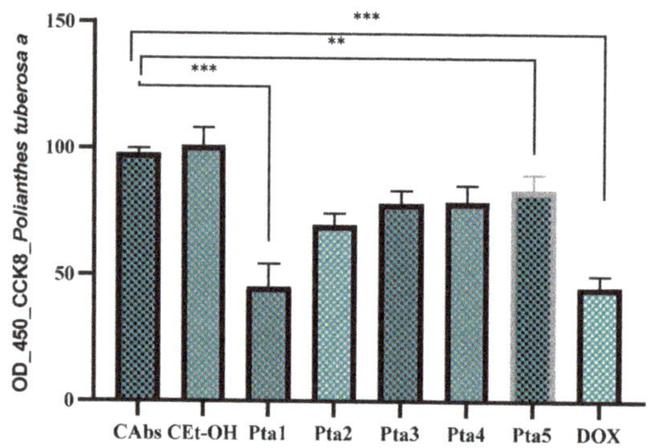

Figure 1. The effect of *A. amica* aerial parts extract on the growth inhibition of DLD cells was evaluated by incubating the cells with different concentrations of the extract (Pta5—0.130 µmol GAE/µL, Pta4—0.260 µmol GAE/µL, Pta3—0.390 µmol GAE/µL, Pta2—0.520 µmol GAE/µL, Pta1—0.650 µmol GAE/µL) for 24 h. Cell proliferation was measured using the CCK-8 assay. Doxorubicin was utilized as a positive control, and statistical significance was determined with ** $p \leq 0.01$, and *** $p \leq 0.001$. The IC_{50} value through dose-response analysis revealed an inhibitory concentration of 0.580 µmol GAE/µL.

The potential antiproliferative activity of Ptb (assessed at five different concentrations, between 0.087 and 0.435 µmol GAE/µL) was evaluated by measuring the absorbance of the chromogenic reaction following the CCK-8 assay. Our results indicate a dose-dependent cytotoxic effect of the Ptb extract, as evidenced by the progressive decrease in absorbance values with increasing extract concentration (Figure 2). The highest concentration (Ptb1) exhibited the most pronounced reduction in absorbance, suggesting significant cytotoxicity and reduced cell viability. Conversely, the lowest concentration (Ptb5) displayed a comparatively higher absorbance, indicating a weaker cytotoxic effect, though still lower than the negative control (CAbs). The ethanol control (CEtOH) showed absorbance values similar to the negative control, suggesting that ethanol alone does not significantly impact cell viability. The positive control (DOX) demonstrated the lowest absorbance values, confirming its expected strong cytotoxic activity. These findings suggest that the extract exerts a dose-dependent inhibitory effect on cell viability, with higher concentrations eliciting greater cytotoxicity.

2.4. Antimicrobial Activity Assays

2.4.1. The Agar-Well Diffusion Method

The *A. amica* extracts' (Pta and Ptb) *in vitro* antimicrobial activity was initially investigated by a screening test, namely the agar-well diffusion method. The obtained results are expressed as diameters of the inhibition zone and are displayed in Table 3.

Table 3. *A. amica* extracts' (Ptb, Pta) *in vitro* antimicrobial activity based on the agar-well diffusion method results.

Tested Samples	Diameters of Inhibition Zone (mm)							
	MSSA	MRSA	*Bacillus cereus*	*Enterococcus faecalis*	*Listeria monocytogenes*	*Escherichia coli*	*Pseudomonas aeruginosa*	*Candida albicans*
Ptb	13.25 ± 0.43 [a]	10.75 ± 0.83 [a]	12.75 ± 0.43 [a]	0	14.25 ± 0.83 [a]	10.5 ± 0.5 [a]	0	19 ± 0.71 [b]
Pta	13.50 ± 0.50 [a]	11.50 ± 0.50 [a]	11 ± 0.71 [a]	0	14.45 ± 0.43 [a]	10.25 ± 0.43 [a]	0	18.5 ± 0.50 [b]
Gentamicin	19 ± 0.00	17 ± 0.25	20 ± 0.00	10 ± 0.00	22 ± 0.50	19 ± 0.00	18 ± 0.00	-
Fluconazole	-	-	-	-	-	-	-	21 ± 0.00

Note: MSSA—methicillin-susceptible *Staphylococcus aureus*, and MRSA—methicillin-resistant *Staphylococcus aureus*. Values represent means of duplicate determinations ($n = 2$) ± standard deviations. Lowercase letters in the same column point out significant differences: [a] $p < 0.05$ (extract vs. gentamicin); [b] $p > 0.05$ (extract vs. fluconazole). Gentamicin (10 µg/disk) and fluconazole (25 µg/disk) were included as positive controls.

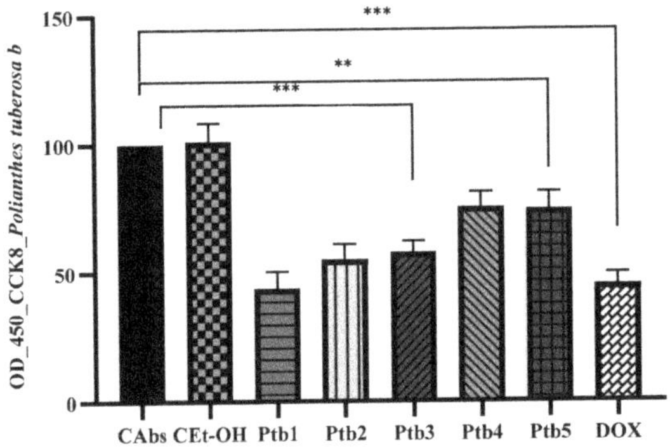

Figure 2. The effect of *A. amica* b extract on the growth inhibition of DLD cells was evaluated by incubating the cells with different concentrations of the extract (Ptb5—0.087 μmol GAE/μL, Ptb4—0.174 μmol GAE/μL, Ptb3—0.261 μmol GAE/μL, Ptb2—0.348 μmol GAE/μL, Ptb1—0.435 μmol GAE/μL) for 24 h. Cell proliferation was measured using the CCK-8 assay. Doxorubicin was utilized as a positive control, and statistical significance was determined with ** $p \leq 0.01$, and *** $p \leq 0.001$. The IC_{50} value through dose-response analysis revealed an inhibitory concentration of 0.285 μmol GAE/μL.

Both *A. amica* extracts (Pta and Ptb) were able to inhibit several tested bacterial strains; still, the *in vitro* antimicrobial potential was found to be significantly lower compared to the positive control gentamicin ($p < 0.05$ for both extracts vs. gentamicin when tested against both Gram-positive and Gram-negative strains). Furthermore, no inhibitory effect was noticed against *Enterococcus faecalis* and *Pseudomonas aeruginosa* (Table 3). Nevertheless, an interesting *in vitro* antifungal activity was indicated when comparing the obtained inhibition zone diameters with fluconazole ($p > 0.05$ for both extracts vs. fluconazole). This notable exception was further confirmed by the results of the broth microdilution method (Table 4) and the antibiofilm assay (Table 5).

Table 4. *A. amica* extracts' (Ptb, Pta) *in vitro* antimicrobial activity based on the broth microdilution assay.

Samples	Microorganisms															
	MSSA		MRSA		*Bacillus cereus*		*Enterococcus faecalis*		*Listeria monocytogenes*		*Escherichia coli*		*Pseudomonas aeruginosa*		*Candida albicans*	
	MIC	MBC	MIC	MBC	MIC	MBC	MIC	MBC	MIC	MBC	MIC	MBC	MIC	MBC	MIC	MFC
Ptb	1.3	2.6	1.3	2.6	1.3	2.6	>2.6	>2.6	1.3	1.3	2.6	> 2.6	>2.6	>2.6	0.325	0.325
Pta	1.3	2.6	1.3	2.6	2.6	2.6	2.6	>2.6	1.3	1.3	2.6	2.6	>2.6	>2.6	0.325	0.325
Gentamicin MIC (mg/L)	3		4		3		3		3		4		-		-	
Fluconazole MIC (mg/L)	-		-		-		-		-		-		-		8	

Note: ">D1" = not active at the tested concentration; MIC: minimum inhibitory concentration (μmol GAE/mL)/; MBC: minimum bactericidal concentration (μmol GAE/mL); MFC: minimum fungicidal concentration (μmol GAE/mL).

Table 5. *A. amica* extracts' (Ptb, Pta) antibiofilm activity.

Samples	Inhibitory Activity Against Biofilm					
	MSSA		*Listeria monocytogenes*		*Candida albicans*	
	T0	T24	T0	T24	T0	T24
Ptb	+	++	+	++	++	++
Pta	+	++	+	++	++	++
Gentamicin	+	++	+	++	-	-
Fluconazole	-	-	-	-	++	++

Note: - = not active; + means poor and ++ means good antibiofilm activity.

2.4.2. The Broth Microdilution Method

The values obtained when performing the broth microdilution method are displayed in Table 4. As indicated by the preliminary screening results, both *A. amica* extracts were able to exhibit a certain degree of *in vitro* antimicrobial efficacy manifested as both inhibitory (MIC) and antimicrobicidal (MBC) activity against the tested organisms. The established values of MIC and MBC generally involved either the lowest tested dilution (against MRSA, *Bacillus cereus*, and *Escherichia coli*) or the second most concentrated form of the extracts (against *Listeria monocytogenes* and MSSA). While such high concentrations of the tested products were required against bacterial strains, *Candida albicans* efficacy of both extracts was further underlined by the four-fold lower MIC and MFC values.

2.4.3. The Antibiofilm Assay

Given the results of the previous two assays, the two *A. amica* extracts' antibiofilm properties were further investigated towards three organisms: two Gram-positive bacteria (MSSA and *Listeria monocytogenes*) and a fungus (*Candida albicans*), with the results presented in Table 5. Using a previously described protocol, inhibition percentage values (%) were calculated and interpreted as good (above 50%, ++), poor (0–50%, +), and no inhibition or enhancement of biofilm development and growth (<0) [37]. Thus, considering these calculated values (%), good antibiofilm activity (above 50%, ++) was observed for both extracts when tested against all organisms for the T24 stage (destruction of 24 h pre-formed biofilm). Regardless of the extract and bacterial strain, the effect on the T0 stage (biofilm attachment) was recorded as rather poor (0–50%, +) and no inhibition or enhancement of biofilm development and growth (<0) (Table 5).

3. Discussion

A. amica is a species cultivated mainly for its ornamental properties, but it is also endowed with medicinal properties. The extracts from its bulbs, flowers, and aerial parts are traditionally used in some countries (in Bangladesh, Algeria, India, and Mexico) for their diuretic, laxative, emetic, uterine relaxant, and antigonorrhoeic properties, and the healing of pediatric rashes, and scientific studies have shown antifungal, antibacterial, anti-MRSA, nematicidal, pesticidal, mosquito larvicidal antioxidant, and antitumor actions in the liver, pancreatic cancers, and promyelocytic leukemia [13,22–24,31,32,38]. In Romania, tuberose is widely grown as an ornamental plant, but pharmacognostic and pharmacological research regarding it is lacking or very limited.

This research brings novelty by the comparative testing, for the first time in Romania and probably in Europe, of the polyphenolic profile and the antioxidant, antimicrobial, and especially antibiofilm and cytotoxic/antiproliferative properties of ethanolic extracts (70% ethanol in water v/v) obtained from the aerial parts (Pta) and bulbs (Ptb) of *A. amica*. Following spectrophotometric evaluations, it was found that the concentration of polyphenolic active principles is different, and the Pta extract is richer in phenolic compounds (TPC: 44.25 mg/g, TC: 12.55 mg/g, TF: 9.20 mg/g, and CADC: 19.95 mg/g, respectively) than Ptb (TPC: 29.60 mg/g, TC: 9.85 mg/g, FC: 5.65 mg/g, and CADC: 8.08 mg/g, respectively). The amounts of TP, flavonoids, and tannins in the Pta sample proved to be higher compared to a crude methanolic extract obtained from *A. amica* leaves from Algeria [13]. Regarding the presence of caffeic acid derivatives (CADC), the Pta sample presented a higher CADC content (19.95 mg/g) than the Ptb sample (8.08 mg/g), with there being no other specialized published works in this regard.

The main constituents of the volatile components in *A. amica* flower extracts are represented by benzyl benzoate, methyl 2-amino benzoate, methyl isoeugenol, isoeugenol, benzyl salicylate, methyl salicylate, geraniol, and 1,8-cineole. As for the total phenolic content of floral extracts, variations are reported in relation to the type of solvent (0.094, 0.18, 0.14, 0.007, 0.004, and 0.110 mg gallic acid equivalent/mg fresh weight in water, methanol, ethanol, ethyl acetate, hexane, and dichloromethane extracts, respectively) [21].

The antioxidant potential of *A. amica* flower extracts was previously reported [18,21]. The highest values were noticed for the methanol, dichloromethane, and aqueous soluble fractions by DPPH and ABTS assays, while the best results of the FRAP assay were obtained for the aqueous extract [21]. The phenolic content of 113.49 mg of GAE/g of extractives and a significant free radical scavenging activity with IC50 values of 71.23 and 84.27 μg/mL, respectively, characterized a crude methanol extract and its hexane soluble fraction [18].

In our study, the antioxidant activity was evaluated by the *in vitro* DPPH and FRAP methods, demonstrating an important antioxidant potential of the extracts obtained from the aerial parts and bulbs of *A. amica* cultivated in Romania. The aerial parts showed a higher DPPH radical scavenging activity than the bulbs, with the effect being possibly due to higher concentrations of caffeic acid (8890 μg/g) and chlorogenic acid (4000 μg/g) in the Pta extract compared to Ptb (4230 μg/g caffeic acid and 1501 μg/g chlorogenic acid). Furthermore, the electron-donating capacity of the *A. amica* extracts was measured by the FRAP method. The capacity to reduce Ferric ions (Fe^{3+}) to ferrous ions (Fe^{2+}) was higher for the aerial parts (79.75 ± 1.80 μM TE/g) than for the bulbs (29.62 ± 0.37 μM TE/g). Therefore, the positive correlation between antioxidant activity and total polyphenolic compound content ($p < 0.001$) suggests that these secondary active principles act as reducing agents and hydrogen donors, and may exert an important antioxidant capacity, as also reported in other publications [13,18].

The cytotoxicity assay results indicate that both extracts (Pta and Ptb) exhibit dose-dependent inhibitory effects on the human colorectal adenocarcinoma cell line DLD-1 cell viability, with higher concentrations leading to increased cytotoxicity. While Ptb shows slightly enhanced cytotoxicity compared to Pta, this difference is relatively modest, particularly at lower concentrations. The absorbance values for Ptb1 are notably lower than those for Pta1, suggesting a stronger inhibitory effect at the highest concentration. However, at the lowest tested concentration (Ptb5 vs. Pta5), the difference in viability reduction is only approximately 10%, indicating that the two extracts have comparable effects at lower doses. The CEtOH exhibits similar values to the negative control, confirming that ethanol itself does not influence cell viability. As expected, the positive control demonstrates the most significant cytotoxic activity. These findings suggest that while Ptb may have a slightly higher potency than Pta, the overall differences are not substantial, particularly at lower concentrations. Further investigations, including chemical composition analysis and mechanistic studies, are necessary to determine whether Ptb possesses a higher concentration of bioactive compounds or if other factors contribute to its slightly stronger cytotoxic effect. Cytotoxicity against tumoral cells was previously reported only by a limited number of studies [20,27,28]. In fact, gold nanoparticles obtained from the floral aqueous extract of *A. amica* (PtubAuNPs) manifested *in vitro* toxicity towards the MCF 7 breast cancer cell line at a concentration of 100 μg/mL and induced these cells' apoptosis in a dose-dependent manner [28]. Furthermore, silver nanoparticle products (PTAgNPs) obtained from *A. amica* aqueous root extract at concentrations of 5, 10, 15, 20, 25, 30, 35, 40, 45, and 50 μg/mL produced Vero cell line morphologic alterations such as cell shrinkage and the loss of colony formation ability [20]. The cytotoxicity of PTAgNPs was demonstrated also on the A-431 melanoma cell line based on the results of a complex protocol that included the MTT assay, cell cycle analysis, the Comet assay, and reactive oxygen species (ROS) level determination: dose-dependent toxicity with an IC_{50} of 56.54 μg/mL, S phase arrest, as well as A431 cell death by inducing ROS [27]. Moreover, using the brine shrimp lethality bioassay, Rumi et al. reported strong cytotoxic activity with an LC_{50} value of 3.56 and

9.31 μg/mL, respectively, for a crude methanol extract and its carbon tetrachloride soluble fraction compared to standard vincristine sulfate (VS) [18]. According to the literature data, polyphenolic compounds present in high concentrations in analyzed extracts such as salicylic acid, chlorogenic acid, caffeic acid, chrysin, luteolin, and hyperoside have shown promising potential in the prevention and treatment of colorectal cancer, through multiple biological mechanisms. They act by inhibiting cell proliferation, inducing apoptosis, blocking cyclin-dependent kinases, reducing inflammation, and modulating signaling pathways involved in carcinogenesis such as Wnt/PI3K/AKT, NF-κB, and PD-L1 [39–43]. Among them, hyperoside stands out for its solid preclinical results in lung cancer, as well as its potential applicability in colorectal cancer [40]. Also, metabolites of salicylic acid have shown notable inhibitory effects on tumor cell division [7]. Overall, these natural compounds may constitute valuable adjuvants in oncological therapies, but rigorous clinical studies are needed to validate their efficacy and safety. Also, optimizing bioavailability and identifying effective doses remain essential challenges for their therapeutic implementation. The presence of these compounds in extracts along with other polyphenols was an important premise in testing the extracts on DLD-1 cell lines and also supported the explanation of antitumor mechanisms.

The testing protocol employed in our study indicated notable anti-*Candida albicans* activity as well as moderate efficacy against *Listeria monocytogenes*, and *Staphylococcus aureus* was displayed by both extracts against planktonic cells and biofilms. In relation to the antimicrobial potential, our results are in agreement with those reported by Rumi et al. Using the disk diffusion test, these authors found mild to moderate activity (zone of inhibition = 9.0–15.0 mm) against *Bacilllus megaterium*, *B. cereus*, *Sarcina lutea*, *Salmonella paratyphi*, and *Vibrio mimicus* for a crude methanol extract obtained from *A. amica* tubers originating from Bangladesh [18]. More recently, *Staphylococcus aureus* efficacy was reported for crude extracts obtained from the leaves of *A. amica* cultivated in Bangladesh, with inhibition zone diameters ranging from 7 to 21 mm, and the ethyl acetate crude extract demonstrated the highest antagonism against MRSA with a minimum inhibitory concentration value of 6.25 μg/mL and a zone of inhibition of 21 mm [31].

The disk diffusion method proved for tuberose flower extracts obtained by green technology, namely supercritical carbon dioxide (SC-CO_2) extraction and solvent extractions (using *n*-hexane and petroleum ether of b.p. 60–80 °C), *in vitro* efficacy against *Staphylococcus aureus*, *Pseudomonas aeruginosa*, *Escherichia coli*, and *Candida albicans* and no activity against *H. pylori* and *V. cholerae*. Still, the authors described significant differences when comparing the MIC values obtained for the two extracts against different bacterial strains, namely the *n*-hexane extract had the highest activity against *P. aeruginosa*, while the petroleum ether extract was the most active against *C. albicans*, but with no anti-*Staphylococcus aureus* potency. Moreover, these two extracts' GC/MS analysis indicated the presence of *p*-xylene, limonene, methyl eugenol, and salicyl hydrazide, all active compounds well known for their antimicrobial potential [30]. The flower, stem, and tuber of *A. amica*-derived extracts prepared by maceration with different polarity solvents (*n*-hexane, ethyl acetate, and methanol) were reported active against *Staphylococcus epidermidis* and *Propionibacterium acnes* [32]. *A. amica* leaves' crude ethanolic extract at concentrations of 100, 250, and 500 μg/mL (disk diffusion) inhibited the growth of *Bacillus subtilis*, *Staphylococcus aureus*, *Micrococcus luteus*, *Escherichia coli*, *Pseudomonas aeruginosa*, and *Salmonella typhi*, with the Gram-positive bacteria more susceptible compared to the Gram-negative ones. This extract was also found with significant antifungal activity against *Aspergillus niger* and *Candida albicans* [38].

Methanolic extracts prepared from the bulbs of cultivated *P. tuberosa* "Double" and six wild species: *P. platyphylla*, *P. montana*, *P. clivicola*, *P. howardii*, *P. pringlei*, and the genera

Manfreda and *Prochnyanthes* from Mexico were able to inhibit two phytopathogens; the *Prochnyanthes* extract displayed the greatest anti-*P. aeruginosa* activity, with a percent of inhibition (PI) of 24 ± 2.34%, while the most effective against *D. dadantii* were *P. ringlei* (PI = 21.8 ± 2.39%), *Prochnyanthes* (PI = 15.5 ± 0.86%), *Manfreda* sp. (PI = 14.4 ± 1.7%), and *P. montana* (PI = 9 ± 1.16%) [44].

Previous studies reported data on the *in vitro* antifungal activity of the absolute of tuberose and its constituents against *Aspergillus niger* and *Candida albicans* [38] and *Colletotrichum gloeosporioides* [33]. The mycelial growth on a potato–dextrose–agar medium was only slightly inhibited by the tuberose absolute at a concentration of 500 mg/L, while significant activity was recorded for three constituents present in the absolute: geraniol, indole, and methyl anthranilate [33].

However, assessed by the disk diffusion method, the bulbs and bulbils alkaloid extracts obtained from *A. amica* cultivated in Algeria did not show *in vitro* inhibitory effects at a concentration of 50 mg/mL against several reference strains (*Escherichia coli* ATCC 25922, *Pseudomonas aeruginosa* ATCC 27853, *Staphylococcus aureus* ATCC 25923, *Enterococcus faecalis* ATCC 29212, *Staphylococcus aureus* resistant to Methicillin (MRSA), and *Candida albicans* ATCC 90028). Furthermore, the authors found only moderate antioxidant activity with the bulbs and bulbils alkaloid extracts exhibiting an antiradical effect with IC_{50} = 0.231 ± 0.017 mg/mL and 0.233 ± 0.093 mg/mL, respectively, compared to vitamin C with IC_{50} = 0.0194 ± 0.0002 mg/mL [13].

Based on the most recent studies on *A. amica*, green synthesized zinc oxide [29], silver [20,27], and gold [28] nanoparticle flower extract biosynthesis was associated with interesting antimicrobial and antiproliferative properties. Green synthesized *A. amica* flower concentrate zinc oxide nanoparticles (ZnONPs) inhibited the growth of selected bacterial and fungal pathogens and suppressed biofilm formation in *Pseudomonas aeruginosa* in a dose-dependent manner [29]. Gold nanoparticles obtained from the floral aqueous extract of *A. amica* (PtubAuNPs) with concentrations of 25 μg/mL, 50 μg/mL, and 75 μg/mL were able to inhibit *in vitro* the *E. coli* and *S. aureus* growth. Higher values of inhibition zone were recorded in the case of the Gram-negative strain due to the nanoparticles' increased ability to penetrate this specific type of bacterial wall. The silver nanoparticles' products (PTAgNPs) are reported to display a relatively more intense and broader *in vitro* antimicrobial efficacy compared to the gold ones [20,27]. Reference strains of *Staphylococcus aureus*, *Bacillus subtilis*, *Escherichia coli*, *Pseudomonas aeruginosa*, and *Klebsiella pneumoniae* were susceptible towards silver nanoparticles obtained from *A. amica* aqueous root extract.

Major compounds identified in the chemical composition of *A. amica* extracts evaluated in the present study, namely chlorogenic acid, caffeic acid, and salicylic acid, were previously reported with antimicrobial and antibiofilm properties through several mechanisms, including disrupting bacterial cell membranes, potentially affecting quorum sensing systems [45]. Caffeic acid's antimicrobial and antibiofilm activity is linked to its ability to act as a cell permeabilizer, causing membrane alterations and potassium leakage. It can also affect phenylalanine ammonia-lyase (PAL) and peroxidase (POD) activity, enzymes that are involved in bacterial defense mechanisms [46]. Salicylic acid's antimicrobial effects are thought to be related to its ability to interact with cell membrane proteins, disrupting chemiosmotic control and leading to cell death. It can also affect bacterial metabolism and potentially interfere with biofilm matrix formation [46]. Luteolin, a flavonoid, was found to exhibit both antimicrobial and antibiofilm properties by interfering with bacterial cell membranes, inhibiting biofilm formation, and potentially disrupting bacterial metabolism and energy production. It has been shown to be effective against various bacteria, including *Staphylococcus aureus* and *Listeria monocytogenes*, and can enhance antibiotic diffusion within biofilms [47]. Hypericin exhibits antimicrobial activity primarily through photodynamic inactivation and by inhibiting bacterial resistance mechanisms. It acts as a photosensitizer, generating reactive oxygen species (ROS) upon exposure to light, which damages and kills microbial cells. Additionally, hypericin can reduce the expression of genes involved in bacterial resistance, potentially enhancing the effectiveness of certain antibiotics [48]. Chrysin exerts antimicrobial activity affecting cell membrane integrity, inhibiting bacterial quorum sensing, and potentially enhancing the effectiveness of other antibiotics. It can also act as an antioxidant and anti-inflammatory agent, which can indirectly support antimicrobial efforts by reducing inflammation and oxidative stress [49].

As described above, several mechanisms support the herbal extract isolated compound's ability to inhibit biofilm formation. Whole herbal extracts represent complex mixtures of compounds and could prove to have a similar or greater efficacy considering their potential to target multiple pathways associated with biofilm formation: bacterial adhesion, quorum sensing, and structural and functional components of the biofilm. Our present study employed a standardized qualitative protocol aimed at assessing the antibiofilm activity, thus further studies aimed at a multi-target approach and employing parameters such as subinhibitory concentrations (Sub-MICs) of extracts are intended to prove the mechanisms' hypothesis. Such information is required to provide a comprehensive scientific base supporting *A. amica* antibiofilm potential applications.

4. Materials and Methods

4.1. Vegetal Material

The vegetal material was represented by the aerial parts and bulbs of *A. amica* (Pta and Ptb) from cultures located in Hoghilag village, Sibiu County, Romania (46°14′ N, 24°37′ E). The aerial parts were harvested during the flowering period (August 2022). During the yellowing of the leaves (autumn), the bulbs were dug up from the soil and cleaned. The vegetal products were dried in the shade in the laboratory. The species was identified at the Department of Pharmaceutical Botany of the "Iuliu Hațieganu" University of Medicine and Pharmacy in Cluj-Napoca, Romania, and was assigned voucher allocation number 202 [50–53].

4.2. Chemical Agents

Chemical reagents and solvents were purchased from several chemical companies. Thus, reagents aluminum chloride, Arnow and Folin–Ciocâlteu, ethanol, methanol, sodium acetate, hydrochloric and acetic acids, DPPH, acetonitrile, and ammonium acetate were purchased from MercK KGaA, Darmstadt (Germany). Ferric chloride III, hydroxide, and sodium carbonate were purchased from Alfa Aesar, USA, Germany, and 6-hydroxy-2,5,7,8-tetramethylchroman-2-carboxylic acid (Trolox) and 2,4,6-tripyridyl-s-triazine (TPTZ) from GmbH Steinheim (Steiheim, Germany). All references used in the LC–MS analysis were obtained from Phytolab (Vestenbergsgreuth, Germany). The spectrophotometer was Agilent Cary 60 UV–vis, (Santa Clara, CA, USA). Specific reagents for cytotoxicity and antimicrobial assays were obtained as follows: the DLD-1 cell line (CCL-221™) was purchased from the American Type Culture Collection, the RPMI-1640 medium was purchased from Gibco Life Technologies (Paisley, UK), fetal bovine serum and glutamine were purchased from Sigma-Aldrich (St. Louis, MO, USA), gentamicin and fluconazole were purchased from Oxoid Ltd., Hampshire, UK, the McFarland standard was purchased from Marcy l'Etoile (France), agar plates were purchased from Mueller–Hinton (MH), and Sabouraud dextrose was purchased from Darmstadt (Germany).

4.3. Extraction Method

The aerial parts (Pta) and bulbs (Ptb) were ground using a knife mill—Grindomix GM 200 (RETSCH GmbH, Haan, Germany) —and then extracted with 70% ethanol (70% ethanol in water v/v). The extracts were prepared hot, at 60 °C on a water bath, for 40 min, using 10 g of plant product and 100 mL of 70% ethanol. They were filtered into a volumetric flask and made up to 100 mL with 70% ethanol. Subsequently, the solutions were filtered through paper filters and centrifuged, and the supernatant solutions were collected and kept in the refrigerator until used for analysis [53–56]. Regarding the standardization of the extraction method, the extraction yield was calculated for each sample as follows: 27% for the Pta sample (g dry extract/g vegetal material), and for the Ptb sample, 40% (g dry extract/g vegetal material).

4.4. Total Polyphenolic Content (TPC) Quantification Method

The determination of the total polyphenolic content (TPC) of the two samples (Pta and Ptb) was spectrophotometrically performed, using the Folin–Ciocâlteu reagent in the presence of 29% sodium carbonate solution [52,54,57]. Absorbance was measured at 760 nm, and the results were calculated in mg gallic acid equivalents (GAE)/g dry plant material, using a gallic acid calibration curve (R^2 = 0.999).

4.5. Flavonoid Content (FC) Quantification Method

The flavonoid content in the two *A. amica* extracts was evaluated by a spectrophotometric method using aluminum chloride (25%) and sodium acetate (10%), and the absorbance was measured at 430 nm. The final results were expressed in mg of rutin equivalents (RE)/g dry plant material, after establishing a calibration curve of the standard, rutin (R^2 = 0.992) [52,54,58].

4.6. Caffeic Acid Derivatives (CADC) Quantification Method

To evaluate the content of caffeic acid derivatives in the two *A. amica* samples, the Arnow reagent (mixture of sodium nitrite and sodium molybdate) was used, the absorbance was measured at 500 nm, and the results were calculated using a caffeic acid calibration curve (R^2 = 0.988). CADC values were expressed in mg caffeic acid equivalents (CAE)/g dry plant material [52,54,58].

4.7. Tannin Content (TC) Quantification Method

To evaluate the tannin content in the *A. amica* extracts, the indirect spectrophotometric method and the Folin–Ciocâlteu reagent were used, according to the European Pharmacopoeia, with some minor modifications [57]. First, the total polyphenol content (TPC) was determined. In the second step, after removing the insoluble tannin–protein complexes obtained with the hide powder reagent, the total content of non-tannin polyphenols (polyphenols not adsorbed on hide powder; TP non-TC) was determined. The tannin content (TC) was expressed as the difference between TPC and TP non-TC, in mg gallic acid equivalent (GAE)/g dry plant material [59,60].

4.8. LC–MS Analysis

The LC–MS analysis was performed using a Shimadzu Nexera I LC–MS-8045 (Kyoto, Japan) UHPLC system, equipped with a quaternary pump, an autosampler, an ESI probe, and a quadrupole rod mass spectrometer. Separation was carried out on a Luna C18 reversed-phase column (Phenomenex-Torrance, Torrance, CA, USA); the mobile phase consisted of a gradient of analytical-grade methanol, ultrapurified water, and formic acid. Detection was performed using a quadrupole rod mass spectrometer with electrospray ionization (ESI), both in negative and positive multiple reaction monitoring (MRM) ion modes. Identification was carried out via a comparison of the retention times, MS spectra, and the transitions between the separate compounds and references. Identification and quantification were based on the main transition from the MS spectra of each individual compound. Calibration curves (R^2 = 0.9964–0.9999) were plotted for the quantification of compounds and references. The samples tested were injected in triplicate [51,53].

4.9. Antioxidant Assays

The estimation of the antioxidant potential of the *A. amica* extracts was established by two tests: the DPPH free radical method and the Ferric ion reducing antioxidant power (FRAP) assay.

4.9.1. DPPH Test

The evaluation of the general antioxidant activity is based on the ability of the scavenger polyphenolic molecule to donate an electron to a stable organic radical of the purple DPPH• type, with its transformation into yellow $DPPH_2$. The color change is spectrophotometrically measured at 517 nm using an ultraviolet–visible (UV/vis) spectrophotometer [61]. Method in brief: 2 mL of each extract (Pta and Ptb) of different concentrations were treated with 2 mL of methanolic DPPH solution (0.1 g/L). After half an hour, the absorbances of the controls (Ac) and samples (As) were determined at 517 nm. The percentages of inhibition of DPPH radicals were calculated, and the results were expressed as IC_{50} (mg/mL), i.e., the concentration of plant product that scavenges 50% of DPPH radicals [54].

4.9.2. FRAP Test

The FRAP method developed by Benzie and Strain was adapted by Pulido R. et al. to measure the antioxidant capacity of plant extracts by reducing Fe3+ to Fe2+ in the presence of the 2,4,6-tripyridyl-s-triazine (TPTZ) radical [62,63]. The absorbance of the blue solution was measured at 450 nm. The results were expressed in µM Trolox equivalents/g dry plant material, using Trolox as a reference (R^2 = 0.994) [54].

4.10. Cell Line and Cytotoxicity Assay

Colorectal adenocarcinoma cell line DLD-1 (CCL-221^{TM}) was cultured in an RPMI-1640 medium (Gibco Life Technologies, Paisley, UK) supplemented with 10% fetal bovine serum (Sigma-Aldrich, St. Louis, MO, USA), 1% glutamine (Sigma-Aldrich, St. Louis, MO, USA), and 1% antibiotics-antimycotics (Gibco Life Technologies, Paisley, UK). Cells were incubated in a humidified atmosphere at 37 °C and 5% CO_2. The antiproliferative potential of *A. amica* extracts (Pta, Ptb) was evaluated using the CCK-8 assay. DLD-1 cells (1×10^4 cells/well) were treated with Pta and Ptb extracts at five distinct concentrations, expressed in µmol GAE/µL (considering the total polyphenol concentration, expressed in mg GAE/g). The concentrations used for Pta were calculated as 0.650 µmol GAE (25 µL extract—Pta1), 0.520 µmol GAE (20 µL extract—Pta2), 0.390 µmol GAE (15 µL extract—Pta3), 0.260 µmol GAE (10 µL—Pta4), and 0.130 µmol GAE (5 µL—Pta1), while for Ptb, the corresponding concentrations were 0.435 µmol GAE (25 µL extract—Ptb1), 0.348 µmol GAE (20 µL extract—Ptb2), 0.261 µmol GAE (15 µL extract—Ptb3), 0.174 µmol GAE (10 µL extract—Ptb4), and 0.087 µmol GAE (5 µL extract—Ptb5). The positive control consisted of doxorubicin (reference compound) at a concentration of 20 µg/mL, and the negative control was represented by cells cultured in a standard medium. In addition, the inhibitory effect of the solvent used for the preparation of the hydroalcoholic extracts was also assessed. After incubation for 24 h, CCK-8 solution (Sigma-Aldrich, St. Louis, MO, USA) was added to each well, and the cells were further incubated for 4 h at 37 °C in the dark. CCK-8, which contains a water-soluble tetrazolium salt, is reduced by viable cells to generate a colored formazan dye. The intensity of the chromogenic reaction correlates directly with the number of viable cells in the sample [64,65]. The absorbance of each well was subsequently measured at 450 nm using a microplate reader (Bio-Rad, Hercules, CA, USA). All experiments were conducted in triplicate and the results were expressed as the mean ± SD. The cell survival percentage was calculated based on the optical densities, with values compared to the optical density of the control group [54,66,67]. IC_{50} values were calculated by plotting cell viability against the logarithm of extract concentrations and fitting the data with a nonlinear sigmoidal dose-response curve using a four-parameter logistic model.

4.11. Antimicrobial Activity Assays

4.11.1. Agar-Well Diffusion Method

The antimicrobial potential of *A. amica* extracts (Pta, Ptb) was initially evaluated based on the agar-well diffusion method [68] according to EUCAST (European Committee on Antimicrobial Susceptibility Testing) criteria [69]. The reference strains included were as follows: *Staphylococcus aureus* ATCC 25923 (methicillin-susceptible *S. aureus*, MSSA), *Staphylococcus aureus* ATCC 700699 (methicillin-resistant *S. aureus*, MRSA), *Bacillus cereus* ATCC 14579, *Enterococcus faecalis* ATCC 29219, *Escherichia coli* ATCC 25922, *Pseudomonas aeruginosa* ATCC 27853, and *Candida albicans* DSMZ 1386. The microbial reference strains were purchased from Oxoid Ltd. (Hampshire, UK). The assay included both positive and negative controls for the antimicrobial activity: gentamicin (10 μg) and fluconazole (25 μg) (Oxoid Ltd., Hampshire, UK) and 70% ethanol in water *v*/*v*, respectively. Each microbial strain was prepared as turbidity 0.5 McFarland (1.0×10^6 CFU/mL) standard (bio-Meriuex, Marcy l'Etoile, France) equivalent inoculum and cultured onto suitable agars: Mueller–Hinton (MH) and Sabouraud dextrose (SD) (Merck, Darmstadt, Germany), for bacteria and *C. albicans*, respectively. Further, 6 mm diameter wells were cut into the inoculated agar plates to add in three wells for the volume of 50 μL for each tested product (extracts, negative control). Thus, considering the total polyphenol concentration, for this assay, the Pta and Ptb extract concentrations were calculated as 1.3 μmol GAE/μL and 0.870 μmol GAE/μL, respectively. After this step, the agar plates were incubated at 37 °C for 24 h for bacteria and 48 h for *C. albicans*, when the growth inhibition zone diameters were measured, and their corresponding values were recorded in mm. This *in vitro* test was performed in duplicate [70].

4.11.2. Broth Microdilution Method

The *A. amica* extracts (Pta, Ptb) were further *in vitro* investigated using the broth microdilution method to determine the minimum inhibitory (MIC), bactericidal (MBC), and fungicidal (MFC) concentrations. This test involved preparing two-fold serial dilutions of each tested herbal extract using 100 μL of a specific broth (MH and SD, respectively) and sterile flat-bottomed 96-well microtiter plates (Deltalab, Barcelona, Spain). These dilutions were mixed with a volume of 5.0 μL microorganism inoculum to be incubated at 37 °C for 24 h for bacteria and 48 h for *C. albicans*; at the end of the incubation, the wells were visually examined against the controls (the two types of broths used to culture bacterial and fungal species (MH and SD, respectively)), considering the absence of turbidity indicative of *in vitro* inhibitory activity and recording its corresponding highest dilution as the MIC value. Further, a volume of 10.0 μL was sampled from each well, inoculated onto specific agar plates, and cultured for 24 h and 48 h for bacteria and *C. albicans*. MBC and MFC values were recorded for if the colonies were absent. Positive and negative controls were tested as well, namely gentamicin 50 mg/mL (Sigma-Aldrich, St. Louis, MO, USA), fluconazole (10–1000 μM) (Sigma-Aldrich, St. Louis, MO, USA), and 70% ethanol in water *v*/*v* with the evaluation made in duplicates for each tested product [70].

4.11.3. Antibiofilm Assay

The *A. amica* extracts' (Pta, Ptb) *in vitro* antimicrobial potential was also verified against the biofilm formation based on previous protocols [37,70,71]. Two stages, namely a biofilm attachment (T0) and a 24 h pre-formed biofilm (T24), were considered. In the case of T0, microbial inoculums were prepared using the same technique as described for the agar-well diffusion method [68] to be inoculated with equal volumes (100 μL) of each tested product in sterile flat-bottomed 96-well microtiter plates without shaking for 24 h at 37 °C. For T24, the plates were inoculated with each microbial inoculum and incubated for 24 h to obtain the formed biofilm stage that was further exposed to equal volumes of each extract. Similar to the other employed methods, this assay also included controls represented by organisms + specific broth, bacteria + MH broth + gentamicin, and *C. albicans* + SD broth + fluconazole. The antibiofilm activity for both stages was recorded based on the crystal violet staining (CVS) results. After 24 h incubation, for each well of the plate, the supernatant was removed, and the plates were washed three times using sterile distilled water. Following drying, the adhered cells were fixed with 96% methanol (150 μL) and further stained with a 0.1% crystal violet solution (100 μL) (Sigma-Aldrich, St. Louis, MO, USA). After 20 min at room temperature, these plates were washed with sterile distilled water and further treated with 150 μL of 100% ethanol. Finally, after gentle shaking, the optic density (OD) was recorded at 490 nm using a microplate reader Sunrise™ (Tecan, Männedorf, Switzerland), and the results were expressed as a percentage of inhibition based on the following equation: Inhibition (%) = $(OD_{control} - OD_{extract})/OD_{control} \times 100$ [54]. These values (%) allowed the interpretation of the *in vitro* antibiofilm activity as good (above 50%, ++), poor (0–50%, +), and no inhibition or enhancement of biofilm development and growth (<0) [37].

4.12. Statistical Analysis

The described methods were performed in triplicate, and the results were presented as the mean standard using the Excel software package 2016. The results were subjected to one-way ANOVA for statistical analysis with the following values: $p < 0.05$, threshold value for statistical significance; $p < 0.001$, very significant; $0.001 < p < 0.05$, significant; and $p > 0.05$, insignificant.

5. Conclusions

Our results showed important polyphenol content and caffeic acid as the main compound for tested *A. amica* ethanolic extracts. Significantly higher amounts of polyphenol compounds and also of antioxidant activity were found for the aerial parts extract compared to the bulbs one. Notable anti-*Candida albicans* activity and moderate efficacy against *Listeria monocytogenes* and *Staphylococcus aureus* were displayed by both extracts against planktonic cells and biofilm. A dose-dependent cytotoxicity towards the colorectal adenocarcinoma cell line was recorded as well. This study brings novelty to the scientific literature by characterizing the phenolic profile and the antimicrobial, antibiofilm, antioxidant, and antiproliferative activities of the ethanolic extracts obtained from *A. amica*, thus highlighting this herbal species's medicinal potential.

Supplementary Materials: The following supporting information can be downloaded at: https://www.mdpi.com/article/10.3390/antibiotics14070638/s1, Figure S1: LC-MS chromatogram and mass fragmentation pattern obtained for the identification of caffeic acid in the Ptb sample; Figure S2. LC-MS chromatogram and mass fragmentation pattern obtained for the identification of chlorogenic acid in the Ptb sample; Figure S3. LC-MS chromatogram and mass fragmentation pattern obtained for the identification of carnosol in the Ptb sample; Figure S4. LC-MS chromatogram and mass fragmentation pattern obtained for the identification of chrysin in the Ptb sample; Figure S5. LC-MS chromatogram and mass fragmentation pattern obtained for the identification of hyperoside in the Ptb sample; Figure S6. LC-MS chromatogram and mass fragmentation pattern obtained for the identification of luteolin-7-O-glucoside in the Ptb sample; Figure S7. LC-MS chromatogram and mass fragmentation pattern obtained for the identification of caffeic acid in the Pta sample; Figure S8. LC-MS chromatogram and mass fragmentation pattern obtained for the identification of chlorogenic acid in the Pta sample; Figure S9. LC-MS chromatogram and mass fragmentation pattern obtained for the identification of salicylic acid in the Pta sample; Figure S10. LC-MS chromatogram and mass fragmentation pattern obtained for the identification of carnosol in the Pta sample; Figure S11. LC-MS chromatogram and mass fragmentation pattern obtained for the identification of chrysin in the Pta sample; Figure S12. LC-MS chromatogram and mass fragmentation pattern obtained for the identification of hyperoside in the Pta sample; Figure S13. LC-MS chromatogram and mass fragmentation pattern obtained for the identification of isoquercetin in the Pta sample; Figure S14. LC-MS chromatogram and mass fragmentation pattern obtained for the identification of luteolin-7-O-glucoside in the Pta sample; Figure S15. LC-MS chromatogram and mass fragmentation pattern obtained for the identification of naringenin in the Pta sample; Figure S16. LC-MS chromatogram and mass fragmentation pattern obtained for the identification of quercetin in the Pta sample.

Author Contributions: Conceptualization, D.B., D.H., I.I., M.N., and E.P.; methodology, D.B., D.H., M.N., T.B., E.P., and M.A.S.; validation, D.B., D.H., I.I., M.N., T.B., E.P., and B.T.; formal analysis, D.B., D.H., M.N., T.B., E.P., B.T., S.-A.B., and R.F.B.; investigation, D.B., D.H., I.I., M.N., T.B., R.F.B., E.P., and S.-A.B.; resources, D.B., D.H., I.I., M.N., T.B., E.P., B.T., M.A.S., and I.O.; writing—original draft preparation, D.B., D.H., I.I., E.P., and M.N.; writing—review and editing, I.O., D.H., I.I., M.N., T.B., E.P., B.T., S.-A.B., R.F.B., and D.B.; visualization, D.B., I.O., D.H., I.I., M.N., T.B., and E.P.; supervision, D.B., M.N., and D.H.; project administration, D.B., I.O., I.I., M.N., T.B., E.P., and D.H.; funding acquisition, D.B. All authors have read and agreed to the published version of the manuscript.

Funding: This research was funded by the "Iuliu Haţieganu" University of Medicine and Pharmacy in Cluj-Napoca, Romania, grant number 32153/8/16.12.2024.

Institutional Review Board Statement: Not applicable.

Informed Consent Statement: Not applicable.

Data Availability Statement: The original contributions presented in this study are included in the article/Supplementary Material; further inquiries can be directed to the corresponding author.

Conflicts of Interest: The authors Timea Bab and Ramona Flavia Burtescu are employed by the company PlantExtrakt Ltd. The remaining authors declare that the research was conducted in the absence of any commercial or financial relationships that could be construed as a potential conflict of interest. The funders had no role in the design of this study, in the collection, analyses, or interpretation of the data, in the writing of the manuscript, or in the decision to publish the results.

References

1. Ahmed, S.K.; Hussein, S.; Qurbani, K.; Ibrahim, R.H.; Fareeq, A.; Mahmood, K.A.; Mohamed, M.G. Antimicrobial Resistance: Impacts, Challenges, and Future Prospects. *J. Med. Surg. Public Health* **2024**, *2*, 100081. [CrossRef]
2. Liu, H.Y.; Prentice, E.L.; Webber, M.A. Mechanisms of Antimicrobial Resistance in Biofilms. *npj Antimicrob. Resist.* **2024**, *2*, 27. [CrossRef] [PubMed]
3. Oliveira, M.; Antunes, W.; Mota, S.; Madureira-Carvalho, Á.; Dinis-Oliveira, R.J.; Dias da Silva, D. An Overview of the Recent Advances in Antimicrobial Resistance. *Microorganisms* **2024**, *12*, 1920. [CrossRef]
4. Zouine, N.; Ghachtouli, N.E.; Abed, S.E.; Koraichi, S.I. A Comprehensive Review on Medicinal Plant Extracts as Antibacterial Agents: Factors, Mechanism Insights and Future Prospects. *Sci. Afr.* **2024**, *26*, e02395. [CrossRef]
5. Lahiri, D.; Dash, S.; Dutta, R.; Nag, M. Elucidating the Effect of Anti-Biofilm Activity of Bioactive Compounds Extracted from Plants. *J. Biosci.* **2019**, *44*, 52. [CrossRef]
6. Ahmadian, M.; Ahmadi, N.; Babaei, A.; Naghavi, M.R.; Ayyari, M. Comparison of Volatile Compounds at Various Developmental Stages of Tuberose (*Polianthes tuberosa* l. Cv. Mahallati) Flower with Different Extraction Methods. *J. Essent. Oil Res.* **2018**, *30*, 197–206. [CrossRef]
7. Srinivasa Suryakoppa, K.; Appadurai, R.; Byrappa, K.; Khan, M.H.M. Phytochemical Analysis of UV Active and Inactive Bioactive Compounds Present in *Polianthes tuberosa* (Linn.) Flower. *J. Sep. Sci.* **2021**, *44*, 3376–3385. [CrossRef]
8. Thiede, J. Agave (Agavaceae). In *Monocotyledons*; Springer: Berlin/Heidelberg, Germany, 2020.

9. Thiede, J.; Eggli, U. Agavaceae. In *Monocotyledons*; Springer: Berlin/Heidelberg, Germany, 2020.
10. Thiede, J.; Govaerts, R. New Combinations in *Agave* (Asparagaceae): *A. Amica*, *A. Nanchititlensis*, and *A. Quilae*. *Phytotaxa* **2017**, *306*, 237–240. [CrossRef]
11. Sadhukhan, R.; Chowdhuri, T.K.; Datta, S.K. *Tuberose (Polyanthes tuberosa Linn./Agave amica)*; Springer: Berlin/Heidelberg, Germany, 2022. [CrossRef]
12. Salomé-Abarca, L.F.; Márquez-López, R.E.; López, M.G. *Agave amica* a Potential Model for the Study of Agavins Metabolism. *Sci. Rep.* **2023**, *13*, 19888. [CrossRef]
13. Barghout, N.; Chebata, N.; Moumene, S.; Khennouf, S.; Gharbi, A.; El Hadi, D. Antioxidant and Antimicrobial Effect of Alkaloid Bulbs Extract of *Polianthes tuberosa* L. (Amaryllidaceae) Cultivated in Algeria. *J. Drug Deliv. Ther.* **2020**, *10*, 173–180. [CrossRef]
14. Barghout, N.; Khennouf, S.; Yekrelef, A.; El-Hadi, D. Polyphenols from *Polianthes tuberosa* L.(Amaryllidaceae) Leaves and Their Antioxidant Properties. *AgroBiologia* **2018**, *8*, 902–912.
15. Bharathi, T.U.; Lallawmzuali, R.; Kirthishree, S.P. Diversity in Flower Morphology of the Single-Type Tuberose (*Agave amica* (Medik.) Thiede & Govaerts). *Genet. Resour. Crop Evol.* **2024**, *71*, 4239–4254. [CrossRef]
16. Copetta, A.; Marchioni, I.; Mascarello, C.; Pistelli, L.; Cambournac, L.; Dimita, R.; Ruffoni, B. *Polianthes tuberosa* as Edible Flower: In Vitro Propagation and Nutritional Properties. *Int. J. Food Eng.* **2020**, *6*, 57–62. [CrossRef]
17. Kutty, N.N.; Mitra, A. Profiling of Volatile and Non-Volatile Metabolites in *Polianthes tuberosa* L. Flowers Reveals Intraspecific Variation among Cultivars. *Phytochemistry* **2019**, *162*, 10–20. [CrossRef]
18. Rumi, F.; Kuddus, M.; Das, S. Evaluation of Antioxidant, Cytotoxic, Antimicrobial, Membrane Stabilizing and Thrombolytic Activities of Polianthes Tuberose Linn. *Br. J. Pharm. Res.* **2014**, *4*, 2106–2115. [CrossRef]
19. Lim, T.K. *Edible Medicinal And Non-Medicinal Plants—Volume 7: Flowers*; Springer: Berlin/Heidelberg, Germany, 2014; Volume 7. [CrossRef]
20. Kumaran, S.; Santhiyaa, R.V.; Prakaesh, U.; Sivasankari, B.; Kokila, D. Biosynthesis of Silver Nanoparticles Using Aqueous Flower Extracts of *Polianthes tuberosa* and Their Antibacterial and Cytotoxicity Activity. *Int. J. Res. Anal. Rev.* **2019**, *5*, 407–414.
21. Maiti, S.; Moon, U.R.; Bera, P.; Samanta, T.; Mitra, A. The in Vitro Antioxidant Capacities of *Polianthes tuberosa* L. Flower Extracts. *Acta Physiol. Plant.* **2014**, *36*, 2597–2605. [CrossRef]
22. Mimaki, Y.; Yokosuka, A.; Sashida, Y. Steroidal Glycosides from the Aerial Parts of *Polianthes tuberosa*. *J. Nat. Prod.* **2000**, *63*, 1519–1523. [CrossRef] [PubMed]
23. Rahmatullah, N.R.; Jannat, K.; Islam, M.; Rahman, T.; Jahan, R.; Rahmatullah, M. A Short Review of *Polianthes tuberosa* L. Considered a Medicinal Plant in Bangladesh. *J. Med. Plants Stud.* **2019**, *7*, 1–4.
24. Lydia, J.; Kannan, M.; Aruna, P.; Jeyakumar, P.; Vetrivelkaalai, P.; Muralidharan, B. Determination of Bioactive Compounds in *Agave amica* L. Cv Arka Prajwal. *Biol. Forum–Int. J.* **2022**, *14*, 366–372.
25. Fragoso-Jiménez, J.C.; Tapia-Campos, E.; Estarron-Espinosa, M.; Barba-Gonzalez, R.; Castañeda-Saucedo, M.C.; Castillo-Herrera, G.A. Effect of Supercritical Fluid Extraction Process on Chemical Composition of *Polianthes tuberosa* Flower Extracts. *Processes* **2019**, *7*, 60. [CrossRef]
26. Yadav, R.; Mohapatra, D.; Kate, A.; Giri, S.K.; Modhera, B. Sequential Ultrasound-Microwave-Assisted Extraction of Tuberose (*Polianthes tuberosa* L.) Concrete: The Effect of Processing Parameters on Yield, Volatile Metabolite Profiles, and Functional Groups. *Biomass Convers. Biorefinery* **2023**. [CrossRef]
27. Alghuthaymi, M.A.; Patil, S.; Rajkuberan, C.; Krishnan, M.; Krishnan, U.; Abd-Elsalam, K.A. *Polianthes tuberosa*-Mediated Silver Nanoparticles from Flower Extract and Assessment of Their Antibacterial and Anticancer Potential: An In Vitro Approach. *Plants* **2023**, *12*, 1261. [CrossRef]
28. Alghuthaymi, M.A.; Rajkuberan, C.; Santhiya, T.; Krejcar, O.; Kuča, K.; Periakaruppan, R.; Prabukumar, S. Green Synthesis of Gold Nanoparticles Using *Polianthes tuberosa* L. Floral Extract. *Plants* **2021**, *10*, 2370. [CrossRef] [PubMed]
29. Christy, J.J.A.; Begum, S.B.; Revathy, M.; Harisma, B.R.; Murugappan, R.M. Antimicrobial and Anti-Inflammatory Efficiency of Green Synthesized Zinc Oxide Nanoparticles Using *Polianthes tuberosa* Flower Concentrate. *J. Environ. Biol.* **2024**, *45*, 243–252. [CrossRef]
30. Ghosh, P.K.; Bhattacharjee, P.; Das, S. Antimicrobial Activity of Supercritical Carbon Dioxide Extracts of Tuberose (*Polianthes tuberosa* Linn.) Flowers against Common Pathogens. *Int. J. Pharm. Sci. Res.* **2014**, *5*, 1279–1289. [CrossRef]
31. Sundar, R.D.V.; Arunachalam, S. Anti-MRSA Activity of Pollianthes Tuberosa Leaf Extracts. *Bangladesh J. Pharmacol.* **2022**, *17*, 11–13. [CrossRef]
32. Setiani, N.A.; Aulifa, D.L.; Selynita; Septiningsih, E. Phytochemical Screening and Antibacterial Activity of Flower, Stem, and Tuber of *Polianthes tuberosa* L. Against Acne-Inducing Bacteria. *Adv. Heal. Sci. Res.* **2020**, *26*, 92–95. [CrossRef]
33. Nidiry, E.S.J.; Babu, C.S.B. Antifungal Activity of Tuberose Absolute and Some of Its Constituents. *Phyther. Res.* **2005**, *19*, 447–449. [CrossRef]
34. Toma, F.; Georgescu, M.I.; Petra, S.; Dobrescu, E. Some Aspects Concerning the Rest Period of Tuberose Bulbs. *Agric. Agric. Sci. Procedia* **2015**, *6*, 179–183. [CrossRef]

35. Moldovan, I.; Cantor, M.; Sabo, R.A.; Lukacs, L.; Somsai, P.A. Research Concerning the Influence of Culture Substrate on the Main Characteristics of Tuberose. *Curr. Trends Nat. Sci.* **2018**, *7*, 54–64.
36. Brînză, M.; Draghia, L.; Ciobănică, M.; Chelaru, E.L.; Bernardis, R. The Influence of Bulb Size on the Morphological and Ornamental Characters of the *Polyanthes tuberosa* L. In Proceedings of the Conferința "Horticultură, Viticultură şi Vinificaţie, Silvicultură şi Grădini Publice, Protecţia Plantelor", Chișinău, Moldova, 1–2 October 2018; pp. 534–539.
37. Olawuwo, O.S.; Famuyide, I.M.; McGaw, L.J. Antibacterial and Antibiofilm Activity of Selected Medicinal Plant Leaf Extracts Against Pathogens Implicated in Poultry Diseases. *Front. Vet. Sci.* **2022**, *9*, 820304. [CrossRef] [PubMed]
38. Ramamoorthy, J.; Venkataraman, S.; Meera, R.; Chidambaranathan, N.; Devi Devisree, P. Phyto-Physico Chemical Investigation, Anti-Inflammatory and Antimicrobial Activities of Pollianthes Tuberosa Linn. *Res. J. Pharm. Technol.* **2009**, *2*, 738–742.
39. Neamțu, A.A.; Maghiar, T.A.; Turcuș, V.; Maghiar, P.B.; Căpraru, A.M.; Lazar, B.A.; Dehelean, C.A.; Pop, O.L.; Neamțu, C.; Totolici, B.D.; et al. A Comprehensive View on the Impact of Chlorogenic Acids on Colorectal Cancer. *Curr. Issues Mol. Biol.* **2024**, *46*, 6783–6804. [CrossRef]
40. Zhang, W.; Wang, R.; Guo, R.; Yi, Z.; Wang, Y.; Wang, H.; Li, Y.; Li, X.; Song, J. The Multiple Biological Activities of Hyperoside: From Molecular Mechanisms to Therapeutic Perspectives in Neoplastic and Non-Neoplastic Diseases. *Front. Pharmacol.* **2025**, *16*, 1538601. [CrossRef] [PubMed]
41. Rauf, A.; Wilairatana, P.; Joshi, P.B.; Ahmad, Z.; Olatunde, A.; Hafeez, N.; Hemeg, H.A.; Mubarak, M.S. Revisiting Luteolin: An Updated Review on Its Anticancer Potential. *Heliyon* **2024**, *10*, e26701. [CrossRef]
42. Naumowicz, M.; Kusaczuk, M.; Zając, M.; Jabłońska-Trypuć, A.; Mikłosz, A.; Gál, M.; Worobiczuk, M.; Kotyńska, J. The Influence of the PH on the Incorporation of Caffeic Acid into Biomimetic Membranes and Cancer Cells. *Sci. Rep.* **2022**, *12*, 3692. [CrossRef]
43. Salari, N.; Faraji, F.; Jafarpour, S.; Faraji, F.; Rasoulpoor, S.; Dokaneheifard, S.; Mohammadi, M. Anti-Cancer Activity of Chrysin in Cancer Therapy: A Systematic Review. *Indian J. Surg. Oncol.* **2022**, *13*, 681–690. [CrossRef]
44. Garcia-Ochoa, E.Y.; León-Morales, J.; Lugo-Cervantes, E.C.; Tapia-Campos, E. In Vitro Biological Activity of Metabolic Extracts of Wild and Cultivated Species of the Genus Polianthes. *Acta Hortic.* **2020**, *1288*, 149–152. [CrossRef]
45. Wang, L.; Cao, X.; Pei, H.; Liu, P.; Song, Y.; Wu, Y. Anti-Biofilm Activity of Chlorogenic Acid against Pseudomonas Using Quorum Sensing System. *Foods* **2023**, *12*, 3601. [CrossRef]
46. Davidova, S.; Galabov, A.S.; Satchanska, G. Antibacterial, Antifungal, Antiviral Activity, and Mechanisms of Action of Plant Polyphenols. *Microorganisms* **2024**, *12*, 2502. [CrossRef] [PubMed]
47. Qian, W.; Liu, M.; Fu, Y.; Zhang, J.; Liu, W.; Li, J.; Li, X.; Li, Y.; Wang, T. Antimicrobial Mechanism of Luteolin against Staphylococcus Aureus and Listeria Monocytogenes and Its Antibiofilm Properties. *Microb. Pathog.* **2020**, *142*, 104056. [CrossRef]
48. Wang, G.; Li, L.; Wang, X.; Li, X.; Zhang, Y.; Yu, J.; Jiang, J.; You, X.; Xiong, Y.Q. Hypericin Enhances β-Lactam Antibiotics Activity by Inhibiting SarA Expression in Methicillin-Resistant Staphylococcus Aureus. *Acta Pharm. Sin. B* **2019**, *9*, 1174–1182. [CrossRef] [PubMed]
49. Chang, A.; He, Q.; Li, L.; Yu, X.; Sun, S.; Zhu, H. Exploring the Quorum Sensing Inhibition of Isolated Chrysin from Penicillium Chrysogenum DXY-1. *Bioorg. Chem.* **2021**, *111*, 104894. [CrossRef]
50. Hanganu, D.; Niculae, M.; Ielciu, I.; Olah, N.-K.; Munteanu, M.; Burtescu, R.; Ştefan, R.; Olar, L.; Pall, E.; Andrei, S.; et al. Chemical Profile, Cytotoxic Activity and Oxidative Stress Reduction of Different *Syringa vulgaris* L. Extracts. *Molecules* **2021**, *26*, 3104. [CrossRef]
51. Ielciu, I.; Niculae, M.; Pall, E.; Barbălată, C.; Tomuţă, I.; Olah, N.K.; Burtescu, R.F.; Benedec, D.; Oniga, I.; Hanganu, D. Antiproliferative and Antimicrobial Effects of *Rosmarinus officinalis* L. Loaded Liposomes. *Molecules* **2022**, *27*, 3988. [CrossRef]
52. Aitynova, A.; Sevastre, B.; Ielciu, I.; Hanganu, D.; Olah, N.K.; Ibragimova, N.; Shalakhmetova, T.; Benedec, D.; Lyu, M.; Krasnoshtanov, A.; et al. Hepatoprotective Activity and Oxidative Stress Reduction of an *Arctium tomentosum* Mill. Root Extract in Mice with Experimentally Induced Hepatotoxicity. *Livers* **2024**, *4*, 696–710. [CrossRef]
53. Ielciu, I.; Filip, G.A.; Sevastre-Berghian, A.C.; Bâldea, I.; Olah, N.K.; Burtescu, R.F.; Toma, V.A.; Moldovan, R.; Oniga, I.; Hanganu, D. Effects of a *Rosmarinus officinalis* L. Extract and Rosmarinic Acid in Improving Streptozotocin-Induced Aortic Tissue Damages in Rats. *Nutrients* **2025**, *17*, 158. [CrossRef]
54. Benedec, D.; Oniga, I.; Hanganu, D.; Vlase, A.M.; Ielciu, I.; Crisan, G.; Fit, N.; Niculae, M.; Bab, T.; Pall, E.; et al. Revealing the Phenolic Composition and the Antioxidant, Antimicrobial and Antiproliferative Activities of Two Euphrasia Sp. Extracts. *Plants* **2024**, *13*, 1790. [CrossRef]
55. Sevastre-Berghian, A.C.; Ielciu, I.; Mitre, A.O.; Filip, G.A.; Oniga, I.; Vlase, L.; Benedec, D.; Gheldiu, A.-M.; Toma, V.A.; Mihart, B.; et al. Targeting Oxidative Stress Reduction and Inhibition of HDAC1, MECP2, and NF-KB Pathways in Rats With Experimentally Induced Hyperglycemia by Administration of Thymus Marshallianus Willd. Extracts. *Front. Pharmacol.* **2020**, *11*, 581470. [CrossRef]
56. Ielciu, I.; Filip, G.A.; Oniga, I.; Olah, N.K.; Bâldea, I.; Olteanu, D.; Burtescu, R.F.; Turcuș, V.; Sevastre-Berghian, A.C.; Benedec, D.; et al. Oxidative Stress and DNA Lesion Reduction of a Polyphenolic Enriched Extract of Thymus Marschallianus Willd. In Endothelial Vascular Cells Exposed to Hyperglycemia. *Plants* **2021**, *10*, 2810. [CrossRef]

57. EDQM Council of Europe. *European Pharmacopoeia*, 11th ed.; EDQM Council of Europe: Strasbourg, France, 2025.
58. Editura, Medicală. *Romanian Pharmacopoeia*, 10th ed.; Editura, Medicală: Bucureşti, Romania, 2005.
59. Moldovan, M.L.; Iurian, S.; Puscas, C.; Silaghi-Dumitrescu, R.; Hanganu, D.; Bogdan, C.; Vlase, L.; Oniga, I.; Benedec, D. A Design of Experiments Strategy to Enhance the Recovery of Polyphenolic Compounds from Vitis Vinifera By-Products through Heat Reflux Extraction. *Biomolecules* **2019**, *9*, 529. [CrossRef] [PubMed]
60. Galvão, M.A.M.; Arruda, A.O.d.; Bezerra, I.C.F.; Ferreira, M.R.A.; Soares, L.A.L. Evaluation of the Folin-Ciocalteu Method and Quantification of Total Tannins in Stem Barks and Pods from Libidibia Ferrea (Mart. Ex Tul) L. P. Queiroz. *Braz. Arch. Biol. Technol.* **2018**, *61*, e18170586. [CrossRef]
61. Mishra, K.; Ojha, H.; Chaudhury, N.K. Estimation of Antiradical Properties of Antioxidants Using DPPH- Assay: A Critical Review and Results. *Food Chem.* **2012**, *130*, 1036–1043. [CrossRef]
62. Benzie, I.; Strain, J. The Ferric Reducing Ability of Plasma (FRAP) as a Measure of "Antioxidant Power": The FRAP Assay. *Anal. Biochem.* **1996**, *239*, 70–76. [CrossRef] [PubMed]
63. Pulido, R.; Bravo, L.; Saura-Calixto, F. Antioxidant Activity of Dietary Polyphenols as Determined by a Modified Ferric Reducing/Antioxidant Power Assay. *J. Agric. Food Chem.* **2000**, *48*, 3396–3402. [CrossRef]
64. Lupșe, I.; Pall, E.; Tudoran, L.B.; Bulboacă, A.E.; Ciurea, A.; Micu, I.C.; Roman, A.; Delean, A.G.; Muntean, A.; Soancă, A. Cytotoxic Effects on Gingival Mesenchymal Stromal Cells and Root Surface Modifications Induced by Some Local Antimicrobial Products Used in Periodontitis Treatment. *Materials* **2021**, *14*, 5049. [CrossRef]
65. Pall, E.; Roman, A.; Olah, D.; Beteg, F.I.; Cenariu, M.; Spînu, M. Enhanced Bioactive Potential of Functionalized Injectable Platelet-Rich Plasma. *Molecules* **2023**, *28*, 1943. [CrossRef]
66. Păltinean, R.; Ielciu, I.; Hanganu, D.; Niculae, M.; Pall, E.; Angenot, L.; Tits, M.; Mocan, A.; Babotă, M.; Frumuzachi, O.; et al. Biological Activities of Some Isoquinoline Alkaloids from Fumaria Schleicheri Soy. Will. *Plants* **2022**, *11*, 1202. [CrossRef]
67. Simea, S.; Ielciu, I.; Hanganu, D.; Niculae, M.; Pall, E.; Burtescu, R.F.; Olah, N.; Cenariu, M.; Oniga, I.; Benedec, D.; et al. Evaluation of the Cytotoxic, Antioxidative and Antimicrobial Effects of Dracocephalum Moldavica L. Cultivars. *Molecules* **2023**, *28*, 1604. [CrossRef]
68. Niculae, M.; Spinu, M.; Sandru, C.D.; Brudasca, F.; Cadar, D.; Szakacs, B.; Scurtu, I.; Bolfa, P.; Mates, C.I. Antimicrobial Potential of Some Lamiaceae Essential Oils against Animal Multiresistant Bacteria. *Lucr. Ştiinţifice Med. Vet.* **2009**, *42*, 170–175.
69. European Committee on Antimicrobial Susceptibility Testing, (EUCAST). *Antimicrobial Susceptibility Testing EUCAST Disk Diffusion Method*; EUCAST: Växjö, Sweden, 2020.
70. Benedec, D.; Oniga, I.; Hanganu, D.; Tiperciuc, B.; Nistor, A.; Vlase, A.M.; Vlase, L.; Pușcaș, C.; Duma, M.; Login, C.C.; et al. Stachys Species: Comparative Evaluation of Phenolic Profile and Antimicrobial and Antioxidant Potential. *Antibiotics* **2023**, *12*, 1644. [CrossRef] [PubMed]
71. Meccatti, V.M.; Santos, L.F.; Carvalho, L.S.d.; Souza, C.B.; Carvalho, C.A.T.; Marcucci, M.C.; Hasna, A.A.; de Oliveira, L.D. Antifungal Action of Herbal Plants' Glycolic Extracts against Candida Species. *Molecules* **2023**, *28*, 2857. [CrossRef] [PubMed]

antibiotics

MDPI

Article

Thai Medicinal Flowers as Natural Antioxidants and Antibacterial Agents Against Pathogenic Enteric Bacteria: A Comparative Study of *Mesua ferrea*, *Mammea siamensis*, and *Clitoria ternatea*

Sureeporn Suriyaprom [1,2], Nitsanat Cheepchirasuk [2,3], Pornpimon Ngamsaard [2], Varachaya Intachaisri [1,2], Angkhana Inta [2] and Yingmanee Tragoolpua [2,4,*]

1 Office of Research Administration, Chiang Mai University, Chiang Mai 50200, Thailand; sureeporn.suriyaprom@cmu.ac.th (S.S.); varachaya.int@cmu.ac.th (V.I.)

2 Department of Biology, Faculty of Science, Chiang Mai University, Chiang Mai 50200, Thailand; nitsanat_cheep@cmu.ac.th (N.C.); promjinnapimon@gmail.com (P.N.); angkhana.inta@cmu.ac.th (A.I.)

3 Doctor of Philosophy Program in Applied Microbiology (International Program), Faculty of Science, Chiang Mai University, Chiang Mai 50200, Thailand

4 Natural Extracts and Innovative Products for Alternative Healthcare Research Group, Faculty of Science, Chiang Mai University, Chiang Mai 50200, Thailand

* Correspondence: yingmanee.t@cmu.ac.th

Abstract: Thai medicinal flowers, namely *Mesua ferrea* L. (Bunnak), *Mammea siamensis* T. Anderson (Saraphi), and *Clitoria ternatea* (Anchan) have long been valued for their traditional medicinal. This study investigated their phytochemical composition and bioactivities, with a particular focus on antioxidant and antibacterial properties. Methods: Ethanolic flower extracts were analyzed by high-performance liquid chromatography (HPLC) and liquid chromatography–mass spectrometry (LC–MS). Antioxidant activities were determined by DPPH, ABTS, and FRAP assays. Antibacterial activity against *Escherichia coli*, *E. coli* O157:H7, *Salmonella* Typhi, *Shigella dysenteriae*, and *Vibrio cholerae* were assessed by agar well diffusion, broth dilution methods, and time–kill assays. Biofilm formation, biofilm disruption, and bacterial adhesion to Caco-2 cells were evaluated. Morphological changes in *E. coli* O157:H7 were examined using scanning electron microscopy (SEM), and leakage of intracellular contents (DNA, RNA, proteins) were quantified. Results: HPLC analysis revealed the highest level of gallic acid in *M. ferrea* and quercetin in *M. siamensis*. LC–MS analysis identified fifteen putative metabolites across the flower extracts, including quercetin, kaempferol, catechin, and luteolin derivatives, with species-specific profiles. *C. ternatea* extract exhibited the greatest total flavonoid content and antioxidant activity. Among the extracts, *M. ferrea* exhibited the strongest inhibitory effect, with inhibition zone of 13.00–15.00 mm and MIC/MBC values of 31.25–62.5 mg/mL. All extracts exhibited time-dependent bactericidal activity, significantly inhibited biofilm formation, disrupted established biofilms, and reduced bacterial adhesion to intestinal epithelial cells. SEM revealed membrane disruption in *E. coli* O157:H7 and leakage of intracellular components. Conclusions: Thai medicinal flower extracts, particularly *M. ferrea*, possess strong antioxidant and antibacterial activities. Their ability to inhibit biofilm formation, interfere with bacterial adhesion, and disrupt bacterial membranes highlights their potential as natural alternatives for preventing or controlling enteric bacterial infections.

Keywords: Thai medicinal flowers; antibacterial; antioxidant; biofilm; bacterial adhesion; enteric pathogens

https://doi.org/10.3390/antibiotics14101038

1. Introduction

Enteric bacterial pathogens are major contributors to foodborne gastroenteritis in humans, representing a significant global public health issue [1]. Notable pathogens in this category include *Salmonella* spp., *Escherichia coli*, *Listeria monocytogenes*, *Vibrio cholerae*, *Norovirus*, and *Campylobacter* [2]. The widespread and often inappropriate use of antibiotics in humans, animals, and agriculture has exacerbated the prevalence of infectious diseases and accelerated the emergence of antimicrobial resistance, particularly in developing countries [1]. This growing burden of drug-resistant infections, caused by both pathogenic and opportunistic microorganisms, presents a serious threat to public health and poses an urgent challenge to healthcare systems worldwide.

In response to the challenges posed by antimicrobial resistance, there has been growing interest in natural antimicrobial agents as alternatives to synthetic chemicals. These natural agents are often perceived by consumers as safer and more environmentally friendly. Among natural sources, flowering plants and their extracts, widely used to enhance the aroma, flavor, and color of food, have gained considerable attention for their potential as natural food preservatives [3]. Numerous studies have highlighted the bioactive properties of these plant-derived substances, particularly their antimicrobial activities, alongside other health-promoting benefits [4]. Herbal plants have long been utilized in folk medicine across various cultures and offer a diverse range of health benefits. Natural extracts from different parts of medicinal plants, such as fruits, seeds, leaves, bark, flowers, and roots, are recognized for their potential applications in nutritional enhancement, food preservation, and therapeutic use [5]. These plant-based compounds often exhibit a broad spectrum of pharmacological activities, including antimicrobial, anti-inflammatory, antioxidant, and anticancer effects. These effects made them invaluable for developing functional foods and alternative medicines [5,6]. In Southeast Asia, flowering medicinal plants have long held a central place in traditional healing practices. Among these, three Thai native species, *Mesua ferrea* Linn. (Bunnak), *Mammea siamensis* T. Anderson (Saraphi), and *Clitoria ternatea* L. (Anchan or butterfly pea), are particularly notable for their historical use in folk remedies and their promising pharmacological properties.

Mesua ferrea Linn., commonly known as Bunnak, is a perennial evergreen tree in the Clusiaceae family, found predominantly in Thailand, Malaysia, India, Sri Lanka, and parts of southern China such as Yunnan and Guangxi. Both flowers and leaves are traditionally used for medicinal purposes [7]. Phytochemical analyses have identified a range of bioactive compounds in the flowers, including coumarins, flavonoids, triterpenoids, and mangiferic acids [8]. Modern pharmacological studies support its antibacterial, anti-inflammatory, anticholinesterase, antioxidant, and anti-platelet aggregation effects [7], highlighting its potential for development into therapeutic agents and functional foods. Another important species, *Mammea siamensis* T. Anderson (Saraphi), is a small evergreen tree from the Calophyllaceae family, native to Thailand and neighboring countries such as Laos, Cambodia, Vietnam, and Myanmar. Its fragrant yellow or white blossoms have been used in Thai traditional medicine as a heart tonic, fever reducer, and appetite stimulant [9,10]. *Clitoria ternatea* L., widely known as butterfly pea or Anchan, is a climbing plant from the Fabaceae family commonly found throughout tropical and subtropical regions. Its roots, flowers, leaves, and seeds are traditionally used in Indian and Southeast Asian folk medicine and recognized for their diverse therapeutic properties [11]. The vivid blue flowers are particularly rich in flavonols, such as quercetin, myricetin, and kaempferol derivatives. Anthocyanins including various ternatin compounds (A1–A3, B1–B4, C1–C4, and D1–D3) have also been identified in the flowers. These compounds contribute to wide applications of the flowers in traditional remedies, cosmetics, and as a natural food colorant in culinary practices [12].

To explore the potential applications of these flowers, their extracts were analyzed for bioactive components, including total phenolics and flavonoids, and the evaluation of antioxidant capacity. Quantitative measurements of gallic acid and quercetin were carried out using high-performance liquid chromatography (HPLC), while liquid chromatography–mass spectrometry (LC–MS) was employed to identify and profile a broader range of phenolic and flavonoid metabolites. The antibacterial activity of these flowers was examined against pathogenic enteric bacteria, emphasizing the inhibition of bacterial growth, suppression of biofilm formation, and reduction in bacterial adhesion to intestinal epithelial cells. Furthermore, structural changes in Escherichia *coli* O157:H7 and the release of intracellular substances, such as DNA, RNA, and proteins, were investigated to clarify the underlying antibacterial mechanisms.

2. Results

2.1. Flower Extraction

Dried flowers of *M. ferrea*, *M. siamensis*, and *C. ternatea* were extracted using 95% ethanol. The percentage yields of the flower extracts are presented in Table 1. Among the three, the *M. ferrea* extract exhibited the highest yield at 18.91%, followed by *C. ternatea* (17.51%) and *M. siamensis* (16.03%).

Table 1. Percentage yields of flower extracts.

Flower Extracts	Yield (%)
Mesua ferrea	18.91
Mammea siamensis	16.03
Clitoria ternatea	17.01

2.2. Phytochemical Compounds in Flower Extracts

Phytochemical compounds in the flower extracts were quantified using high-performance liquid chromatography (HPLC). The levels of gallic acid and quercetin in *M. ferrea*, *M. siamensis*, and *C. ternatea* extracts were analyzed in this study (Figure 1). The *M. ferrea* extract contained the highest amount of gallic acid (16.956 $\pm$ 0.059 mg/g extract), followed by *M. siamensis* (0.921 $\pm$ 0.015 mg/g extract). Similarly, the quercetin content was significantly higher in the *M. siamensis* extract (0.678 $\pm$ 0.025 mg/g extract), followed by *M. ferrea* (0.260 $\pm$ 0.027 mg/g extract) (Table 2 and Figures 2 and 3). In contrast, neither gallic acid nor quercetin was detected in the *C. ternatea* extract under the HPLC conditions used (Figure 4). Nevertheless, comprehensive chemical profiles of other bioactive compounds in the flower extracts were also demonstrated using LC–MS analysis.

To further verify and expand the chemical profiling results, liquid chromatography–mass spectrometry (LC–MS) analysis was performed. The total ion chromatograms revealed a diverse array of metabolites eluting at different retention times across the three flower extracts. A total of fifteen putative metabolites were identified with high confidence (identification scores $\geq$ 1.50), as summarized in Figure 5. Among these, kaempferol exhibited the highest identification score (1.73), indicating strong reliability in compound assignment. Other detected major phenolic compounds included catechol, 3,4-dihydroxybenzoic acid, catechin, quercetin, rutin, and luteolin-6-C-glucoside. Interestingly, LC–MS analysis also confirmed the presence of HPLC-quantified compounds, namely gallic acid and quercetin, in *M. ferrea* and *M. siamensis*, while their signals were negligible in *C. ternatea* (Table S1). Additionally, species-specific metabolites were observed. Mesuaferrone A and nootkatone were abundant in *M. ferrea* whereas trigonelline and kaempferol derivatives predominated in *C. ternatea*.

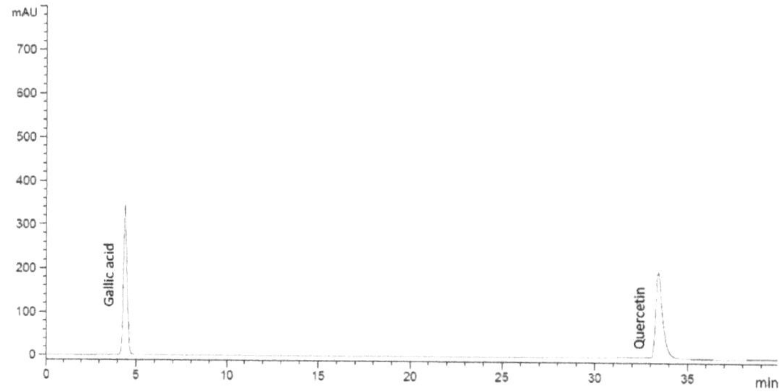

Figure 1. HPLC chromatograms of the standard mixture containing gallic acid and quercetin.

Table 2. Phytochemicals compounds of flower extracts by HPLC analysis.

Flower Extracts	Gallic Acid (mg/g Extract)	Quercetin (mg/g Extract)
Mesua ferrea	16.956 ± 0.059 [b]	0.260 ± 0.027 [a]
Mammea siamensis	0.921 ± 0.015 [a]	0.678 ± 0.025 [b]
Clitoria ternatea	ND	ND

ND; non-detected. Statistical significance was denoted by different letters ([a,b]) indicating a significant difference ($p < 0.05$).

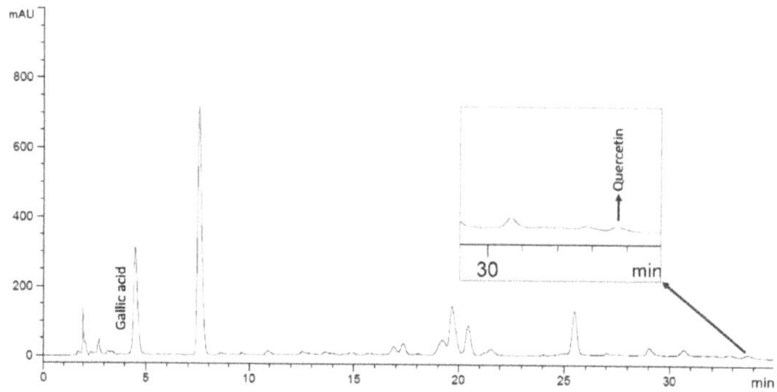

Figure 2. HPLC chromatograms of *M. ferrea* extract, indicating peaks corresponding to gallic acid and quercetin.

2.3. *Total Phenolic and Flavonoid Contents in Flower Extracts*

The bioactive compounds in the flower extracts were identified by determining their total phenolic and flavonoid contents, as presented in Table 3. The *M. ferrea* extract exhibited the highest total phenolic content (50.09 ± 1.01 milligram gallic acid equivalents per gram of extract; mg GAE/g extract), followed by *M. siamensis* (26.02 ± 0.62 mg GAE/g extract) and *C. ternatea* (10.66 ± 0.85 mg GAE/g extract). In contrast, the highest total flavonoid content was observed in the *C. ternatea* extract (19.37 ± 0.91 milligram quercetin equivalent per gram of extract; mg QE/g extract), followed by *M. siamensis* (16.51 ± 0.01 mg QE/g extract) and *M. ferrea* (12.48 ± 0.48 mg QE/g extract). All differences were statistically significant.

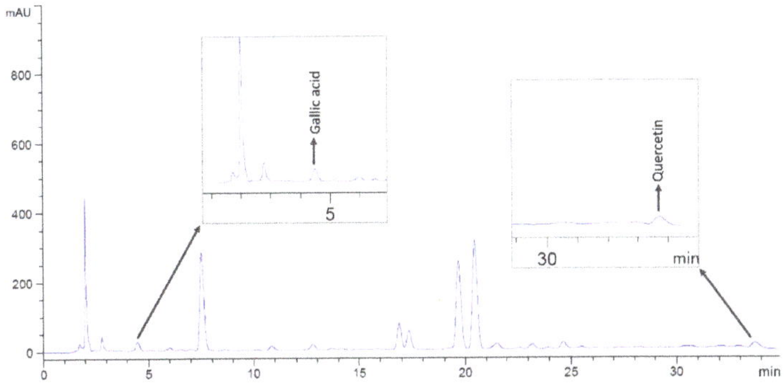

Figure 3. HPLC chromatograms of *M. siamensis* extract, indicating peaks corresponding to gallic acid and quercetin.

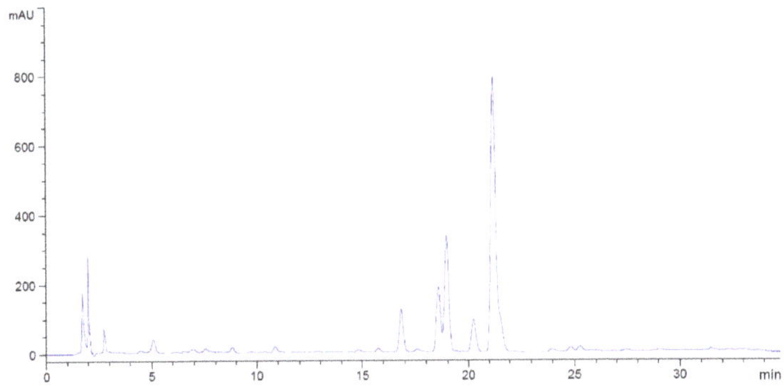

Figure 4. HPLC chromatograms of *C. ternatea* extract. Gallic acid and quercetin were not detected in this study.

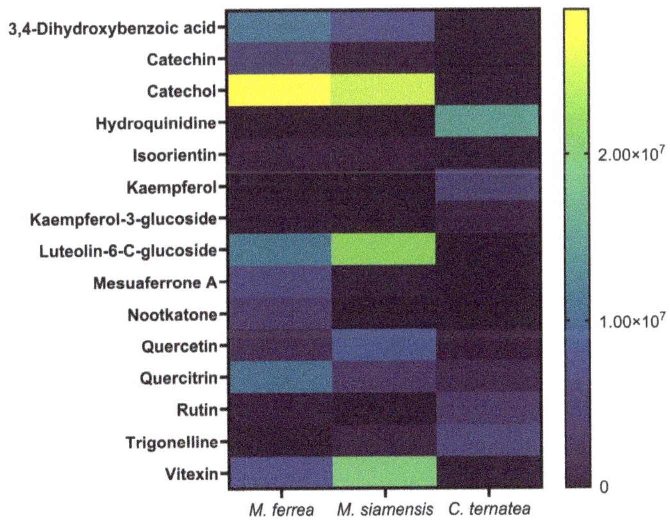

Figure 5. Heat map of LC–MS–identified metabolites in flower extracts.

Table 3. Total phenolic and flavonoid content in flower extracts.

Flower Extracts	Total Phenolic Content (mg GAE/g Extract)	Total Flavonoid Content (mg QE/g Extract)
Mesua ferrea	50.09 ± 1.01 [c]	12.48 ± 0.48 [a]
Mammea siamensis	26.02 ± 0.62 [b]	16.51 ± 0.01 [b]
Clitoria ternatea	10.66 ± 0.85 [a]	19.37 ± 0.91 [c]

Statistical significance was denoted by different letters ([a–c]) indicating a significant difference ($p < 0.05$).

2.4. Antioxidant Activities of Flower Extracts

Antioxidant properties of the flower extracts were determined using DPPH, ABTS, and FRAP assays. IC_{50} values, which indicate the concentration of extract that needed to neutralize 50% of free radicals, were calculated for both DPPH and ABTS assays. Among all tested extracts, *C. ternatea* showed the strongest antioxidant activity, with IC_{50} values of 0.30 mg/mL (DPPH) and 1.00 mg/mL (ABTS) (Table 4). The activities were also compared to reference standards, gallic acid for DPPH and Trolox for ABTS. In these comparisons, *C. ternatea* extract presented the highest antioxidant capacity, with 16.48 mg GAE/g extract in the DPPH assay and 159.46 mg TEAC/g extract in the ABTS assay. In contrast, the highest FRAP value was observed in *M. ferrea* extract, equivalent to 186.49 mg $FeSO_4$/g extract.

Table 4. Evaluation of antioxidant activities of flower extracts using DPPH, ABTS, and FRAP assays.

Flower Extracts	DPPH		ABTS		FRAP
	IC_{50} (mg/mL)	Antioxidant Activity (mg GAE/g Extract)	IC_{50} (mg/mL)	Antioxidant Activity (mg TEAC/g Extract)	Antioxidant Activity (mg $FeSO_4$/g Extract)
Mesua ferrea	2.09 ± 0.06 [c]	2.31 ± 0.14 [a]	10.96 ± 1.04 [c]	15.78 ± 1.59 [a]	186.49 ± 9.36 [c]
Mammea siamensis	0.47 ± 0.02 [b]	10.20 ± 0.25 [b]	2.87 ± 0.28 [b]	60.29 ± 5.81 [b]	81.03 ± 4.02 [b]
Clitoria ternatea	0.30 ± 0.03 [a]	16.48 ± 2.00 [c]	1.00 ± 0.14 [a]	159.46 ± 3.21 [c]	37.45 ± 3.11 [a]

Statistical significance was denoted by different letters ([a–c]) indicating a significant difference ($p < 0.05$).

2.5. Antibacterial Activities of Flower Extracts

The antibacterial potential of the flower extracts (500 mg/mL) was examined against major foodborne and enteric pathogens, including *E. coli*, *E. coli* O157:H7, *S.* Typhi DMST 22842, *S. dysenteriae* DMST 1511, and *V. cholerae*, using the agar well diffusion assay. Inhibition zone diameters were recorded to evaluate the effectiveness of each extract. Among the tested samples, *M. ferrea* extract demonstrated strong antibacterial activity, with inhibition zones ranging from 13.00 to 15.00 mm (Table 5 and Figure 6). The *M. siamensis* extract inhibited *E. coli*, *S.* Typhi, and *S. dysenteriae*, with inhibition zones ranging from 10.00 to 11.00 mm. In contrast, *C. ternatea* extract showed no inhibitory effect against any of the tested bacteria. The positive control (gentamicin at 1 mg/mL) produced inhibition zones of 26.67–29.33 mm, while no clear zones were observed in the negative control (99.9% DMSO).

The antibacterial potency of the flower extracts was assessed by determining the minimum inhibitory concentration (MIC) and minimum bactericidal concentration (MBC) using the broth dilution technique. All tested bacterial species were susceptible to the extracts, with MIC and MBC values ranging between 31.25 and 125 mg/mL (Table 6). The *M. ferrea* extract exhibited the highest activity against *S.* Typhi, *S. dysenteriae*, and *V. cholerae*, with both MIC and MBC values of 31.25 mg/mL. In contrast, the *C. ternatea* extract showed the lowest activity against all tested bacteria, with MIC and MBC values of 125 mg/mL. Gentamicin, used as a positive control, effectively inhibited the growth of all tested bacteria, with MIC and MBC values of 0.0078 mg/mL.

Table 5. Antibacterial activity of flower extracts against some pathogenic enteric bacteria.

Flower Extracts (500 mg/mL)	Inhibition Zone Diameter (mm)				
	E. coli	*E. coli* O157:H7	*S.* Typhi	*S. dysenteriae*	*V. cholerae*
Mesua ferrea	13.00 ± 0.87 [cd]	13.33 ± 1.15 [cd]	15.00 ± 0.00 [d]	14.00 ± 2.65 [d]	14.83 ± 1.04 [d]
Mammea siamensis	10.00 ± 0.00 [b]	0.00 [a]	11.00 ± 1.73 [bc]	10.67 ± 1.53 [bc]	0.00 [a]
Clitoria ternatea	0.00 [a]	0.00 [a]	0.00 [a]	0.00 [a]	0.00 [a]
Gentamicin (1 mg/mL)	26.67 ± 0.58 [e]	29.33 ± 0.58 [f]	26.67 ± 0.58 [e]	28.33 ± 0.58 [ef]	26.67 ± 0.58 [e]

Data in the table are presented as mean ± standard deviation (n = 3) for flower extracts at 500 mg/mL. Inhibition zone diameters (mm) were recorded using the agar well diffusion assay. Statistical significance was evaluated at $p < 0.05$, with distinct letters ([a–f]) denoting significant differences among groups. Gentamicin (1 mg/mL) served as the positive control.

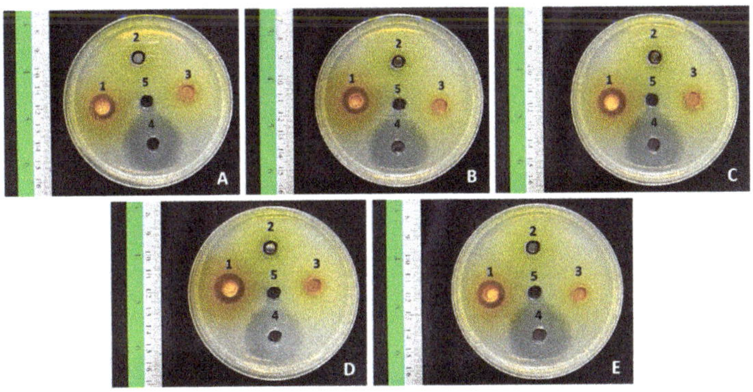

Figure 6. Antibacterial activity of (1) *M. ferrea* extract, (2) *C. ternatea* extract, (3) *M. siamensis* extract, (4) gentamicin (1 mg/mL), and (5) DMSO (99.9%) against pathogenic enteric bacteria: (**A**) *E. coli*, (**B**) *E. coli* O157:H7, (**C**) *S.* Typhi, (**D**) *S. dysenteriae*, and (**E**) *V. cholerae*.

Table 6. Evaluation of antibacterial effects of flower extracts based on MIC and MBC.

Flower Extracts	Concentration of Flower Extracts (mg/mL)									
	E. coli		*E. coli* O157:H7		*S.* Typhi		*S. dysenteriae*		*V. cholerae*	
	MIC	MBC	MIC	MBC	MIC	MBC	MIC	MBC	MIC	MBC
Mesua ferrea	62.5	62.5	62.5	62.5	31.25	31.25	31.25	31.25	31.25	31.25
Mammea siamensis	62.5	62.5	62.5	62.5	62.5	62.5	31.25	31.25	62.5	62.5
Clitoria ternatea	125	125	125	125	125	125	125	125	125	125
Gentamicin	0.0078	0.0078	0.0078	0.0078	0.0078	0.0078	0.0078	0.0078	0.0078	0.0078

The bactericidal efficacy of the flower extracts was assessed by monitoring the time required to eradicate the tested bacteria. The time to achieve no viable cell count varied among the bacterial species and flower extracts (Figure 7). *M. ferrea* extract completely eradicated *E. coli* O157:H7 within 24 h, while *M. siamensis* extract completely inhibited *S.* Typhi within 12 h. *C. ternatea* extract achieved the same result for *S.* Typhi within 24 h. Furthermore, *C. ternatea* extract completely inhibited *S. dysenteriae* after 24 h, and *M. siamensis* extract completely killed *V. cholerae* within 24 h. The flower extracts exhibited the slowest killing rate against *E. coli*, taking 24 h to inhibit more than 60% of the bacteria (Table S2).

2.6. Antibiofilm Activity of Flower Extracts

The antibiofilm activity of the flower extracts was evaluated using the crystal violet assay to assess their potential to inhibit biofilm formation. The inhibitory effects on both bacterial adherence during biofilm development and on pre-formed (established) biofilms are illustrated in Figure 8 and Tables S3 and S4. The extracts were tested at their respective MIC values, and the results are expressed as percentage inhibition relative to the untreated control group. Inhibition of bacterial adherence during biofilm formation ranged from 53.18% to 100% (Figure 8A). *M. siamensis* and *C. ternatea* extracts at their MIC values showed the highest efficacy against *E. coli*, *S. dysenteriae*, and *V. cholerae*, with inhibition rates between 93.38% and 100%. Similarly, *M. ferrea* extract effectively inhibited biofilm formation in *S.* Typhi, with inhibition rates of 96.20%. In contrast, it showed the lowest inhibitory effect on *E. coli* O157:H7, with inhibition rates of approximately 53.18%.

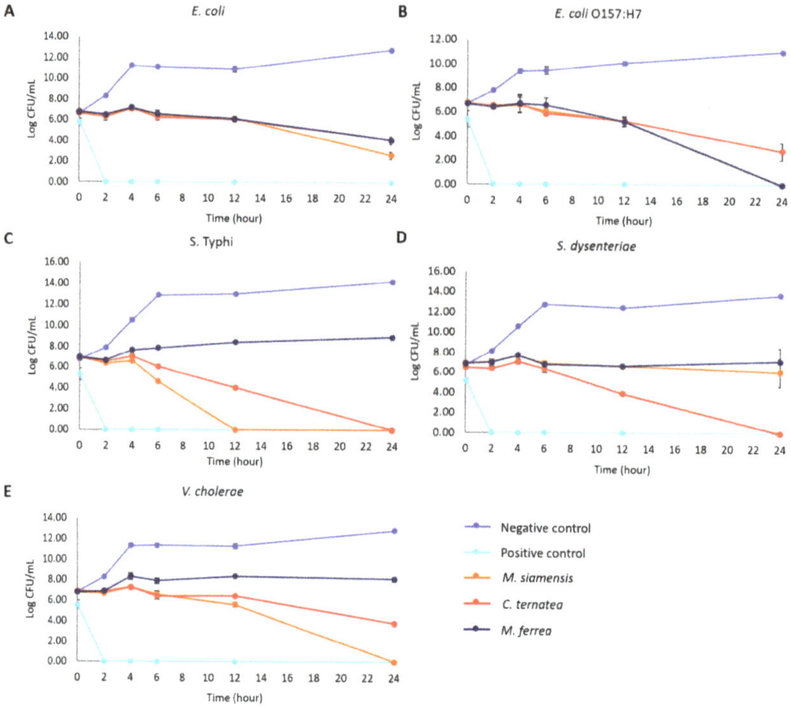

Figure 7. Effect of flower extracts (at 1 MIC) against (**A**) *E. coli*, (**B**) *E. coli* O157:H7, (**C**) *S.* Typhi, (**D**) *S. dysenteriae*, and (**E**) *V. cholerae*. Time–kill data are presented as mean ± SD (n = 3). Gentamicin at 1 mg/mL was used as the positive control, and untreated bacteria served as the negative control.

Regarding the inhibition of established biofilms, the flower extracts exhibited inhibition rates ranging from 5.25% to 100% (Figure 8B). *M. siamensis* extract demonstrated the highest inhibitory activity against established biofilms formed by *E. coli* O157:H7, with inhibition rates of 95.90%, and showed complete inhibition (100%) of *V. cholerae* biofilm formation. Additionally, *M. ferrea* extract exhibited significant activity against *E. coli* O157:H7 biofilm formation, with an inhibition rate of 96.90%. Although *C. ternatea* extract showed a strong inhibitory effect against *E. coli* (82.90%) and *S.* Typhi (93.14%) biofilm formation, it exhibited the lowest activity against *S. dysenteriae*, with an inhibition rate of only 5.25%. Gentamicin, used as the positive control, inhibited bacterial adherence during biofilm formation and against established biofilms, with inhibition rates ranging from 94.71% to 98.43% and 18.21% to 88.81%, respectively.

2.7. Antibacterial Adhesion Activity of Flower Extracts on Caco-2 Cells

The ability of the flower extracts to inhibit bacterial adhesion was assessed using Caco-2 cells. Before conducting the adhesion assay, cytotoxicity of the extracts toward Caco-2 cells was examined by MTT assay (Figure S1 and Table S5). A vehicle control (1% DMSO) was also included, no significant difference in cell viability was observed compared with the untreated control, indicating that DMSO at this concentration had no cytotoxic effect. At concentrations of 0.08 mg/mL (*M. ferrea*), 0.0024 mg/mL (*M. siamensis*), and 0.63 mg/mL (*C. ternatea*), cell viability remained comparable to the untreated control. The IC_{50} values and the corresponding selectivity indices (SI), calculated as the ratio of IC_{50} to the minimum inhibitory concentration (MIC), are presented in Table S6. These data quantitatively confirm the safety of the extracts, demonstrating that the concentrations used in the adhesion assay were well below the cytotoxic threshold.

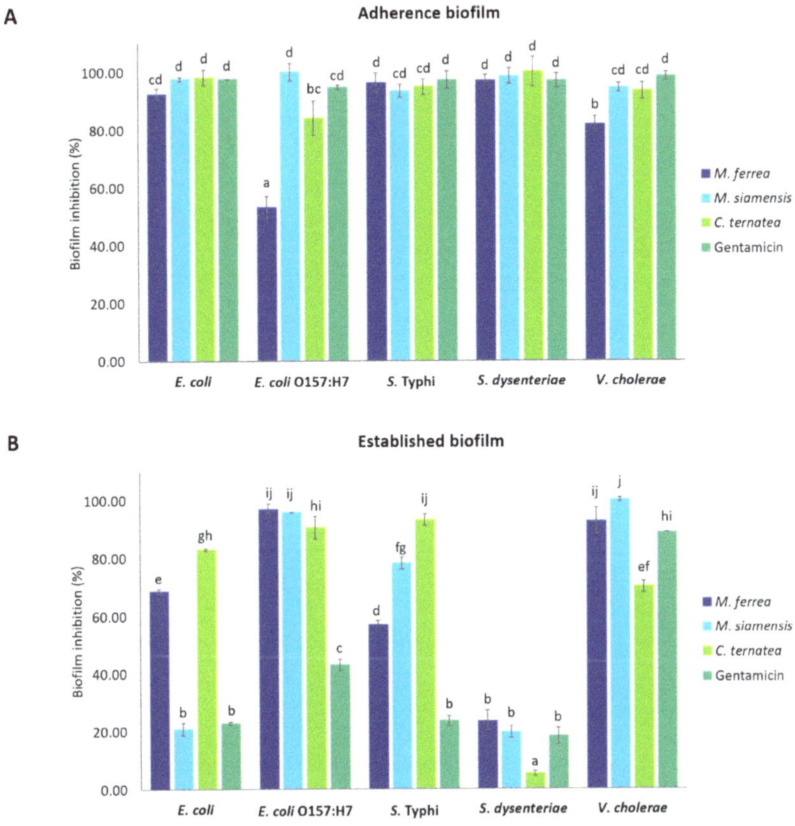

Figure 8. Inhibition of (**A**) adherence biofilm and (**B**) established biofilm by flower extracts against selected pathogenic enteric bacteria. Data are presented as mean ± standard deviation (n = 3). Statistical significance was indicated by different letters (a–j), representing significant differences ($p < 0.05$). Statistical analysis was performed using one-way ANOVA followed by Tukey's post hoc test.

The antibacterial adhesion activity of the flower extracts was expressed as a percentage relative to the untreated control. Inhibition of bacterial adhesion ranged from 57.41% to 83.14%, as shown in Table 7 and Figure 9. Among the tested extracts, *M. siamensis* and *C. ternatea* demonstrated the highest activity against the adhesion of *S.* Typhi to Caco-2 cells, with inhibition rates of 57.41% and 55.18%, respectively. *M. ferrea* exhibited inhibitory activity against *Escherichia coli* (20.49%); however, it showed the lowest efficacy in reducing *S.* Typhi adhesion, with only 3.14% inhibition. *M. siamensis* demonstrated the highest inhibitory effect on *V. cholerae* adhesion, while *C. ternatea* was most effective against *E. coli* O157:H7. In contrast, all flower extracts exhibited relatively low inhibitory activity against the adhesion of *S. dysenteriae*.

Table 7. Inhibition of bacterial adhesion to intestinal epithelial cells by flower extracts.

Flower Extracts	Inhibition of Bacterial Adhesion (%)				
	E. coli	***E. coli* O157:H7**	***S.* Typhi**	***S. dysenteriae***	***V. cholerae***
Mesua ferrea	20.49 ± 1.78 [de]	24.08 ± 0.36 [e]	3.14 ± 1.22 [a]	5.13 ± 0.52 [ab]	12.43 ± 1.41 [bc]
Mammea siamensis	16.06 ± 0.95 [cd]	24.78 ± 3.61 [e]	57.41 ± 0.21 [g]	6.41 ± 0.26 [ab]	24.29 ± 6.06 [e]
Clitoria ternatea	16.11 ± 3.53 [cd]	38.61 ± 0.39 [f]	55.18 ± 0.54 [g]	4.76 ± 2.07 [ab]	17.86 ± 1.01 [cde]

Statistical significance was denoted by different letters ([a–g]) indicating a significant difference ($p < 0.05$).

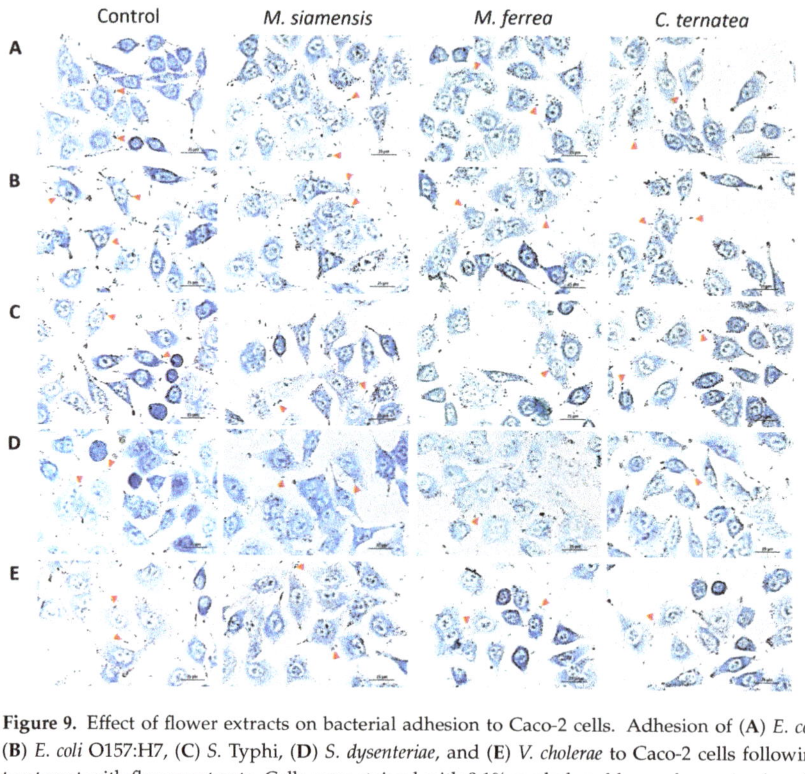

Figure 9. Effect of flower extracts on bacterial adhesion to Caco-2 cells. Adhesion of (**A**) *E. coli*, (**B**) *E. coli* O157:H7, (**C**) *S.* Typhi, (**D**) *S. dysenteriae*, and (**E**) *V. cholerae* to Caco-2 cells following treatment with flower extracts. Cells were stained with 0.1% methylene blue and examined under a light microscope. The red triangles indicate representative adhered bacterial cells. Images were captured at 400× magnification.

2.8. *Effects of Flower Extracts on the Cellular Structure of* E. coli *O157:H7*

The morphology of *E. coli* O157:H7 treated with flower extracts at 1 and $\frac{1}{2}$ MIC was examined using scanning electron microscopy (SEM), revealing marked morphological alterations. As shown in Figure 10, untreated *E. coli* O157:H7 cells exhibited a normal, rod-shaped structure with smooth and intact surfaces. In contrast, cells treated with flower extracts displayed significant morphological disruptions, including distorted shapes, wrinkled and damaged surfaces, and irregular, shriveled, and rugged appearances. These morphological changes suggest that the flower extracts induce structural damage to the bacterial cell envelope, indicating their potential antibacterial mechanism of action.

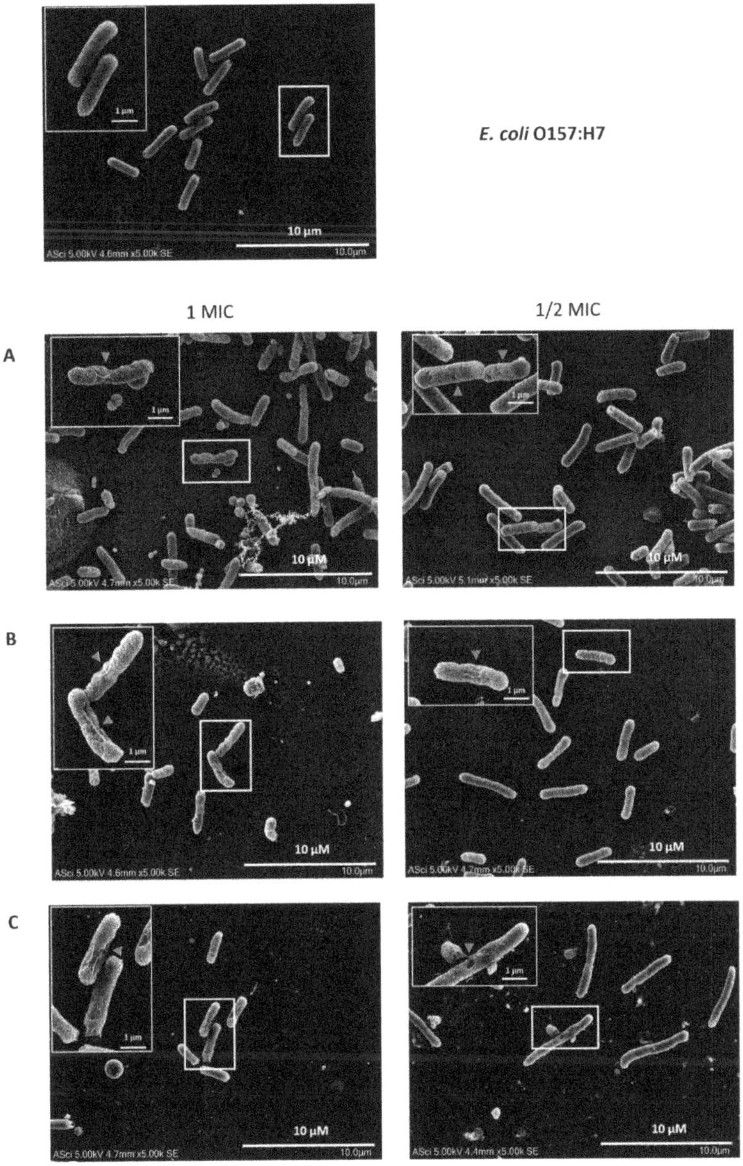

Figure 10. Morphological changes in *E. coli* O157:H7 following treatment with (**A**) *M. ferrea*, (**B**) *M. siamensis*, and (**C**) *C. ternatea*, as observed under a scanning electron microscope. The red triangles indicate representative distorted and damaged bacterial cells.

The influence of flower extracts on the integrity of bacterial cell membranes was investigated by measuring the release of DNA, RNA, and proteins from *E. coli* O157:H7 into the extracellular medium. As shown in Figure 11, the levels of extracellular DNA and

RNA increased with increasing concentrations of the flower extracts. A similar trend was observed for protein leakage (Figure 11C). These results indicate that the leakage of DNA, RNA, and protein was concentration-dependent, with higher concentrations of the extracts causing more pronounced effects.

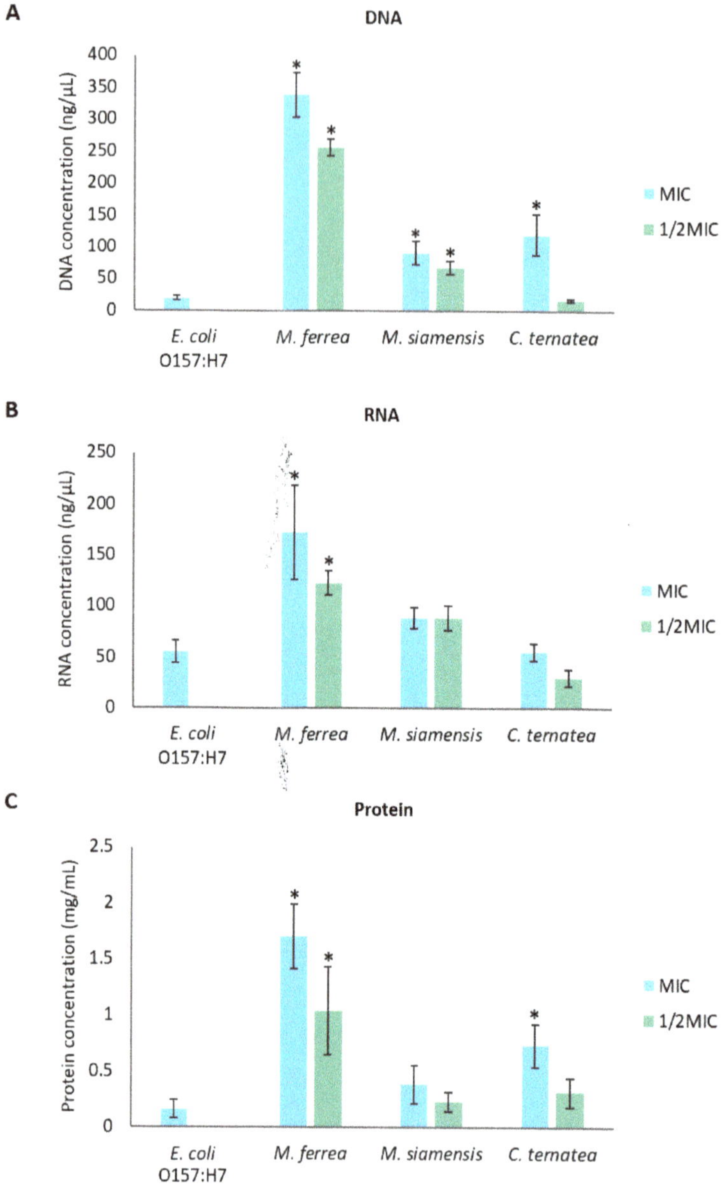

Figure 11. Effects of flower extracts on the leakage of (**A**) DNA, (**B**) RNA, and (**C**) protein from *E. coli* O157:H7. * Indicates significant difference ($p < 0.05$) compared with the untreated *E. coli* O157:H7 control group.

3. Discussion

Several phytochemical studies have demonstrated that Thai flowers such as *Mesua ferrea, Mammea siamensis,* and *Clitoria ternatea* contain a diverse array of bioactive constituents, including flavonoids, phenolic acids, and other polyphenolic compounds. These phytochemicals are known to contribute to the antioxidant, antibacterial, and anti-inflammatory properties, supporting their traditional medicinal uses and highlighting their therapeutic potential. In this study, the HPLC analysis revealed varying levels of gallic acid and quercetin among the flower extracts, which were selected as representative markers of phenolic acids and flavonoids, respectively. These two compounds were chosen as standards because they are well-characterized and commonly used as indicators of total phenolic and flavonoid content, allowing for reliable comparison across samples [13]. Although gallic acid and quercetin do not fully capture the phytochemical complexity of the extracts, they provide suitable indicators for linking chemical composition with the observed antioxidant and antibacterial activities [14]. *M. ferrea* exhibited the highest gallic acid content (16.956 ± 0.059 mg/g extract), which aligns with previous findings reporting gallic acid and quercetin levels of 13.17 mg/g extract and 0.34 mg/g extract, respectively [15]. Additionally, other phenolic compounds such as coumaric acid, catechin, rutin, and ferulic acid have also been identified in *M. ferrea* ethanolic extracts, highlighting its rich polyphenolic profile. In contrast, gallic acid and quercetin were not detected in the *C. ternatea* extract, which differs from earlier reports identifying these compounds in the same species [16]. However, our findings are consistent with those of Escher et al. (2020), who also failed to detect quercetin using HPLC-PAD-UV [11]. Interestingly, despite the non-detection of gallic acid and quercetin, *C. ternatea* extract exhibited the highest total flavonoid content and demonstrated strong antioxidant activity in both DPPH and ABTS assays. Previous studies have primarily focused on the antioxidant activity of *C. ternatea,* with reported IC_{50} values in the DPPH assay ranging from 0.08 to 4 mg/mL [12], further supporting its potent antioxidant capacity. The discrepancy between the absence of detectable gallic acid and quercetin and the observed antioxidant activity may be explained by the presence of other flavonoid compounds that were either not specifically targeted or present at concentrations below the detection limit of HPLC [17]. Moreover, total flavonoid and antioxidant assays detect a broader range of phenolic and flavonoid structures, many of which may not be individually identified by standard HPLC methods [18]. The discrepancy between the HPLC profile and antioxidant activity suggests that other bioactive flavonoids may contribute to the functional properties of the extract.

To further explore this, LC–MS analysis was conducted to obtain a more comprehensive chemical profile of the extracts. The LC–MS results confirmed the chemical diversity among the three species and revealed several metabolites. Specifically, in *C. ternatea*, compounds such as kaempferol derivatives, trigonelline, and hydroquinidine were identified, which have been reported to possess both antioxidant and antibacterial properties [19–22]. These results are consistent with *C. ternatea* extract exhibiting the highest total flavonoid content and strongest antioxidant activity. The additional peaks observed in the HPLC chromatogram of *C. ternatea*, particularly between 15 and 25 min, may correspond to these flavonoid and phenolic compounds detected by LC–MS. Therefore, these findings suggest that the biological activities of *C. ternatea* are attributed to the synergistic effects of diverse flavonoid and phenolic constituents, rather than the presence of gallic acid or quercetin alone.

Regarding *M. ferrea*, the extract exhibited the highest total phenolic content among the tested samples. HPLC analysis identified *p*-coumaric acid and gallic acid as the predominant phenolic acids, while rutin was the major flavonoid detected in the extract [15]. The elevated total phenolic content may be largely attributed to the abundant presence of gallic acid and quercetin, both of which are well-documented for their potent antioxidant properties [23]. This is consistent with the FRAP assay results, where *M. ferrea* extract demonstrated the highest antioxidant activity, further supporting the role of these phenolic compounds in contributing to its antioxidant potential. Several studies have shown that *p*-coumaric acid, gallic acid, and ferulic acid exhibit strong antioxidant activity across various assays, including DPPH, ABTS, FRAP, and ORAC [24]. Collectively, these compounds enhance the free radical scavenging ability of the extract, supporting the observed antioxidant effects.

Among the tested flower extracts, *M. ferrea* demonstrated the most potent antibacterial activity, effectively inhibiting all tested pathogenic enteric bacteria, including *E. coli*, *E. coli* O157:H7, *S.* Typhi, *S. dysenteriae*, and *V. cholerae*. In comparison, *M. siamensis* extract showed inhibitory activity against *E. coli*, *S.* Typhi, and *S. dysenteriae*, while *C. ternatea* extract exhibited no inhibitory effect against any of the tested bacteria under the same experimental conditions using the agar well diffusion method. Nevertheless, all tested extracts demonstrated bactericidal activity in broth microdilution assays, with MIC and MBC values being equal, ranging from 31.25 to 125 mg/mL. These findings contrast with a previous study, which reported that ethanolic extracts of *M. ferrea* and *M. siamensis* at 500 mg/mL inhibited *Staphylococcus aureus*, but not *E. coli*, *S. dysenteriae*, or *S.* Typhimurium when tested using the agar disk diffusion method [25].

Although *C. ternatea* extract exhibited no antibacterial activity against enteric pathogens in the agar well diffusion assay, it demonstrated clear inhibitory effects in the broth microdilution assay, as indicated by measurable MIC and MBC values. This discrepancy suggests that its antibacterial efficacy may be method-dependent, potentially influenced by factors such as compound diffusion efficiency and solubility. These limitations are well-documented in agar-based methods, where compounds with low diffusion capacity may not produce observable inhibition zones despite possessing antimicrobial activity [26]. Since agar diffusion assays rely heavily on the ability of compounds to permeate through the agar matrix, substances with high molecular weight, low aqueous solubility, or lipophilic characteristics may fail to diffuse adequately, thus underestimating their true antibacterial potential [27]. In contrast, the broth microdilution method enables uniform distribution of test substances in a liquid medium, ensuring consistent exposure between bacteria and the compounds throughout incubation [28]. This allows for more accurate detection of inhibitory activity, particularly in the case of slow-acting or low-potency antimicrobials, and offers greater sensitivity and quantification capacity compared to diffusion-based methods [29]. These findings emphasize the importance of employing complementary methodologies when evaluating the antibacterial potential of plant-derived compounds and highlight the limitations of relying solely on agar-based techniques.

Moreover, time–kill assays further substantiated the antibacterial efficacy of the flower extracts by enabling real-time monitoring of bacterial growth [29]. The extracts exhibited time-dependent bactericidal activity, as evidenced by extended lag phases and slowed log phases in all tested enteric pathogens. The prolonged lag phase reflects the bacterial adaptation period following exposure to antimicrobial stress, during which macromolecular damage may be repaired and essential cellular components synthesized for growth resumption [4]. This suggests that the flower extracts may interfere with key biosynthetic processes involved in macromolecule production. Additionally, the varied responses among bacterial species may be attributed to differences in the composition and structure of their lipopolysaccharide (LPS) layers. Despite all being Gram-negative, variations in the LPS structure can result in differential susceptibility to antibacterial agents [30].

Biofilms are organized bacterial communities encased in a matrix of self-secreted extracellular polymeric substances (EPS) [31]. The development of a biofilm starts when free-floating (planktonic) bacterial cells temporarily adhere to surfaces, either biotic or abiotic surfaces, and progresses to stable attachment and maturation of the community [32]. In the present study, the flower extracts were found to significantly inhibit both the initial adhesion phase and the development of mature biofilms in pathogenic enteric bacteria, with inhibition rates ranging from 53.18% to 100% for initial adherence and 5.25% to 96.90% for established biofilms. The higher efficacy observed against the early adhesion stage may be due to the increased susceptibility of planktonic cells, which are more accessible and vulnerable to antibacterial agents than cells embedded within mature biofilms [33]. Thus, planktonic cells are primarily responsible for initiating the biofilm matrix, targeting them can effectively disrupt biofilm development at its earliest stage [34]. In contrast, mature biofilms present a substantial challenge due to their complex EPS matrix and extended growth periods, which confer high resistance to antimicrobial agents [35]. The biofilm formation process involves not only cell adhesion via pili but also sophisticated cell-to-cell communication through quorum sensing (QS), which facilitates the coordination and maintenance of biofilm communities [36]. Disrupting QS pathways and adhesion mechanisms has been suggested as an effective strategy to prevent biofilm establishment, thereby offering a promising approach for novel therapeutic development [4]. In addition to interfering with adhesion and QS signaling, plant extracts may also reduce the availability of essential nutrients, both inorganic and organic, thereby impairing bacterial attachment and subsequent growth [37].

The antibacterial and anti-biofilm effects of Thai flower extracts are largely attributed to their phytochemical composition. Medicinal plants generally contain diverse secondary metabolites, many of which serve as natural defense agents against microbial pathogens [36]. According to previous studies, *M. ferrea* flowers were found to harbor a variety of bioactive compounds, including alkaloids, glycosides, tannins, phenolics, coumarins, sterols, xanthones, volatile oils, triterpenoids, resins, and saponins. Specific constituents such as α-copaene, germacrene D, β-amyrin, and β-sitosterol were detected, along with newly identified compounds including mesuanic acid, mesuferrols A and B, mesuaxanthones A and B, mesuaferrins A–C, mesuaferrones A and B, mesuarin, and mesuol [38]. Similarly, the phytochemical profile of *M. siamensis* flowers included predominant compounds such as coumarins and mammeasins A and B [39], as well as xanthones, triterpenes, steroids [40], and flavonoids [41]. Notably, vitexin and isovitexin were identified as functional components common to both *M. ferrea* and *M. siamensis* [7]. For *C. ternatea*, anthocyanins, particularly ternatins, were identified as the major phytochemicals [11], in addition to flavonoids such as kaempferol, quercetin, and myricetin [12]. Other constituents included saponins, tannins, glycosides, triterpenoids, steroids, and alkaloids [42]. LC–MS profiling confirmed the chemical diversity among the three Thai flower extracts. *M. ferrea* extract showed biflavonoids such as mesuaferrone A. *M. siamensis* extract contained catechol and luteolin-6-C-glucoside, while *C. ternatea* extract enriched with trigonelline, kaempferol derivatives, and hydroquinidine. These distinct phytochemical profiles contribute to the differences in antioxidant, anti-bacteria, and antibiofilm activities.

Moreover, the compounds are known to exert anti-biofilm effects through multiple mechanisms including inhibition of substrate availability, disruption of cell walls and membranes, interference with bacterial adhesion molecules, protein binding, and interaction with DNA, which may also contribute to antiviral effects [43]. Certain compounds, such as terpenoids, may promote detachment of planktonic cells from the biofilm by compromising membrane integrity, leading to effective eradication of biofilm-associated cells. Terpenoids are believed to alter the fatty acid composition of bacterial membranes, reducing surface hydrophobicity and thereby interfering with biofilm formation [36]. In comparison, the standard antibiotic gentamicin, used as a positive control, exhibited potent antibacterial and anti-biofilm activities at a lower dose. This superior efficacy can be attributed to its purified form, which contrasts with the complex mixtures of bioactive compounds present in crude plant extracts [36]. However, the diverse mechanisms of action demonstrated by the phytochemicals in the plant extracts underscore their potential as complementary or alternative therapeutic agents in the treatment of biofilm-related infections. DMSO was used as a negative control to dissolve the plant extracts, which were initially prepared at 500 mg/mL in 99.9% DMSO. In the various assays, including agar well diffusion, MIC, MBC, time–kill, and biofilm experiments, the final DMSO concentrations varied according to the dilutions of the extracts. Although DMSO at high concentrations may exhibit inhibitory effects on certain bacteria or biofilm formation [44], the specific DMSO concentrations used in each treatment are provided in Table S4 and Figure S2 to account for these potential effects.

The use of Caco-2 cells provides a physiologically relevant model to assess the impact of flower extracts on bacterial adherence to host tissues. Since adhesion is a key virulence factor, evaluating bacterial attachment to intestinal epithelial cells offers valuable insights into host–pathogen interactions [45]. Bacterial adhesion to intestinal epithelial cells represents an essential initial step in the establishment of infection. In the present work, we examined the potential of flower extracts to disrupt this early adhesion process, which is known to contribute significantly to pathogenesis [46]. Pathogenic enteric bacteria were able to adhere to Caco-2 cells due to the presence of adhesin genes. Both fimbrial and non-fimbrial adhesins are essential mediators of bacterial attachment to intestinal cells, contributing to the development of diarrheal diseases [47].

Treatment with several flower extracts significantly reduced the adhesion of pathogenic enteric bacteria to Caco-2 cells. This anti-adhesion effect can be attributed to the phytochemicals present in the extracts, particularly polyphenols. These compounds have been shown to bind to bacterial adhesins and cell walls, disrupt bacterial membranes, inhibit enzymatic activity, and chelate essential metal ions [48]. Additionally, polyphenolic compounds such as rutin, ferulic acid, and gallic acid can interfere with bacterial adhesion by modulating the expression of adhesin-related genes or by directly binding to surface proteins, thereby preventing bacterial attachment to host cell receptors [49,50]. Moreover, these compounds may alter host cell receptors or associated signaling pathways, further hindering bacterial recognition and attachment [51,52]. Further investigations are needed to elucidate the specific mechanisms underlying these effects.

The increasing resistance of bacteria, including *E. coli* O157:H7, to multiple antibiotics poses a significant challenge for treatment and demonstrates the need for alternative antimicrobial strategies [53]. Flower extracts were effective in disrupting biofilms and reducing the adhesion of various pathogenic enteric bacteria at their MIC values. These effects were further confirmed by SEM analysis of *E. coli* O157:H7, which revealed significant damage to the bacterial cell walls. In the presence of flower extracts, *E. coli* O157:H7 cells exhibited notable morphological alterations, including irregular, shriveled, and rugged surfaces, indicating a loss of structural integrity. These observations suggest that flower extracts can induce substantial disruption of the cell wall, ultimately leading to bacterial cell damage. Similarly, Zulkamal et al. [54] reported that *C. ternatea* aqueous extract induced morphological changes in *Streptococcus mutans*, resulting in the loss of the original bacterial shape and severe structural damage, including crumpling, cavitation, and irregular forms, ultimately leading to cell death. *M. ferrea* leaf and fruit extracts caused direct alterations in the morphology of *S. aureus*, affecting membrane permeability and integrity. Additionally, exposure to high levels of gallic acid induced irregularities in cell morphology in *E. coli* 3110 [55]. Although the exact mechanisms remain unclear, it has been suggested that the active components of the extracts may disrupt the bacterial cytoplasmic membrane by binding to the cell surface and penetrating the phospholipid bilayer, leading to damage to both the cell wall and membrane [56].

To further investigate the effects of flower extracts on bacterial membrane permeability, the levels of extracellular DNA, RNA, and proteins were measured in *E. coli* O157:H7. An increase in the leakage of these intracellular components indicated compromised membrane integrity following exposure to the extracts [57]. Damage to bacterial membranes leads to the release of essential ions such as K^+ and $(PO_4)^{3-}$, along with larger molecules like DNA and RNA [58]. The leakage of nucleic acids and proteins serves as a critical marker of membrane disruption. Previous studies have demonstrated that flavonoids and gallic acid can induce membrane perforation, inhibit nucleic acid synthesis, interfere with energy metabolism, and ultimately compromise the cell integrity of *E. coli* and *S.* Typhimurium [59,60]. Flavonoids interact with DNA primarily through non-covalent bonds, and the overall charge of the flavonoid–DNA complex can be affected by the type of interaction and the surrounding pH. Depending on these factors, the complex may carry a positive or negative charge [61]. This interaction can result in drastic impacts on bacterial DNA, potentially inhibiting bacterial viability [62]. Moreover, hyperoside and quercetin disrupted the cell membrane integrity of *Pseudomonas aeruginosa* and *Shewanella putrefaciens*, causing ion leakage and increased medium conductivity. SEM analysis revealed morphological damage, including shrinkage and pore formation. Both compounds inhibited motility, EPS production, and biofilm formation. They also interacted with or degraded bacterial DNA, with molecular docking suggesting binding to AT-rich minor grooves, contributing to bacterial death [63]. Therefore, it is plausible that the antibacterial activity of the flower extracts against *E. coli* O157:H7 may be attributed to flavonoids, including hyperoside and quercetin, which promote membrane damage and cellular leakage. However, several limitations must be considered in this study. First, the precise molecular mechanisms underlying the antibacterial and anti-biofilm activities of the flower extracts remain unclear. Future studies should focus on identifying and characterizing specific active compounds and their interactions with bacterial targets. Additionally, while this study used in vitro models, further research using in vivo models would provide more physiologically relevant data on the effectiveness of these extracts in real-world applications. Finally, the stability and bioavailability of the flower extracts need to be evaluated to assess their potential for use in clinical or pharmaceutical settings. Further investigations into these areas will provide a more comprehensive understanding of the therapeutic potential of these natural compounds.

4. Materials and Methods

4.1. Plant Material

The dried flowers of *M. ferrea* L. (Bunnak), *M. siamensis* T. Anderson (Saraphi), and *C. ternatea* (Anchan) were purchased from the Lanna Herbal Shop in Chiang Mai Province, northern Thailand. The plant materials were taxonomically identified by Dr. Wittaya Pongamornkul, an ethnobotanist and botanist at the Research and Conservation Department, Queen Sirikit Botanic Garden Herbarium (QSBG Herbarium). Voucher specimens of *M. ferrea* L. (WP9500), *M. siamensis* T. Anderson (WP9501), and *C. ternatea* (WP9502) were deposited at the Queen Sirikit Botanic Garden Herbarium (QSBG), Chiang Mai, Thailand.

4.2. Preparation of Flower Samples and Extraction Procedure

The extraction procedure was performed using the maceration technique [64]. The dried flowers of *M. ferrea* L., *M. siamensis*, and *C. ternatea* (Figure 12) were ground into a fine powder. For the extraction, 200 g of flower powder was mixed with 2 L of 95% ethanol (1:10 *w*/*v*) in a 2.5 L container. The mixture was stirred on an orbital shaker (IKA, Staufen, Germany) at 150 rpm at room temperature for 72 h. After extraction, the solution was filtered through Whatman No. 1 filter paper. The filtrate was then concentrated under reduced pressure at 45 °C using a rotary evaporator (Heidolph, Schwabach, Germany) and freeze-dried (LABCONCO, Kansas City, MO, USA). The resulting crude extracts were weighed to determine yield and dissolved in 99.9% DMSO (RCI Labscan, Bangkok, Thailand) to a final concentration of 500 mg/mL for subsequent experiments.

4.3. Identification of Bioactive Compounds in Flower Extracts by HPLC Analysis

The levels of gallic acid and quercetin in the flower extracts were analyzed using high-performance liquid chromatography (HPLC) following slight modifications of published protocols [65,66]. An Agilent 1260 Infinity II system (Santa Clara, CA, USA) with a C-18 column (4.6 × 150 mm, 5 µm; GL Sciences, Tokyo, Japan) maintained at 40 °C was used. The mobile phase consisted of 1% formic acid in water (solvent A) and acetonitrile (solvent B). A gradient program was applied: from 0 to 30 min, solvent A decreased linearly from 90% to 70%, and from 30 to 40 min, it increased from 70% to 95%. The flow rate was 1.0 mL/min, with detection at 254 nm.

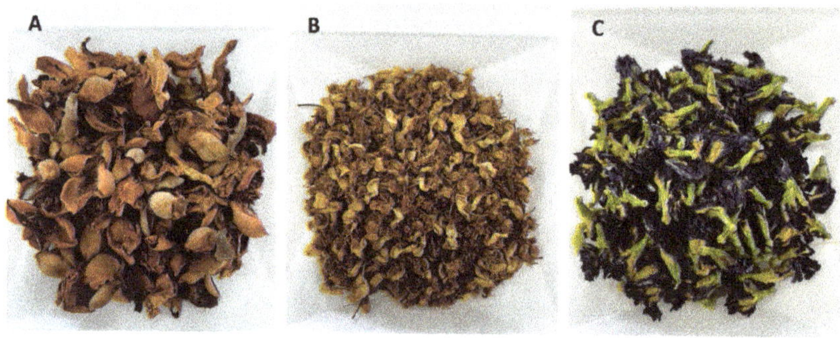

Figure 12. Dried flowers of (**A**) *Mesua ferrea* L. (Bunnak), (**B**) *Mammea siamensis* T. Anderson (Saraphi), and (**C**) *Clitoria ternatea* (Anchan).

Prior to injection, the flower extracts were dissolved in a methanol–water mixture (6:4, *v*/*v*) and filtered through a 0.45 µm sterile syringe filter. Standard solutions of gallic acid (Sigma-Aldrich, St. Louis, MO, USA) and quercetin (Merck, Billerica, MA, USA) were also prepared under the same conditions. Compounds in the extracts were identified by comparing their retention times with those of the standards (Figure 13), and concentrations were calculated using standard calibration curves.

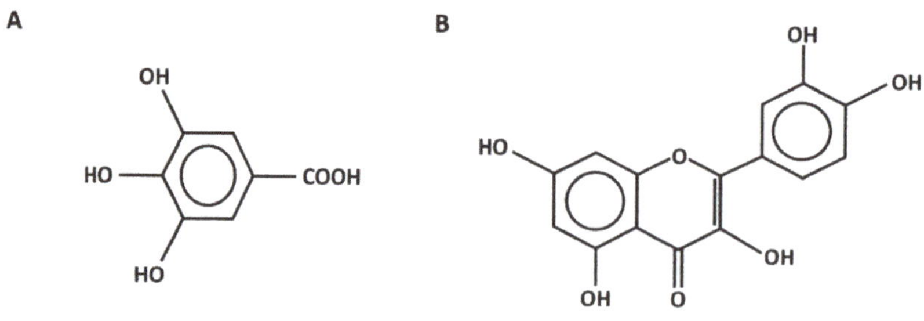

Figure 13. Chemical structure of (**A**) gallic acid and (**B**) quercetin.

4.4. Identification of Bioactive Compounds in Flower Extracts by LC-MS Analysis

The chemical constituents of the flower extracts were characterized using liquid chromatography coupled with quadrupole time-of-flight mass spectrometry (LC-QTOF-

MS; 6545XT AdvanceBio LC/Q-TOF, Agilent Technologies, Santa Clara, CA, USA) [67]. Chromatographic separation was performed on an InfinityLab Poroshell 120 EC-C18 column (2.1 × 100 mm, 2.7 µm, 120 Å; Agilent Technologies, Santa Clara, CA, USA) maintained at 50 °C. The mobile phase consisted of 0.1% (*v*/*v*) formic acid in water (solvent A) and 0.1% (*v*/*v*) formic acid in acetonitrile (solvent B). A gradient elution was applied over 20 min at a flow rate of 0.4 mL/min, with an injection volume of 10 µL.

Mass spectrometry detection was conducted under both positive and negative electrospray ionization (ESI) modes using high-resolution settings. The ion source parameters were as follows: drying gas at 325 °C and 13 L/min, sheath gas at 275 °C and 12 L/min, nebulizer pressure at 45 psi, and capillary voltage of +4000 V/−3000 V. The instrument was set to scan over an m/z range of 40–1700 (MS1) and 25–1000 (MS2), with collision energies of 20 eV in positive and 10 eV in negative modes.

Data acquisition and processing were performed using MS-DIAL version 5.3 (RIKEN CSRS, Yokohama, Kanagawa, Japan). Sulfadimethoxine served as the internal standard for normalization, and metabolite identification was based on comparison with the in-house ESI(+/−) MS/MS spectral library of authentic standards provided in MS-DIAL.

4.5. *Determination of Total Phenolic and Flavonoid Contents in Flower Extracts*

The Folin–Ciocalteu assay [68] was employed to determine total phenolic content in the flower extracts. In brief, 20 µL of each extract was added to 10 µL of 50% Folin–Ciocalteu reagent, followed by 100 µL of distilled water and 20 µL of 95% ethanol. After allowing the reaction to proceed for 5 min, 20 µL of 5% (*w*/*v*) sodium carbonate was incorporated. The mixture was incubated for 1 h at room temperature, after which the absorbance was measured at 725 nm using a microplate spectrophotometer (DYNEX SPECTRA MR, Chantilly, VA, USA). Phenolic content was quantified against a gallic acid standard curve (10–100 µg/mL) and expressed as milligrams of gallic acid equivalents per gram of extract (mg GAE/g extract).

The aluminum chloride colorimetric assay [69] was employed to quantify total flavonoids in the flower extracts. Briefly, 20 µL of each extract was combined with 4 µL of 10% aluminum chloride, 60 µL of methanol, 4 µL of 1 M potassium acetate, and 112 µL of distilled water. The reaction mixture was allowed to incubate for 30 min at room temperature, and absorbance was measured at 415 nm using a microplate spectrophotometer. Total flavonoid content was calculated using a quercetin standard curve (7.81–125 µg/mL) and expressed as mg quercetin equivalents per gram of extract (mg QE/g extract).

4.6. *Antioxidant Activities*

4.6.1. DPPH Radical Scavenging Assay

The DPPH assay [70] was used to evaluate the radical scavenging activity of flower extracts with minor modifications. In brief, a 0.1 mM DPPH solution in methanol (Sigma-Aldrich, St. Louis, MO, USA) was prepared, and 150 µL of this solution was added to 50 µL of each extract. The mixture was gently mixed and incubated in the dark for 20 min at room temperature. Absorbance was read at 517 nm using a microplate spectrophotometer. The antioxidant capacity was expressed as IC_{50}, representing the concentration of extract required to neutralize 50% of DPPH radicals, and was calculated against a gallic acid standard curve (1–8 µg/mL), with results reported as mg GAE/g extract.

4.6.2. ABTS Radical Cation Decolorization Assay

The ABTS•$^+$ assay [71] was used to determine the radical cation decolorization activity of flower extracts with slight modifications. Briefly, ABTS•$^+$ stock solution was prepared by reacting 7 mM ABTS diammonium salt with 2.45 mM potassium persulfate and incubating in the dark for 12–16 h. The solution was then diluted with distilled water to an absorbance of 0.700 ± 0.020 at 734 nm. A total of 195 μL of ABTS•$^+$ solution was added to 5 μL of each extract, mixed gently, and incubated in the dark for 10 min at room temperature. Absorbance was measured at 734 nm using a microplate spectrophotometer. The antioxidant capacity was expressed as IC_{50}, representing the extract concentration required to quench 50% of ABTS•$^+$ radicals, and calculated based on a Trolox standard curve (25–300 μg/mL), with results reported as milligrams of Trolox equivalent antioxidant capacity per gram of extract (mg TEAC/g extract).

4.6.3. Ferric Reducing Antioxidant Power (FRAP) Assay

The FRAP assay [72] was used to evaluate the ferric reducing antioxidant power of flower extracts. The FRAP working reagent was prepared by mixing 10 mL of 300 mM acetate buffer (pH 3.6), 1 mL of 10 mM TPTZ (2,4,6-tri(2-pyridyl)-s-triazine) (Merck, Billerica, MA, USA) in 40 mM HCl, and 1 mL of 20 mM ferric chloride solution. A total of 150 μL of FRAP reagent was added to 5 μL of each extract, mixed gently, and incubated in the dark for 15 min at room temperature. Absorbance was measured at 593 nm using a microplate spectrophotometer. Antioxidant capacity was calculated against a ferrous sulfate ($FeSO_4$) standard curve (10–100 μg/mL) and reported as mg milligrams of ferrous sulfate ($FeSO_4$) equivalent per gram of extract (mg $FeSO_4$/g extract).

4.7. Antibacterial Activities

4.7.1. Bacterial Strains

Pathogenic enteric bacteria used in this study; *E. coli* and *V. cholerae* were obtained from the Division of Microbiology, Department of Medical Technology, Faculty of Associated Medical Sciences, whereas *E. coli* O157:H7 DMST 12743, *S.* Typhi DMST 22842, and *S. dysenteriae* DMST 1511 were obtained from the SCB 2711 Microbiology Laboratory. These strains were utilized to evaluate the antibacterial activity of flower extracts.

4.7.2. Agar Well Diffusion Method

The antibacterial activity of flower extracts was evaluated using the agar well diffusion method [26]. Bacterial strains, including *E. coli*, *E. coli* O157:H7, *S.* Typhi, *S. dysenteriae*, and *V. cholerae*, were adjusted to a turbidity equivalent to a 0.5 McFarland standard (CLSI M07-A10) [28] and then swabbed onto Mueller–Hinton agar (MHA) plates (HiMedia, Mumbai, Maharashtra, India). A hole with an 8 mm diameter was aseptically punched using a sterile cork borer, and 100 μL of the flower extract was transferred into each well. Gentamicin (1 mg/mL; Bio Basic Inc., Markham, ON, Canada) was used as the positive control, and 99.9% DMSO served as the negative control. Plates were incubated at 37 °C for 24 h, and antimicrobial activity was evaluated by measuring the inhibition zone diameters (mm) for each bacterial species. All assays were performed in triplicate.

4.7.3. Minimum Inhibitory Concentration (MIC) and Minimum Bactericidal Concentration (MBC)

The inhibitory effect of flower extracts on bacterial growth was determined using the broth microdilution method, as indicated by the Clinical and Laboratory Standards Institute of the United States of America (CLSI M07-A10) [28]. Flower extracts were diluted in Mueller–Hinton broth (MHB) by two-fold serial dilution and added to each well of a sterile, round-bottom 96-well culture plate (100 μL/well; SPL Life Sciences, Pocheon-si, Gyeonggi-do, Korea). The bacterial suspension, previously adjusted to a 0.5 McFarland standard and containing approximately 1–2 × 10^8 colony-forming units (CFU)/mL, was appropriately diluted to obtain a final concentration of 10^6 CFU/mL. Then, 100 μL of this suspension was inoculated into each well, yielding a final bacterial concentration of approximately 10^5 CFU/mL. The plates were incubated at 37 °C for 24 h. After incubation, the minimum inhibitory concentration (MIC) was determined as the lowest extract concentration that visibly inhibited bacterial growth [73].

To determine the minimum bactericidal concentration (MBC), viable cells from wells showing no visible bacterial growth were streaked onto agar plates, followed by incubation at 37 °C for 24 h. The MBC endpoint was defined as the lowest concentration of the flower extract that resulted in a 99.9% reduction in the bacterial population [73].

4.7.4. Time–Kill Assay

The bactericidal effect of flower extracts was evaluated using a time–kill assay [74]. Bacterial suspension was adjusted to a 0.5 McFarland standard and diluted to approximately 10^5 CFU/mL. The bacterial suspension was treated with flower extracts at concentrations equal to the MIC and incubated at 37 °C. Samples were collected at 0, 2, 4, 6, 12, and 24 h, serially diluted in phosphate-buffered saline (PBS), and 100 µL of each dilution was spread onto Mueller–Hinton agar plates using a glass bead method [75]. Following incubation at 37 °C for 24 h, colonies were counted, and bacterial viability was expressed as $\log_{10}$ CFU/mL relative to the initial inoculum. All experiments were performed in triplicate.

4.8. Anti-Biofilm Assay

4.8.1. Effects on Adherence Biofilm

The effects of flower extracts on adherent biofilms were assessed using the crystal violet staining method [76]. A 100 µL bacterial inoculum was adjusted to an initial optical density at 600 nm (OD_{600}) of 0.01 in BD Bacto™ tryptic soy broth (TSB; Becton, Dickinson and Company, Franklin Lakes, NJ, USA) supplemented with 1% glucose and added to 96-well plates. Then, 100 µL of the flower extract was distributed into the wells at a final concentration of 1 MIC. The plates were incubated at 37 °C for 24 h. After incubation, the medium was discarded, and the wells were washed with PBS. Adherent cells were then fixed and stained with 0.1% crystal violet (PanReac AppliChem, Barcelona, Spain) for 10 min. The wells were washed twice with sterile distilled water to remove excess stains. The plates were air-dried to fix the biofilms, which were subsequently quantified by resolubilizing the crystal violet in 200 µL of 33% acetic acid for 5 min. Absorbance was measured at 595 nm using a microplate spectrophotometer [77]. The inhibitory effect of each extract was expressed as a percentage relative to the untreated control using the following formula:

$$\text{Relative inhibition } (\%) = \frac{\text{Absorbance } \textit{control} - \text{Absorbance } \textit{treatment}}{\text{Absorbance } \textit{control}} \times 100$$

4.8.2. Effects on Established Biofilm

The effects of flower extracts on established biofilms were assessed using the crystal violet staining method [76]. The bacterial inoculum was adjusted to an OD_{600} of 0.01 in BD Bacto™ TBS supplemented with 1% glucose. Then, 200 µL aliquots of the inoculum were transferred to 96-well plates and incubated at 37 °C to allow biofilm formation for 24 h. After removing the planktonic cells, 200 µL of the flower extracts at a concentration of 1 MIC were added to the wells and incubated at 37 °C for 24 h [78]. The biofilms were then fixed and stained with 0.1% crystal violet for 10 min. The wells were washed twice with sterile distilled water to remove excess stains. The plates were air-dried to fix the biofilms, which were subsequently quantified by resolubilizing the crystal violet in 200 µL of 33% acetic acid for 5 min. Biofilm biomass was then measured at 595 nm using a microplate spectrophotometer, following the same procedure as described above in Section 4.8.1.

4.9. Antibacterial Adhesion Activity on Caco-2 Cells

4.9.1. Cell Culture and Cytotoxicity by Flower Extracts

Caco-2 human colorectal carcinoma cells were cultured in Dulbecco's Modified Eagle's Medium (DMEM; Gibco, Grand Island, NY, USA) supplemented with 10% heat-inactivated fetal bovine serum (FBS) and 1% penicillin–streptomycin (100 U/mL penicillin and 100 μg/mL streptomycin; Gibco, Grand Island, NY, USA) at 37 °C in a 5% CO_2 incubator [79]. Cytotoxicity of the flower extracts was assessed using the MTT assay according to standard protocols [80].

4.9.2. Determination of Antibacterial Adhesion by Flower Extracts

The efficacy of the flower extracts in inhibiting bacterial adhesion to Caco-2 cells was determined using the antibacterial adhesion assay [77]. A cell suspension at 1×10^5 cells/mL was transferred to 24-well cell culture plates and incubated in a 5% CO_2 atmosphere at 37 °C for 24 h. Afterward, the medium was discarded, and a bacterial suspension grown to mid-log phase in diluted medium was added to the 24-well plates at a multiplicity of infection (MOI) of 100. Then, flower extracts at a non-toxic concentration were immediately added to the co-culture. After co-culture for 1 h, unbound bacteria were removed by washing with PBS. Adherent bacteria were detached with 0.1% Triton X-100 and quantified by viable plate counting [75]. The relative percentage of antibacterial adhesion was calculated as

$$\text{Relative antibacterial adhesion (\%)} = \frac{\text{N control} - \text{N treatment}}{\text{N control}} \times 100$$

where N control is the number of bacteria attached in the untreated control, and N treatment is the number attached in wells treated with flower extracts.

4.10. Cellular Structure of E. coli O157:H7

4.10.1. SEM Analysis

Morphological changes in *E. coli* O157:H7 after treatment with flower extracts were observed by scanning electron microscopy (SEM) [81,82]. The bacterial inoculum was adjusted to an OD_{600} of 0.01 and added to 24-well plates containing glass coverslips coated with 0.2% gelatin [83]. Flower extracts at a concentration of 1 and 1/2 MIC were added to the wells, and the plates were incubated at 37 °C for 24 h. The bacterial specimens were fixed with 2.5% (v/v) glutaraldehyde, followed by incubation with 1.0% osmium tetroxide (OsO_4). The bacterial samples were progressively dehydrated using increasing concentrations of ethanol, followed by drying at the critical point with CO_2. The dried specimens were attached to aluminum stubs, covered with a thin layer of gold, and visualized using a Hitachi SU3800 scanning electron microscope (Minato-ku, Tokyo, Japan).

4.10.2. Cell Membrane Damage Assays

Assays for nucleic acid (DNA and RNA) and protein leakage were conducted to evaluate the destructive effect of the flower extract on the *E. coli* O157:H7 cell membrane [57,84]. A 10 mL suspension of 24 h *E. coli* O157:H7 culture was centrifuged at 3500 rpm for 15 min. The resulting pellets were washed three times with PBS, after which 10 mL of PBS was added and mixed with the flower extracts at concentrations of 1 and 1/2 MIC. The mixture was incubated at 37 °C for 24 h, followed by centrifugation at 3500 rpm for 15 min. The supernatant was then used to assess cell membrane leakage. DNA and RNA released from the cytoplasm were quantified at 260 nm using a NanoDrop One^c Microvolume UV-Vis Spectrophotometers (Thermo Scientific, Waltham, MA, USA). Protein levels in the collected supernatants were determined using a BCA protein assay kit (EMD Millipore, Burlington, MA,

USA). Absorbance was recorded at 562 nm, and bovine serum albumin (BSA) was employed as the calibration standard.

4.11. Statistical Analysis

All experiments were conducted in triplicate, and the results are presented as mean ± standard deviation (SD). Statistical comparisons among groups were performed using one-way analysis of variance (ANOVA) followed by Tukey's honestly significant difference (HSD) post hoc test. Differences with $p < 0.05$ were considered statistically significant. Analyses were carried out using SPSS software (version 17.0, IBM Corp., Chicago, IL, USA). For the analysis of DNA, RNA, and protein contents, comparisons were performed exclusively between flower extract-treated samples and the untreated *E. coli* O157:H7 control group, without intergroup comparisons among treated samples.

5. Conclusions

This study demonstrated that Thai flower extracts are rich sources of phenolic and flavonoid compounds, which significantly contribute to their antioxidant activities, as evidenced by the DPPH, ABTS, and FRAP assays. LC–MS profiling further revealed the presence of specific bioactive metabolites. Despite only detecting gallic acid and quercetin through HPLC analysis, the crude extracts exhibited marked antibacterial effects against pathogenic enteric bacteria, including inhibition of biofilm formation, eradication of established biofilms, and reduction in bacterial adhesion to intestinal epithelial cells. The antibacterial mechanism, particularly against *E. coli* O157:H7, appears to involve damage to the bacterial cell wall and membrane integrity, as well as interference with nucleic acid and protein synthesis. These findings highlight the promising potential of these flower extracts as natural antibacterial agents. However, further studies, including detailed mechanistic investigations, in vivo evaluations, and assessments of stability and bioavailability, are necessary to fully elucidate their therapeutic potential, safety profiles, and practical applicability in clinical settings.

Supplementary Materials: The following supporting information can be downloaded at https://www.mdpi.com/article/10.3390/antibiotics14101038/s1. Figure S1: Effect of flower extracts on the viability of Caco-2 cells; Figure S2: Effect of DMSO on time kill assay of flower extracts; Table S1: Putative metabolites detected in flower extracts by LC–MS, showing retention time, mass-to-charge ratio, adduct ion, identification score, and relative abundance; Table S2: Inhibition of bacterial growth (%) as determined by time–kill assay; Table S3: Inhibition of adherence biofilm and established biofilm; Table S4: Inhibition of established biofilm; Table S5: Effect of DMSO on antibacterial activity of flower extracts; Table S6. IC50 values and selectivity index (SI) of flower extracts.

Author Contributions: Conceptualization, S.S. and Y.T.; methodology, S.S. and N.C.; formal analysis, S.S.; investigation, S.S., N.C., P.N. and V.I.; resources, A.I.; writing—original draft preparation, S.S.; writing—review and editing, S.S. and Y.T.; supervision, Y.T.; funding acquisition, Y.T. All authors have read and agreed to the published version of the manuscript.

Funding: This research was funded by CMU Proactive Researcher Program, Chiang Mai University, grant number 794/2567.

Institutional Review Board Statement: Not applicable.

Informed Consent Statement: Not applicable.

Data Availability Statement: The original contributions presented in this study are included in the article/Supplementary Materials; further inquiries can be directed to the corresponding author.

Acknowledgments: The authors gratefully acknowledge the support from the CMU Proactive Researcher Program, Chiang Mai University. This research was partially supported by Chiang

Mai University. The authors would also like to thank Wittaya Pongamornkul, ethnobotanist and botanist at the Queen Sirikit Botanic Garden Herbarium (QSBG), for his assistance in the taxonomic identification of the plant materials used in this study. We also thank Pitchayuth Srisai, at the Interdisciplinary Program in Biotechnology, Chiang Mai University, for helpful guidance in the LC–MS analysis.

Conflicts of Interest: The authors declare no conflicts of interest.

Abbreviations

The following abbreviations are used in this manuscript:

ABTS	2,20-azinobis-(3-ethylbenzothiazolin-6-sulfonic acid
DPPH	2,2-diphenyl-1-picrylhydrazil
FRAP	ferric reducing antioxidant power
GAE	gallic acid equivalent
HPLC	high-performance liquid chromatography
LC-MS	liquid chromatography–mass spectrometry
MIC	minimum inhibitory concentration
MBC	minimum bactericidal concentration
QE	quercetin equivalent
TEAC	trolox equivalent antioxidant capacity

References

1. Getie, M.; Abebe, W.; Tessema, B. Prevalence of Enteric Bacteria and Their Antimicrobial Susceptibility Patterns among Food Handlers in Gondar Town, Northwest Ethiopia. *Antimicrob. Resist. Infect. Control* **2019**, *8*, 111. [CrossRef]
2. Wallace, M.J.; Fishbein, S.R.S.; Dantas, G. Antimicrobial Resistance in Enteric Bacteria: Current State and next-Generation Solutions. *Gut Microbes* **2020**, *12*, 1799654. [CrossRef] [PubMed]
3. Purohit, S.R.; Rana, S.S.; Idrishi, R.; Sharma, V.; Ghosh, P. A Review on Nutritional, Bioactive, Toxicological Properties and Preservation of Edible Flowers. *Future Foods* **2021**, *4*, 100078. [CrossRef]
4. Awadelkareem, A.M.; Al-Shammari, E.; Elkhalifa, A.O.; Adnan, M.; Siddiqui, A.J.; Mahmood, D.; Azad, Z.R.A.A.; Patel, M.; Mehmood, K.; Danciu, C.; et al. Anti-Adhesion and Antibiofilm Activity of *Eruca Sativa* Miller Extract Targeting Cell Adhesion Proteins of Food-Borne Bacteria as a Potential Mechanism: Combined in Vitro-in Silico Approach. *Plants* **2022**, *11*, 610. [CrossRef] [PubMed]
5. Liu, X.; Wang, S.; Cui, L.; Zhou, H.; Liu, Y.; Meng, L.; Chen, S.; Xi, X.; Zhang, Y.; Kang, W. Flowers: Precious Food and Medicine Resources. *Food Sci. Hum. Wellness* **2023**, *12*, 1020–1052. [CrossRef]
6. Książkiewicz, M.; Karczewska, M.; Nawrot, F.; Korybalska, K.; Studzińska-Sroka, E. Traditionally Used Edible Flowers as a Source of Neuroactive, Antioxidant, and Anti-Inflammatory Extracts and Bioactive Compounds: A Narrative Review. *Molecules* **2025**, *30*, 677. [CrossRef]
7. Kuerban, M.; Ma, F.; Shan, L.; Wang, Y.; Zhou, G. Comparative Discriminant Analysis of *Mesua Ferrea* L. and Its Adulterants. *Heliyon* **2024**, *10*, e28459. [CrossRef]
8. Chahar, M.K.; Kumar, S.D.S.; Geetha, L.; Lokesh, T.; Manohara, K.P. *Mesua Ferrea* L.: A Review of the Medical Evidence for Its Phytochemistry and Pharmacological Actions. *Afr. J. Pharm. Pharmacol.* **2013**, *7*, 211–219. [CrossRef]
9. Luo, F.; Manse, Y.; Chaipech, S.; Pongpiriyadacha, Y.; Muraoka, O.; Morikawa, T. Structures of Mammeasins P and Q, Coumarin-Related Polysubstituted Benzofurans, from the Thai Medicinal Plant *Mammea Siamensis* (Miq.) T. Anders.: Anti-Proliferative Activity of Coumarin Constituents against Human Prostate Carcinoma Cell Line LNCaP. *Pharmaceuticals* **2023**, *16*, 231. [CrossRef]
10. Sitthisuk, P.; Poorahong, W.; Innajak, S.; Krajarng, A.; Samosorn, S.; Watanapokasin, R. *Mammea Siamensis* Flower Extract-Induced Cell Death Apoptosis in HCT116 Colon Cancer Cells via Vacuolar-Type H+-ATPase Inhibition Associated with GSK-3β/β-Catenin, PI3K/Akt/NF-κB, and MAPK Signaling Pathway. *Pharmaceuticals* **2025**, *18*, 441. [CrossRef]
11. Escher, G.B.; Marques, M.B.; do Carmo, M.A.V.; Azevedo, L.; Furtado, M.M.; Sant'Ana, A.S.; da Silva, M.C.; Genovese, M.I.; Wen, M.; Zhang, L.; et al. *Clitoria Ternatea* L. Petal Bioactive Compounds Display Antioxidant, Antihemolytic and Antihypertensive Effects, Inhibit α-Amylase and α-Glucosidase Activities and Reduce Human LDL Cholesterol and DNA Induced Oxidation. *Food Res. Int.* **2020**, *128*, 108763. [CrossRef]
12. Jeyaraj, E.J.; Lim, Y.Y.; Choo, W.S. Antioxidant, Cytotoxic, and Antibacterial Activities of *Clitoria Ternatea* Flower Extracts and Anthocyanin-Rich Fraction. *Sci. Rep.* **2022**, *12*, 14890. [CrossRef] [PubMed]

13. Sulastri, E.; Zubair, M.; Anas, N.; Abidin, S.; Hardani, R.; Yulianti, R.; Yulianti, R. Total Phenolic, Total Flavonoid, Quercetin Content and Antioxidant Activity of Standardized Extract of Moringa Oleifera Leaf from Regions with Different Elevation. *Phcog J.* **2018**, *10*, s104–s108. [CrossRef]
14. Sulaiman, M.; Ebehairy, L.; Nissapatorn, V.; Rahmatullah, M.; Villegas, J.; Dupa, H.J.; Verzosa, R.C.; Dolma, K.G.; Shabaz, M.; Lanting, S.; et al. Antibacterial Phenolic Compounds from the Flowering Plants of Asia and the Pacific: Coming to the Light. *Pharm. Biol.* **2024**, *62*, 713–766. [CrossRef] [PubMed]
15. Plekratoke, K.; Boonyarat, C.; Monthakantirat, O.; Nualkaew, N.; Wangboonskul, J.; Awale, S.; Chulikhit, Y.; Daodee, S.; Khamphukdee, C.; Chaiwiwatrakul, S.; et al. The Effect of Ethanol Extract from *Mesua Ferrea* Linn Flower on Alzheimer's Disease and Its Underlying Mechanism. *Curr. Issues Mol. Biol.* **2023**, *45*, 4063–4079. [CrossRef]
16. Srichaikul, B. Ultrasonication extraction, bioactivity, antioxidant activity, total flavonoid, total phenolic and antioxidant of *Clitoria ternatea* Linn flower extract for anti-aging drinks. *Pharmacogn. Mag.* **2018**, *14*, 56. [CrossRef]
17. Dong, M.; Llanas, A. The Essence of Modern HPLC: Advantages, Limitations, Fundamentals, and Opportunities. *LCGC Asia Pac.* **2013**, *31*, 472.
18. Oldoni, T.L.C.; Da Silva, R.C.; Carpes, S.T.; Massarioli, A.P.; Alencar, S.M.D. Antioxidant Activity and Development of One Chromatographic Method to Determine the Phenolic Compounds from Agroindustrial Pomace. *Ann. Acad. Bras. Ciênc.* **2020**, *92*, e20181068. [CrossRef]
19. Periferakis, A.; Periferakis, K.; Badarau, I.A.; Petran, E.M.; Popa, D.C.; Caruntu, A.; Costache, R.S.; Scheau, C.; Caruntu, C.; Costache, D.O. Kaempferol: Antimicrobial Properties, Sources, Clinical, and Traditional Applications. *Int. J. Mol. Sci.* **2022**, *23*, 15054. [CrossRef]
20. Kar, A.; Mukherjee, S.K.; Barik, S.; Hossain, S.T. Antimicrobial Activity of Trigonelline Hydrochloride Against Pseudomonas Aeruginosa and Its Quorum-Sensing Regulated Molecular Mechanisms on Biofilm Formation and Virulence. *ACS Infect. Dis.* **2024**, *10*, 746–762. [CrossRef]
21. Sekhar, M.G.; Ramudu Shanmugam, K.; Chakrapani, I.S. Trigonelline, a Fenugreek Bioactive Protects Heart Tissue against Alcohol Intoxication: An in-Vivo Study Focusing on Antioxidant Perspective. *J. Ayurveda Integr. Med.* **2024**, *15*, 100963. [CrossRef]
22. Tatsimo, S.J.N.; Tamokou, J.d.D.; Havyarimana, L.; Csupor, D.; Forgo, P.; Hohmann, J.; Kuiate, J.-R.; Tane, P. Antimicrobial and Antioxidant Activity of Kaempferol Rhamnoside Derivatives from Bryophyllum Pinnatum. *BMC Res. Notes* **2012**, *5*, 158. [CrossRef] [PubMed]
23. Zhang, M.-Q.; Zhang, J.; Zhang, Y.-T.; Sun, J.-Y.; Prieto, M.A.; Simal-Gandara, J.; Putnik, P.; Li, N.-Y.; Liu, C. The Link between the Phenolic Composition and the Antioxidant Activity in Different Small Berries: A Metabolomic Approach. *LWT* **2023**, *182*, 114853. [CrossRef]
24. Velderrain-Rodríguez, G.R.; Torres-Moreno, H.; Villegas-Ochoa, M.A.; Ayala-Zavala, J.F.; Robles-Zepeda, R.E.; Wall-Medrano, A.; González-Aguilar, G.A. Gallic Acid Content and an Antioxidant Mechanism Are Responsible for the Antiproliferative Activity of 'Ataulfo'Mango Peel on LS180 Cells. *Molecules* **2018**, *23*, 695. [CrossRef] [PubMed]
25. Batm, A.S.; Chaiyawatthanananthn, P.; Itharat, A. Antibacterial Activity of Extracts from a Thai Traditional Remedy Called Prasachandaeng and Its Plant Components. *J. Med. Assoc. Thai* **2017**, *100*, S58–S66.
26. Balouiri, M.; Sadiki, M.; Ibnsouda, S.K. Methods for in Vitro Evaluating Antimicrobial Activity: A Review. *J. Pharm. Anal.* **2016**, *6*, 71–79. [CrossRef]
27. Ríos, J.L.; Recio, M.C. Medicinal Plants and Antimicrobial Activity. *J. Ethnopharmacol.* **2005**, *100*, 80–84. [CrossRef]
28. *Methods for Dilution Antimicrobial Susceptibility Tests for Bacteria that Grow Aerobically: M07-A10; Approved Standard*, 10th ed.; Clinical and Laboratory Standards Institute, Ed.; Documents/Clinical and Laboratory Standards Institute; Committee for Clinical Laboratory Standards: Wayne, PA, USA, 2015; ISBN 978-1-56238-987-1.
29. Hossain, T.J. Methods for Screening and Evaluation of Antimicrobial Activity: A Review of Protocols, Advantages, and Limitations. *Eur. J. Microbiol. Immunol.* **2024**, *14*, 97–115. [CrossRef]
30. Reygaert, W.C. An Overview of the Antimicrobial Resistance Mechanisms of Bacteria. *AIMS Microbiol.* **2018**, *4*, 482–501. [CrossRef]
31. Saharan, B.S.; Beniwal, N.; Duhan, J.S. From Formulation to Function: A Detailed Review of Microbial Biofilms and Their Polymer-Based Extracellular Substances. *Microbe* **2024**, *5*, 100194. [CrossRef]
32. Liu, H.Y.; Prentice, E.L.; Webber, M.A. Mechanisms of Antimicrobial Resistance in Biofilms. *npj Antimicrob. Resist.* **2024**, *2*, 1–10. [CrossRef]
33. Vani, S.; Vadakkan, K.; Mani, B. A Narrative Review on Bacterial Biofilm: Its Formation, Clinical Aspects and Inhibition Strategies. *Future J. Pharm. Sci.* **2023**, *9*, 50. [CrossRef]
34. Banerjee, A.; Chowdhury, P.; Bauri, K.; Saha, B.; De, P. Inhibition and Eradication of Bacterial Biofilm Using Polymeric Materials. *Biomater. Sci.* **2023**, *11*, 11–36. [CrossRef] [PubMed]
35. da Silva, R.A.G.; Afonina, I.; Kline, K.A. Eradicating Biofilm Infections: An Update on Current and Prospective Approaches. *Curr. Opin. Microbiol.* **2021**, *63*, 117–125. [CrossRef] [PubMed]

36. Perumal, S.; Mahmud, R. Chemical Analysis, Inhibition of Biofilm Formation and Biofilm Eradication Potential of *Euphorbia Hirta* L. against Clinical Isolates and Standard Strains. *BMC Complement. Altern. Med.* **2013**, *13*, 346. [CrossRef] [PubMed]
37. Sandasi, M.; Leonard, C.M.; Viljoen, A.M. The in Vitro Antibiofilm Activity of Selected Culinary Herbs and Medicinal Plants against *Listeria Monocytogenes*. *Lett. Appl. Microbiol.* **2010**, *50*, 30–35. [CrossRef]
38. Kshirsagar, P.R.; Patil, S.M. Phytochemistry and Pharmacology of *Mesua Ferrea* L. In *Bioactive Compounds in Underutilized Fruits and Nuts*; Murthy, H.N., Bapat, V.A., Eds.; Springer International Publishing: Cham, Switzerland, 2020; pp. 223–256, ISBN 978-3-030-30182-8.
39. Morikawa, T.; Sueyoshi, M.; Chaipech, S.; Matsuda, H.; Nomura, Y.; Yabe, M.; Matsumoto, T.; Ninomiya, K.; Yoshikawa, M.; Pongpiriyadacha, Y.; et al. Suppressive Effects of Coumarins from *Mammea Siamensis* on Inducible Nitric Oxide Synthase Expression in RAW264.7 Cells. *Bioorg Med. Chem.* **2012**, *20*, 4968–4977. [CrossRef]
40. Chaniad, P.; Chukaew, A.; Payaka, A.; Phuwajaroanpong, A.; Techarang, T.; Plirat, W.; Punsawad, C. Antimalarial Potential of Compounds Isolated from *Mammea Siamensis* T. Anders. Flowers: In Vitro and Molecular Docking Studies. *BMC Complement. Med. Ther.* **2022**, *22*, 266. [CrossRef]
41. Tung, N.H.; Uto, T.; Sakamoto, A.; Hayashida, Y.; Hidaka, Y.; Morinaga, O.; Lhieochaiphant, S.; Shoyama, Y. Antiproliferative and Apoptotic Effects of Compounds from the Flower of *Mammea Siamensis* (Miq.) T. Anders. on Human Cancer Cell Lines. *Bioorg Med. Chem. Lett.* **2013**, *23*, 158–162. [CrossRef]
42. Multisona, R.R.; Shirodkar, S.; Arnold, M.; Gramza-Michalowska, A. *Clitoria Ternatea* Flower and Its Bioactive Compounds: Potential Use as Microencapsulated Ingredient for Functional Foods. *Appl. Sci.* **2023**, *13*, 2134. [CrossRef]
43. Shamim, A.; Ali, A.; Iqbal, Z.; Mirza, M.A.; Aqil, M.; Kawish, S.M.; Siddiqui, A.; Kumar, V.; Naseef, P.P.; Alshadidi, A.A.F.; et al. Natural Medicine a Promising Candidate in Combating Microbial Biofilm. *Antibiotics* **2023**, *12*, 299. [CrossRef]
44. Summer, K.; Browne, J.; Hollanders, M.; Benkendorff, K. Out of Control: The Need for Standardised Solvent Approaches and Data Reporting in Antibiofilm Assays Incorporating Dimethyl-Sulfoxide (DMSO). *Biofilm* **2022**, *4*, 100081. [CrossRef]
45. Bonetti, A.; Toschi, A.; Tugnoli, B.; Piva, A.; Grilli, E. A Blend of Selected Botanicals Maintains Intestinal Epithelial Integrity and Reduces Susceptibility to *Escherichia Coli* F4 Infection by Modulating Acute and Chronic Inflammation in Vitro. *Front. Vet. Sci.* **2023**, *10*, 1275802. [CrossRef]
46. Famuyide, I.M.; Aro, A.O.; Fasina, F.O.; Eloff, J.N.; McGaw, L.J. Antibacterial Activity and Mode of Action of Acetone Crude Leaf Extracts of Under-Investigated Syzygium and Eugenia (Myrtaceae) Species on Multidrug Resistant Porcine Diarrhoeagenic Escherichia Coli. *BMC Vet. Res.* **2019**, *15*, 162. [CrossRef] [PubMed]
47. Hernandes, R.T.; Velsko, I.; Sampaio, S.C.F.; Elias, W.P.; Robins-Browne, R.M.; Gomes, T.A.T.; Girón, J.A. Fimbrial Adhesins Produced by Atypical Enteropathogenic *Escherichia Coli* Strains. *Appl. Environ. Microbiol.* **2011**, *77*, 8391–8399. [CrossRef] [PubMed]
48. Tiwari, P.; Kumar, B.; Kaur, M.; Kaur, G.; Kaur, H. Phytochemical Screening and Extraction: A Review. *Int. Pharm. Sci.* **2011**, *1*, 98–106.
49. Borges, A.; Saavedra, M.J.; Simões, M. The Activity of Ferulic and Gallic Acids in Biofilm Prevention and Control of Pathogenic Bacteria. *Biofouling* **2012**, *28*, 755–767. [CrossRef]
50. Ivanov, M.; Novović, K.; Malešević, M.; Dinić, M.; Stojković, D.; Jovčić, B.; Soković, M. Polyphenols as Inhibitors of Antibiotic Resistant Bacteria—Mechanisms Underlying Rutin Interference with Bacterial Virulence. *Pharmaceuticals* **2022**, *15*, 385. [CrossRef]
51. Klančnik, A.; Šimunović, K.; Sterniša, M.; Ramić, D.; Smole Možina, S.; Bucar, F. Anti-Adhesion Activity of Phytochemicals to Prevent *Campylobacter Jejuni* Biofilm Formation on Abiotic Surfaces. *Phytochem. Rev.* **2021**, *20*, 55–84. [CrossRef]
52. Lewis, A.J.; Richards, A.C.; Mendez, A.A.; Dhakal, B.K.; Jones, T.A.; Sundsbak, J.L.; Eto, D.S.; Rousek, A.A.; Mulvey, M.A. Plant Phenolics Inhibit Focal Adhesion Kinase and Suppress Host Cell Invasion by Uropathogenic *Escherichia Coli*. *Infect. Immun.* **2024**, *92*, e00080-24. [CrossRef]
53. Haile, A.F.; Alonso, S.; Berhe, N.; Atoma, T.B.; Boyaka, P.N.; Grace, D. Prevalence, Antibiogram, and Multidrug-Resistant Profile of E. Coli O157: H7 in Retail Raw Beef in Addis Ababa, Ethiopia. *Front. Vet. Sci.* **2022**, *9*, 734896. [CrossRef]
54. Zulkamal, L.M.; Zelan, M.A.H.A.; Aris, F.; Zakaria, N.A.; Yusof, F.Z.M.; Ibrahim, D.; Jalil, M.T.M.J. Aqueous Extract of *Clitoria Ternatea* Attenuates the Growth of *Streptococcus Mutans*. *J. Pure Appl. Microbiol.* **2023**, *17*, 1047–1055. [CrossRef]
55. Liu, L.; Ma, X.; Bilal, M.; Wei, L.; Tang, S.; Luo, H.; Zhao, Y.; Wang, Z.; Duan, X. Toxicity and Inhibition Mechanism of Gallic Acid on Physiology and Fermentation Performance of *Escherichia Coli*. *Bioresour. Bioprocess.* **2022**, *9*, 76. [CrossRef] [PubMed]
56. Aruldass, C.A.; Marimuthu, M.M.; Ramanathan, S.; Mansor, S.M.; Murugaiyah, V. Effects of *Mesua Ferrea* Leaf and Fruit Extracts on Growth and Morphology of *Staphylococcus Aureus*. *Microsc. Microanal.* **2013**, *19*, 254–260. [CrossRef] [PubMed]
57. Tang, C.; Chen, J.; Zhang, L.; Zhang, R.; Zhang, S.; Ye, S.; Zhao, Z.; Yang, D. Exploring the Antibacterial Mechanism of Essential Oils by Membrane Permeability, Apoptosis and Biofilm Formation Combination with Proteomics Analysis against Methicillin-Resistant *Staphylococcus Aureus*. *Int. J. Med. Microbiol.* **2020**, *310*, 151435. [CrossRef]
58. Sulistyani, N.; Mahfudh, N.; Mawardi, R.H.; Zakaria, Z.A. Cell Leakage Mechanism and Time Kill Studies on *Staphylococcus Aureus* after Exposure to Ethanol Leaf Extract of *Muntingia Calabura* L. *Trop. J. Pharm. Res.* **2023**, *22*, 355–362. [CrossRef]

59. He, M.; Wu, T.; Pan, S.; Xu, X. Antimicrobial Mechanism of Flavonoids against *Escherichia Coli* ATCC 25922 by Model Membrane Study. *Appl. Surf. Sci.* **2014**, *305*, 515–521. [CrossRef]
60. Hossain, M.A.; Park, H.-C.; Lee, K.-J.; Park, S.-W.; Park, S.-C.; Kang, J. In Vitro Synergistic Potentials of Novel Antibacterial Combination Therapies against *Salmonella Enterica* Serovar Typhimurium. *BMC Microbiol.* **2020**, *20*, 118. [CrossRef]
61. Yilmaz, H.; Gultekin Subasi, B.; Celebioglu, H.U.; Ozdal, T.; Capanoglu, E. Chemistry of Protein-Phenolic Interactions toward the Microbiota and Microbial Infections. *Front. Nutr.* **2022**, *9*, 914118. [CrossRef]
62. Palamae, S.; Mittal, A.; Buatong, J.; Zhang, B.; Hong, H.; Benjakul, S. Chitooligosaccharide-Catechin Conjugate: Antimicrobial Mechanisms toward *Vibrio Parahaemolyticus* and Its Use in Shucked Asian Green Mussel. *Food Control* **2023**, *151*, 109794. [CrossRef]
63. Tagrida, M.; Palamae, S.; Saetang, J.; Ma, L.; Hong, H.; Benjakul, S. Comparative Study of Quercetin and Hyperoside: Antimicrobial Potential towards Food Spoilage Bacteria, Mode of Action and Molecular Docking. *Foods* **2023**, *12*, 4051. [CrossRef] [PubMed]
64. Abubakar, A.R.; Haque, M. Preparation of Medicinal Plants: Basic Extraction and Fractionation Procedures for Experimental Purposes. *J. Pharm. Bioallied Sci.* **2020**, *12*, 1–10. [CrossRef] [PubMed]
65. Fernandes, F.H.A.; Salgado, H.R.N. Gallic Acid: Review of the Methods of Determination and Quantification. *Crit. Rev. Anal. Chem.* **2016**, *46*, 257–265. [CrossRef] [PubMed]
66. Mansour, F.R.; Abdallah, I.A.; Bedair, A.; Hamed, M. Analytical Methods for the Determination of Quercetin and Quercetin Glycosides in Pharmaceuticals and Biological Samples. *Crit. Rev. Anal. Chem.* **2025**, *55*, 187–212. [CrossRef]
67. Zhong, B.; Robinson, N.A.; Warner, R.D.; Barrow, C.J.; Dunshea, F.R.; Suleria, H.A.R. LC-ESI-QTOF-MS/MS Characterization of Seaweed Phenolics and Their Antioxidant Potential. *Mar. Drugs* **2020**, *18*, 331. [CrossRef]
68. Lamuela-Raventós, R.M. Folin–Ciocalteu Method for the Measurement of Total Phenolic Content and Antioxidant Capacity. In *Measurement of Antioxidant Activity & Capacity*; John Wiley & Sons, Ltd.: Hoboken, NJ, USA, 2018; pp. 107–115, ISBN 978-1-119-13538-8.
69. Matić, P.; Sabljić, M.; Jakobek, L. Validation of Spectrophotometric Methods for the Determination of Total Polyphenol and Total Flavonoid Content. *J. AOAC Int.* **2017**, *100*, 1795–1803. [CrossRef]
70. Gulcin, İ.; Alwasel, S.H. DPPH Radical Scavenging Assay. *Processes* **2023**, *11*, 2248. [CrossRef]
71. Kut, K.; Cieniek, B.; Stefaniuk, I.; Bartosz, G.; Sadowska-Bartosz, I. A Modification of the ABTS• Decolorization Method and an Insight into Its Mechanism. *Processes* **2022**, *10*, 1288. [CrossRef]
72. Benzie, I.F.F.; Devaki, M. The Ferric Reducing/Antioxidant Power (FRAP) Assay for Non-Enzymatic Antioxidant Capacity: Concepts, Procedures, Limitations and Applications. In *Measurement of Antioxidant Activity & Capacity*; John Wiley & Sons, Ltd.: Hoboken, NJ, USA, 2018; pp. 77–106, ISBN 978-1-119-13538-8.
73. Parvekar, P.; Palaskar, J.; Metgud, S.; Maria, R.; Dutta, S. The Minimum Inhibitory Concentration (MIC) and Minimum Bactericidal Concentration (MBC) of Silver Nanoparticles against *Staphylococcus Aureus*. *Biomater. Investig. Dent.* **2020**, *7*, 105–109. [CrossRef]
74. Qureshi, K.A.; Imtiaz, M.; Parvez, A.; Rai, P.K.; Jaremko, M.; Emwas, A.-H.; Bholay, A.D.; Fatmi, M.Q. In Vitro and in Silico Approaches for the Evaluation of Antimicrobial Activity, Time-Kill Kinetics, and Anti-Biofilm Potential of Thymoquinone (2-Methyl-5-Propan-2-Ylcyclohexa-2, 5-Diene-1, 4-Dione) against Selected Human Pathogens. *Antibiotics* **2022**, *11*, 79. [CrossRef]
75. Prusokiene, A.; Hawkins, M.; Nieduszynski, C.A.; Retkute, R. Effectiveness of Glass Beads for Plating Cell Cultures. *Phys. Rev. E* **2021**, *103*, 052410. [CrossRef]
76. Gascón, E.; Merino, N.; Pagán, E.; Berdejo, D.; Pagán, R.; García-Gonzalo, D. Assessment of in Vitro Biofilms by Plate Count and Crystal Violet Staining: Is One Technique Enough? In *Detection and Enumeration of Bacteria, Yeast, Viruses, and Protozoan in Foods and Freshwater*; Magnani, M., Ed.; Springer: New York, NY, USA, 2021; pp. 53–63, ISBN 978-1-0716-1932-2.
77. Molina Bertrán, S.d.C.; Monzote, L.; Cappoen, D.; Escalona Arranz, J.C.; Gordillo Pérez, M.J.; Rodríguez-Ferreiro, A.O.; Chill Nuñez, I.; Novo, C.P.; Méndez, D.; Cos, P.; et al. Inhibition of Bacterial Adhesion and Biofilm Formation by Seed-Derived Ethanol Extracts from *Persea Americana* Mill. *Molecules* **2022**, *27*, 5009. [CrossRef] [PubMed]
78. Jaśkiewicz, M.; Janczura, A.; Nowicka, J.; Kamysz, W. Methods Used for the Eradication of Staphylococcal Biofilms. *Antibiotics* **2019**, *8*, 174. [CrossRef] [PubMed]
79. Wanes, D.; Naim, H.Y.; Dengler, F. Proliferation and Differentiation of Intestinal Caco-2 Cells Are Maintained in Culture with Human Platelet Lysate Instead of Fetal Calf Serum. *Cells* **2021**, *10*, 3038. [CrossRef] [PubMed]
80. Nowak, A.; Zakłos-Szyda, M.; Żyżelewicz, D.; Koszucka, A.; Motyl, I. Acrylamide Decreases Cell Viability, and Provides Oxidative Stress, DNA Damage, and Apoptosis in Human Colon Adenocarcinoma Cell Line Caco-2. *Molecules* **2020**, *25*, 368. [CrossRef]
81. Minuti, A.E.; Labusca, L.; Herea, D.-D.; Stoian, G.; Chiriac, H.; Lupu, N. A Simple Protocol for Sample Preparation for Scanning Electron Microscopic Imaging Allows Quick Screening of Nanomaterials Adhering to Cell Surface. *Int. J. Mol. Sci.* **2023**, *24*, 430. [CrossRef]
82. SEM Preparation. Available online: https://electron-microscopy.hms.harvard.edu/sem-preparation (accessed on 27 March 2025).

83. Fischer, A.H.; Jacobson, K.A.; Rose, J.; Zeller, R. Preparation of Slides and Coverslips for Microscopy. *Cold Spring Harb. Protoc.* **2008**, *2008*, pdb.prot4988. [CrossRef]
84. Bai, J.; Li, J.; Chen, Z.; Bai, X.; Yang, Z.; Wang, Z.; Yang, Y. Antibacterial Activity and Mechanism of Clove Essential Oil against Foodborne Pathogens. *LWT* **2023**, *173*, 114249. [CrossRef]

MDPI AG
Grosspeteranlage 5
4052 Basel
Switzerland
Tel.: +41 61 683 77 34

Antibiotics Editorial Office
E-mail: antibiotics@mdpi.com
www.mdpi.com/journal/antibiotics

www.ingramcontent.com/pod-product-compliance
Lightning Source LLC
Chambersburg PA
CBHW061613210926
51850CB00009B/2309
9783725878178